MACHINERY NOISE AND DIAGNOSTICS

MACHINERY NOISE AND DIAGNOSTICS

RICHARD H. LYON
Massachusetts Institute of Technology

Butterworths
Boston London Durban Singapore Sydney Toronto Wellington

Library of Congress Cataloging-in-Publication Data

Lyon, Richard H.
Machinery noise and diagnostics.

Bibliography: p.
Includes index.
1. Machinery—Noise. I. Title.
TJ179.L96 1986 621.8′11 86-34341
ISBN 0-409-90101-6

Butterworth Publishers
80 Montvale Avenue
Stoneham, MA 02180

10 9 8 7 6 5 4 3 2 1

Printed in the United States of America

CONTENTS

PREFACE

This book is intended to provide engineers with an understanding of how dynamic forces produce structural vibration in machines and how these vibrations are transmitted through the machine and produce radiated sound. The application of these ideas to machinery noise is obvious. If one knows how forces are generated, then modifying the mechanisms may result in less excitation at audible frequencies and therefore generate less noise. Or the transmission of vibration may be reduced by modifying shafts, bearing supports, frames, or other machine elements by changing compliances or mass loadings at critical locations. Finally, the radiation of sound by the structure may be reduced by lowering the vibration amplitudes of housings or panels or by reducing the coupling of these structures to the surrounding air.

The application of these principles described in this book will allow design engineers to develop quieter products or to modify existing products to make them quieter. Engineers who do not have training in acoustics are often surprised by the very small amounts of power carried away by sound radiation, typically a few milliwatts! Also, machine structures behave differently at audible frequencies (200–5000 Hz) than they do at operating speeds. Thus, a housing that seems stiff and heavy may resonate at audible frequencies and be a good sound radiator also. This book is intended to help the engineer develop a feel for these processes and to be able to predict the effects of structural changes on vibration transmission and sound radiation.

The reader will discover that this book differs from many noise control books by the absence of discussions of add-on or "band-aid" noise control devices, such as enclosures, mufflers, or isolators. Our motivation is to design machines that will produce less noise. For machines already in service, quieting often can be accomplished only by add-on devices. Books are available that deal with noise control devices.

The second major concern of this book is the use of vibration or acoustical signals generated by a machine to reveal its operating conditions. The vibrations produced by mechanism forces can be used to reveal faults in the mechanism itself or some change in the vibration path. The fault revealed may be an actual malfunction or perhaps only a change in an operating parameter. In the latter case, the operating condition may be changed by a control system, but if a failure is detected, the machine probably needs to be shut down. Changes in the vibratory path may signal the need for the replacement or repair of structural elements.

This use of vibration to signal changes in mechanisms or structure is termed "machinery diagnostics." It is closely related conceptually to the problem of machinery noise because of the common areas of vibration generation and transmission. Sound radiation is also important in diagnostics if a sound wave is used for the diagnostic signal. But the greatest differences between diagnostics and noise are that diagnostics is generally concerned with a broader range of frequencies and the received signals tend to be electronically processed rather than listened to. Because of this, the book emphasizes various signal processing techniques designed to recover dynamic signatures of interest.

This book is intended to help the engineer who wants to develop a diagnostic system for a particular machine or to understand the basis on which such systems operate. No attempt is made to recount the properties of various commercial systems. Such information is readily obtained from the manufacturers in the form of brochures or seminars. By emphasizing the principles of diagnostics, we show the reader how modern systems can reveal defects and we indicate what diagnostic systems may be able to do in the future.

This book has been developed from the author's short course in machinery noise and diagnostics presented to working engineers at M.I.T. and in Australia, China, and Greece. The course notes were originally prepared from tape recordings of the lectures, and in spite of the numerous revisions they retain a conversational style. This makes the book particularly appropriate for self-study. The notes have also been used for a full semester graduate course at M.I.T. It can also be used as a text for a university course at the senior or graduate level.

Any book dealing with the subjects that this does, and at a rather fundamental level, could become fairly mathematical. I have tried to avoid excessive mathematics by motivating a number of the derivations by physical arguments. In addition, there is a great deal of experimental data presented so that the derivations are buttressed by practical discussions. I hope that readers of a practical bent will benefit from this combined approach and be convinced that theory does have its role in their work. Likewise, there are many people interested in the more theoretical aspects of noise and diagnostics, and I hope they will find this book helpful in relating theory to practical aspects of these fields.

However, there are practices in any field that have important applications that cannot be fully justified from basic principles. These practices are the "art" of that field, and diagnostics work has a substantial amount of art. We shall touch on some of this art in this book, but it is difficult to include much without the book becoming a series of anecdotes or recitals of "so and so did such and such and it seems to work." Since it is very difficult to draw general conclusions from such examples, our use of such material is limited.

CHAPTER 1

Introduction to Machinery Noise and Diagnostics

1.1 RELATIONS BETWEEN MACHINERY NOISE AND DIAGNOSTICS

Machinery noise reduction is carried out at many companies as a product design activity or for plant environmental reasons. The engineers who carry out this work are generally trained in acoustics, and since their interest is in noise at frequencies that are large compared to machine operating frequencies, their interest in vibration is also at high frequencies. Noise reduction engineers can have a variety of educational backgrounds, but electrical engineering, mechanical engineering, and physics are the most common. They are most concerned with impact, gear mesh, flexural waves in structures, sound radiation, and sound transmission.

Vibration monitoring is also used at many companies to anticipate needs for machine repair. The vibration signatures are most often frequency spectra that encompass several multiples of the machine operating speed to reveal malfunctions such as imbalance, misalignment, and bearing race defects, as shown in Figure 1.1. The frequencies involved tend to be lower than for noise problems. Engineers who design and use diagnostic systems tend to have backgrounds in machine dynamics and vibration. Their education is in mechanical design or engineering mechanics; there is little overlap between this group and noise reduction engineers.

Recent trends in noise reduction and diagnostics have brought these two groups closer together in methods and interests, even though the goals of each have remained distinct. This has created an educational challenge—to teach practicing engineers and those in school the common ground between these fields.

When mechanisms operate within a machine, forces that produce vibratory motions are generated. These vibrations are transmitted through the machine and produce exterior surface vibrations and radiated sound. Whether the sound or vibration is of concern as a noise problem or is to be used for fault detection in a diagnostic system, these features of excitation, transmission, and radiation are important. The similarity of these basic aspects of noise problems and diagnostics is evident.

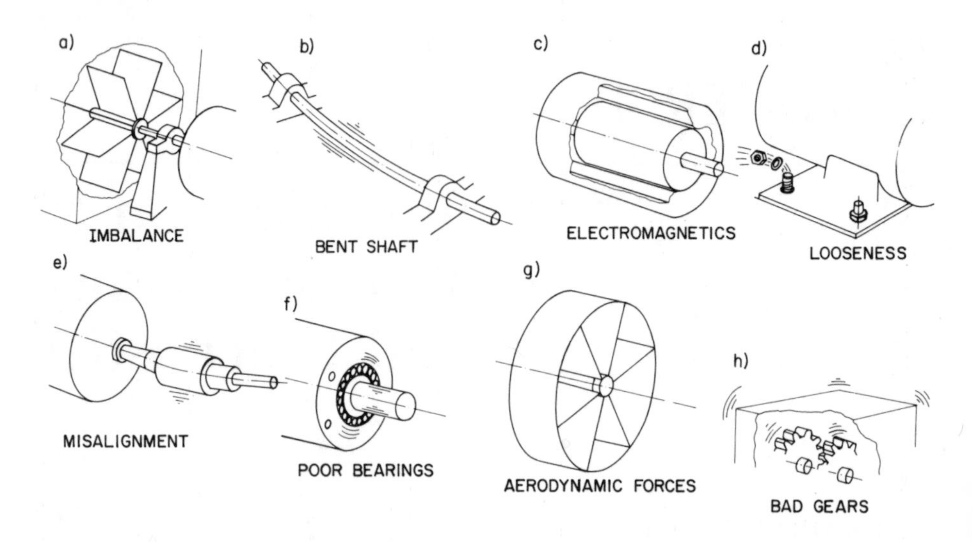

Figure 1.1 A variety of faults or operating conditions in machines can lead to vibration spectra that are usable as "signatures" for that fault. Most practical systems are applied to rotating machinery, but new advanced systems are applied to reciprocating and impacting mechanisms.

Both diagnostics and noise analysis use power spectra of excitation sources as signals of interest. Diagnostics has generally used line spectra of rotating machinery to indicate imbalance, misalignment, and other faults. Because the relative strengths of the spectral lines that correspond to shaft multiples are important, the frequency dependence of the transmission path for such vibrations is also important.

Broadband, full-, or third-octave spectra are of special importance in noise analysis, and they can also be useful for diagnostics. For example, impact and roughness in bearings produce broadband excitations, and simple energy detection in these high-frequency bands can indicate faults and jitter in a once-per-rev impact; the jitter will cause a transition from a line spectrum at low frequencies to a continuous spectrum at high frequencies.

1.2 THE GOALS OF NOISE REDUCTION COMPARED WITH THOSE OF DIAGNOSTICS

The greatest difference between diagnostics and noise reduction lies in their respective goals. A machine operating properly and without faults can still be very noisy, and a machine that has developed a major fault may operate quietly. Signal features that may be important for diagnostics (such as phase) are generally ignored in noise problems. Signal energy in the speech frequency bands, so important in noise problems, has no special role in diagnostics.

The goals of a noise reduction program generally include one or more of the following:

- Reduction of hazard (such as hearing loss or hypertension)
- Reduction in functional impairment (such as speech interference or task performance)
- Increased acceptability of the sound (such as consumer satisfaction) and the related goal
- Enhanced product image

These last two goals are occasionally of interest to manufacturers of consumer products, and represent areas in which noise reduction engineering could be of greater service in product design. Helping manufacturers to meet these goals will require new approaches in product sound evaluation that move away from *A*-weighted intensity measures to evaluation procedures that may be very product specific, as discussed in Appendix B.

The goals of any diagnostic system will include one or more of the following: detecting faults in machine operation, identifying the fault, and anticipating breakdown in machine operation. This last aspect is sometimes referred to as prognostics. A diagnostic system can also be used to determine certain operating characteristics of the machine and thereby supply data that are useful for the control or modification of machine operation.

These goals of diagnostics are usually thought of in positive terms. It is perhaps more appropriate to describe them in a more negative manner—in terms of mistakes. Two kinds of mistakes can occur in diagnostics:

Type I error—There is no fault, but a faulty condition is declared.

Type II error—There is a fault, but a safe condition is declared.

Depending on the situation, the relative penalties for making each kind of mistake can be quite different. Type 1 errors can be very costly if a production process is needlessly interrupted; Type 2 errors can be particularly damaging in security systems. We discuss the relationship of these errors to system design in Section 1.11.

1.3 GENERAL FEATURES OF A NOISE REDUCTION PROGRAM

According to some textbooks, machinery noise control consists of wrapping or enclosing the noisy component, putting a muffler on its exhaust, damping the interior or surface structure, and setting the machine on springs. It is easy to understand the lack of favor toward such an approach by the users or manufacturers of machines. Such "band-aids" are an unproductive extra cost; they reduce accessibility and visibility, and they may cause safety hazards and reduce reliability.

Table 1.1 Comparison of Noise Control and Design for Reduced Noise

Noise Control	*Design for Reduced Noise*
Technology is in place	Principles are understood
Implemented by handbook	Practitioners are scarce
Add-on after product design	Integrate with design cycle
Design relatively independent	Design highly interactive
Costing is straightforward	Costing is difficult
May reduce reliability, maintainability, safety, utility	May enhance function and other attributes

The advantage of this form of noise control is that the technology is in place and fairly easy to apply, as noted in Table 1.1. This has led to the curious assumption that noise reduction always causes unfavorable impacts on other aspects of the machine and makes it more expensive. The idea that a change in design to reduce noise can solve other problems at the same time and may actually make the machine less costly to produce would appear to be a pipe dream. Yet, this can happen if noise reduction is made a part of the basic design and not treated as an afterthought.

What are the requirements for design? First, there must be people who can do it. As noted earlier, mechanical design engineers generally equate balance and vibration with noise—the notion that the vibration that actually radiates audible sound can neither be seen nor felt is not intuitive to most design engineers. We must have the resources and the knowledge base to train mechanical engineers in noise reduction principles.

Second, if the goals regarding sound levels and implementation periods take cognizance of the design cycle and depreciation schedules of industrial equipment, a noise reduction program can be more effectively programmed into the design process. Then design changes for noise reduction can be integrated with aspects such as operator use, maintainability, cost reduction, durability, and aesthetics.

One "disadvantage" of the design approach is that it becomes very difficult to assign specific costs to noise reduction. If a machine cover is perforated, costs are involved to perform this operation and to build the tooling for it. Less noise is radiated by the cover, but ventilation may be increased, less material is used, and "add-on" noise reduction, such as damping and resilient supports for the covers, may be eliminated. The actual manufacturing cost of the new cover could be reduced by the changes, but, in any event, the noise-related contribution to the change might be difficult to single out. Also, since design improvements are transferable across a product line, it may be inappropriate to assess all the costs to a single product.

Such "difficulties," however, really exemplify the benefit of noise reduction by design. Integrated with the design cycle, antinoise features are far less likely to interfere with the maintenance or utilization of the product, and manufacturing costs assessable to noise may not be increased. Such benefits are clearly desirable, but they come at the cost of greater design effort, a longer payout, and much more effort in understanding the noise generation, transmission, and sound radiation processes within the machine.

1.4 NOISE REDUCTION BY DESIGN—SOME PRINCIPLES

Noise control engineers are familiar with the source–path–receiver idea. Table 1.2 shows a similar triad in machinery noise production: generator–transmission path–radiating area. Since much of the work in noise-related design involves dealing with these items, discussing them is worthwhile.

The *generator* is the producer or source of the high-frequency vibrational energy that ends up as radiated sound. In many machines, the generator may not itself be a strong or efficient radiator of sound. For example, the pressure transient within a diesel engine combustion chamber is such a generator, but the high-frequency energy must be transduced and propagated through the engine structure as vibration before it can be radiated as sound by the engine casing. A meshing gear produces high-frequency vibrational energy in its shaft, but the fluctuating forces at the tooth are usually very weak *direct* radiators of sound, and such sound may be blocked in any case by the gear enclosure. Other examples of generators include cams, timing belts, link joints, air impingement, bearings, and fluctuating stresses that occur in materials forming and fracture, as shown in Figure 1.2.

A review of the table shows that the production of high-frequency vibrational energy by a generator is not necessarily a result of its failure to perform its designed function. Also, the amount of audible frequency energy will usually be a very minute part of the energy in the basic process involved in machine operation. Thus,

Table 1.2 Elements Considered in Design

Generator ("It's better to control the noise at the source")
Audible frequency energy small fraction of machine work
Little energy may be radiated directly
Modification may have strong interaction with function
Transmission path
Often made up of assembled parts
Offers good opportunity for redesign
Available methods are damping and mismatch
impedance mismatch
isolators
spermassen (loading masses)
Radiating areas
Reduce vibration by decoupling or damping
Reduce radiation efficiency by "opening up"
Light covers are difficult to damp or isolate

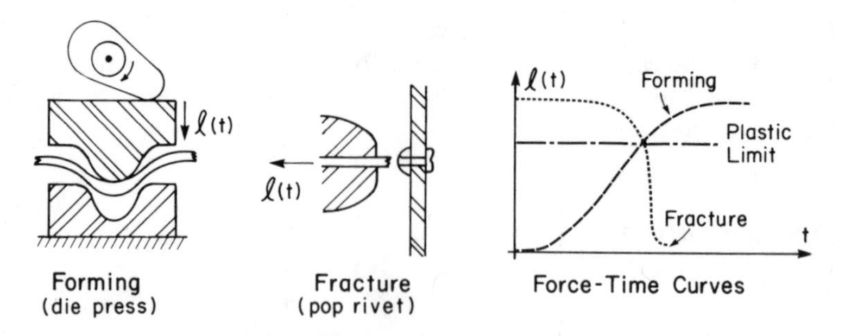

Figure 1.2 Reducing the rapidity of force changes in industrial processes can lead to reduced high-frequency excitation of machine structures.

attempts to modify the production of audible sound by modifying the generator may have serious and unacceptable consequences on machine operation, efficiency, or cost. Thus, the general dictum "it's better to control the noise at its source" may not be applicable in many design situations.

The *transmission* of the high-frequency energy from the place where it is generated to places where it can radiate to the machine exterior may involve shafts, frames, bearing supports, links, castings, or interior acoustical spaces, as shown in Figure 1.3. The designs of these machine elements rarely take high-frequency (audible) energy transmission into account, but because they are mostly assembled items, they are easily modified and offer excellent possibilities for achieving noise reduction.

The noise reduction principles that are applicable to these transmission paths are familiar—reflection of the energy back to the source and its removal by damping—but the systems for achieving these effects may be unconventional. The reflection or impedance discontinuity element may be a resilient bearing support structure or loading masses on a link. The damping elements can be dissipative elements in the joints or modified structural elements with viscoelastic damping materials added. Added damping is often ineffective in machine structures, however, because of the large amount of damping already present from structural junctions.

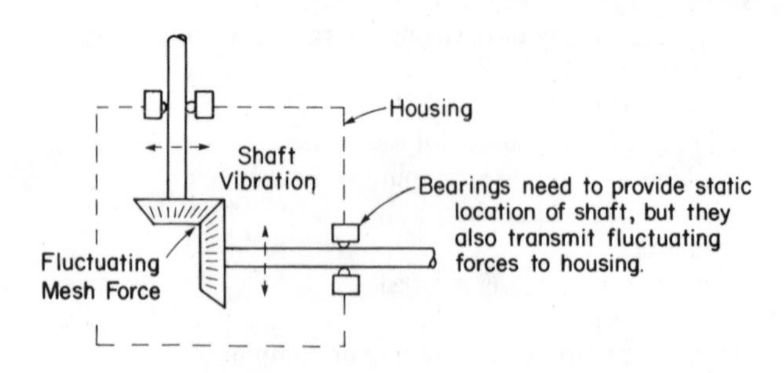

Figure 1.3 Internal dynamical forces generate vibration that is transmitted to radiating surfaces by machine elements.

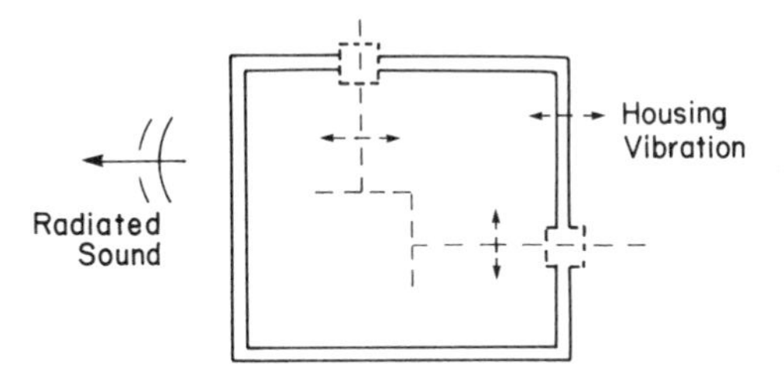

Figure 1.4 Radiated sound is produced by vibration of large, flat surfaces that are often excited by internal mechanical elements.

Major structural castings sometimes act as reservoirs for vibrational energy. Low-order resonances can act as "amplifiers" of vibrational excitation in narrow frequency ranges, and a noise reduction benefit may be gained by moving such a resonance, usually to a higher frequency. This change requires a higher stiffness-to-weight ratio, which may involve a reduction in weight and greater stiffness-producing benefits, using less material and offering better machinability. Such a design goal is familiar to aerospace structural engineers, but may be novel to a machine designer.

The *sound radiating* areas of a machine are usually parts of the structural castings (work surfaces, casing, etc.), external covers, and openings such as air vents. This structure is often excited by the operating mechanisms that it supports or encloses, as shown in Figure 1.4. The treatment of such elements by mufflers, wrappings, damping, and enclosing is a common practice, and if these add-ons are acceptable such treatment can be quite effective. Structural redesign may offer the possibility of cover isolation, designed-in damping, or perforation (or replacement of plate by truss elements), which reduces the ability of the structure to radiate sound. Sometimes a change in the stiffness or weight of a radiating surface can also be useful.

Every approach to noise reduction is sometimes useful and sometimes ineffective. For example, perforation is helpful if a cover is mechanically excited by its mountings, but it may be harmful if the cover is keeping the sound in interior spaces from getting out. It is also possible that the sound radiation from a cover is due to mechanical excitation in some frequency bands and to excitation by internal sound in others. Clearly, perforation of a cover would not be appropriate as the only remedy in such a situation.

1.5 INTEGRATING NOISE REDUCTION INTO THE DESIGN PROCESS

A desirable approach from an engineering viewpoint is to incorporate noise reduction principles into the basic design. Such principles are reasonably well understood and are discussed in this book. Few engineers have experience with such design procedures, however, and companies have difficulty finding people with the necessary experience and knowledge. The acoustical design should be integrated within the design cycle of the product, and this could delay the appearance of a new product with reduced noise. But when it is introduced, the machine may very well be quieter and more efficient, because this type of noise design is highly interactive with other aspects of the design process, which can include surface finish, machining accuracy, and structural and cover design.

The many sources of excitation and radiating surfaces means that to achieve a significant reduction, it may be necessary to attack the noise at many points. This combined effort may include some treatment of sources, some path efforts, and modification of radiating surfaces. Indeed, the work on any particular machine will likely result in changes in all of these areas.

Source control may involve changes in gear form and construction or changing a motor belt drive. Path modifications can involve shaft isolation, stiffening of major castings, and acoustical absorption of interior spaces. Control of radiating surfaces may involve perforating and removing covers or damping and mechanically isolating covers.

Damping is often suggested for reducing sound radiated by a housing. The reader may have been exposed to the demonstration, usually presented by a salesman of damping materials, in which a thin plate is first struck by a small hammer, resulting in a loud, ringing tone. Then with a small amount of damping material added to the plate, the same experiment results in a very dead sound, representing a sizable decrease in radiated sound energy.

There are two things to remember, however, in considering the actual structures to which damping may be applied. The first is that any applied damping must be greater than that provided by the attachment of the panel to the main supporting structure, and in many cases this damping may be fairly sizable. Thus, the benefit gained by adding damping material may not be nearly as great as when the plate is freely suspended.

The second point is that damping material works by dissipating a certain fraction of the potential energy stored in the damping material. In simple bending motions of flat plates, the fraction of energy put into the damping material can be relatively large for a given amount of deflection. When the structure is stiffened by adding supporting ribs or by changing its shape through drawing or rolling operations, the same structural deflection may result in a much larger amount of potential energy stored in the metal of the structure and the same amount of potential energy stored in the damping material. Thus the fractional amount of energy dissipated by the damping material is much reduced for such a stiff structure, and, consequently, there is less reduction of vibrational energy.

Interactions between the addition of a perforated cover, for example, and cooling requirements, surface finish, and the like show that a balanced design will require integration of acoustical and nonacoustical factors. This shows why noise control engineers do not normally have the opportunity to make fundamental changes in machine design and why mechanical design engineers should have more training to recognize those aspects of a design that may affect noise production.

When the noise control engineer becomes part of the design team, it is possible for these interactions to be dealt with in the most beneficial way. In the rest of this section, we describe some interactions between noise design and other design aspects.

A common problem in machine operation is the coincidence of some particular operating frequency and a major structural resonance. Because a machine potentially has a range of operating speeds and associated excitation frequencies and a number of resonances, such coincidences cannot be entirely avoided. Structural changes (such as stiffening or mass addition, for example) will interact with manufacturability, material costs, and weight. If we want to increase the frequency of a particular resonance, this means increasing the stiffness-to-weight ratio of the castings. This is accomplished by adding material in some areas while taking it away in others. The changes can improve manufacturability (less deflection during machining operations) while slightly increasing the weight of the machine.

If perforated or more open construction covers are used, then shock hazard, fluids containment, and materials selection become issues. If an internal cover is removed (for acoustical purposes), then the electrical and combustibility requirements on the outer cover may change. In addition, mold design, materials choice, and color properties must also be considered in order to produce an aesthetically pleasing cover.

The innovative substitution of materials such as plastics for gears and pulleys produces benefits and problems in noise behavior. Plastic gears are much more forgiving in terms of alignment and location, but in some applications they may have load and life limitations not shared by metal gears. Substituting timing belts for geared systems will affect noise output and timing accuracy of machine parts.

We cannot be exhaustive in this account of interactions, but perhaps the point has been made. Design for reduced noise must be carried out by design engineers capable in (or at least highly sensitive to) the noise characteristics of various machine elements. They must be joined by other engineers (we can call them noise control engineers) capable of (or sensitive to) understanding the relation between noise and other features of machine behavior.

1.6 COSTS ASSESSMENT

When a muffler is added to a machine to reduce its noise, the production costs assignable to noise control are fairly easy to quantify. When noise reduction is a part of redesign, it is difficult to assign costs to noise reduction, even as it it also difficult to specify just what fraction of the design effort or product cost is related to structural integrity, lubrication, or finish. Also, a proper acoustical design carried out within the context of other design aspects may actually enhance the basic function of the machine.

Production costs may be divided into "fixed" and "variable" in various mixes according to the accounting practices of the company. Fixed costs include design effort, tooling changes, plant modifications, and the like. Variable costs include labor and materials costs associated with the manufacture of machines. The costs of worker benefits and environmental support (heating, etc.) are included as overhead on the labor component of variable costs.

The costs and benefits of redesign may be given as a fraction of total variable manufacturing costs. Comparison of costs for other products and noise control treatments can be made in a more meaningful way if costs are presented in this way.

Noise Costs

We cannot evaluate all of the costs of noise in terms of product image, lost sales, extra sales effort, and warranty expense. For consumer sewing machines, for example, we can identify the expense of reworking machines to make them meet the radiated noise product specification as monitored by the QC (quality control) department.

At one company, each machine is tested for radiated sound; if the level is above a specified limit, the machine is adjusted until the limit is met. This reworking is quite expensive, amounting on the average to about 1.5% of the variable production cost. Redesign might reduce this cost to about 0.5%. This projected 1% benefit does not include other benefits of noise reduction or the benefits of nonacoustical aspects of the redesign.

Redesign Costs

Redesign costs include the costs of engineering effort, tooling modifications, and changes in production labor, materials, and overhead. The variable costs of production in our example had a net reduction of about 0.5% due to the design changes. This result is a combination of the expected 1% cost reduction from reduced rework with a 0.5% increase in materials and labor costs for producing the new design. The tooling and engineering costs must be distributed over the production run of machines expected to benefit from the design changes. This amounts to about 0.35% for the engineering effort and about 0.65% for the tooling changes.

The net cost impact of the design changes was about a 0.5% increase in the cost to produce a machine. This is a conservative figure, because the number of produced machines that benefit from the changes has been conservatively estimated, and the intangible benefits were not included in the analysis.

1.7 APPROACHES TO DIAGNOSTIC SYSTEM DESIGN

Most current diagnostic and monitoring systems are based on empirical procedures. In this approach, there may be no clear reason why a particular characteristic of the measured signal should be identifiable with a particular fault, or the relationship may be largely intuitive; but as a result of experience and general correlation of machine behavior with noise or vibration, such a relationship can be established. The appropriate signal characteristic is monitored; when it goes beyond certain limits or displays certain trends, a corrective action may be called for. The signals used may be the output of displacement or acceleration sensors, and they may be either broadband noise or narrowband tones. The analysis may include time-domain behavior for which temporal peaks of vibration are detected.

Generally, statistical procedures for estimating probable life or the appearance of a fault are used because there is an inherent variability in the measured signals over a class of machines. A trend may be more significant than an absolute value. A machine component may display a high level of vibration over its entire life, but if the level of vibration does not change very much, the machine operation is stable and can be expected to remain so. On the other hand, a vibration that starts out very weak but increases consistently as the machine is monitored is more likely to indicate a developing fault and a need for corrective action.

In many ways, the empirical approach is similar to that used by physicians, who identify some sound or other symptom with a particular problem after observing many cases. The physician is usually not terribly concerned with a detailed explanation of the relationship between the symptom and the problem but happily accepts the signal's indication of the problem. The empirical approach is strongly based on experience and is generally used by instrument companies when applying and marketing diagnostic systems. Because of its nature it is difficult to give general principles for applying the empirical approach, and discussions of it tend to be recitals of the use of various sensors and instruments and their application of diagnostic and monitoring problems by various users.

The discussion in this book will center on an analytic approach. The analytic approach represents a series of techniques and studies recently developed; few systems in place use its principles. Nevertheless, it appears to be the basis on which future systems will be developed. Of course, there is no hard and fast line between analytic and empirical methods, and even a highly analytic method will have certain elements of empiricism associated with it.

1.8 EXTRACTION OF SOURCE FEATURES OR "SIGNATURES"

The analytic approach is concerned with determining various properties of the event that produces the disturbance in the machine. We generally refer to such events as "sources," since they are the cause of the vibration being measured. The source properties of interest might be the temporal waveform, a frequency spectrum, or some aspect of signal phase. The basic idea is that if we can determine the actual waveform of the event (such as combustion pressure in a cylinder of an engine), we can learn quite a bit about the operation of the machine. If we can recover the combustion pressure, we can determine the nature of the combustion, the power produced by that cylinder, and other quantities.

The recovery of source signatures from output data requires that we know the magnitude and/or phase of the signal much better than we need to for noise analysis.

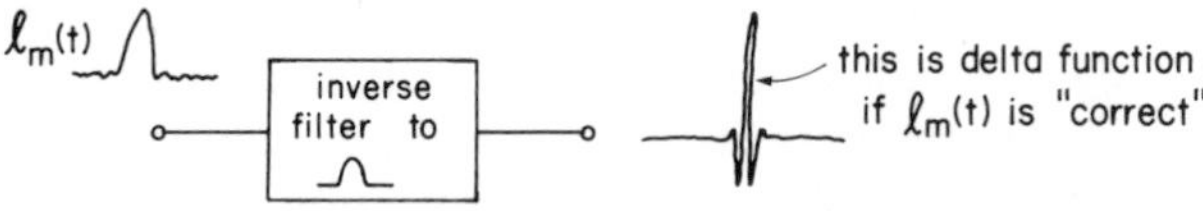

Figure 1.5 An inverse filter will convert a measured output to a delta function impulse for a particular signal—other inputs will cause an output that has a time spread.

The recovery of temporal waveforms requires a detailed knowledge of the phase spectrum, an almost unexplored area in structural transmission. Recent research has shed more light on the phase properties of transfer functions, but it is still not clear how the information needed to recover source signatures is to be obtained.

The need for detailed information on structural transmission raises a related issue of variability. A diagnostic system may be installed in a variety of machines, and the transfer function for vibration and the transducer mounting impedance can vary greatly. Vibration spectra on machines of the same general type, which are different models or from different manufacturers, may have significant differences. This makes the use of vibration criteria based on vibration levels alone unreliable, particularly at frequencies above the first few machine resonances.

Signature recovery may be accomplished with "inverse filters." These filters are intended to take the potentially complicated waveform that results from the action of the source and to convert it into a simpler signal from which one can make judgments about system operation. An inverse filter has a gain that is the negative of the log magnitude signal spectrum, and a phase that is the negative of the signal phase. If the inverse filter is applied to the measured signal, as shown in Figure 1.5, the resulting output will be a delta function. If the measured pulse departs from this, however, then the output of the inverse filter will depart from the delta function and will be "smeared out."

1.9 DETECTION AND CLASSIFICATION OF FAULTS

The extraction of signatures or features of a diagnostic signal is only part of the process. We must determine whether a fault exists (detection), but we must also determine the nature of the fault (classification). Because listening to machines is a traditional method of diagnostics, some diagnostic system designers have sought to "do what the person does" in detecting faults. Although this approach can be successful in particular instances, it may also lead to time-wasting efforts to "diagnose" the listener (to learn what the person is doing in making the diagnostic judgments). In general, it is better to direct this effort toward learning enough about fault-related vibration generation and transmission to be able to design the diagnostic system to detect and recover the appropriate signals.

Another important area is the development of decision models for selecting system parameters. Decision models tend to be empirical, and one has to gain experience regarding the effects of various uncertainties in the signal processing and in the machine operation itself to compare the nature of the variations in a machine in good condition with those associated with a fault. We need to classify variations in the output of the inverse filter to achieve the best identification of machines that are in poor condition or out of their operating parameter envelope.

1.10 MACHINE CONTROL USING DIAGNOSTIC SIGNALS

Diagnostic systems can also be used in machine control. A very good example is the knock detector, which has been incorporated in some automobiles with spark ignition engines. Under conditions of knock, the combustion in the cylinder is very erratic and rapid, much as in a diesel engine, which causes a rapid and erratic rise in combustion chamber pressure. The pressure variations produce vibration of mechanical components, and heat release produces ringing of the combustion chamber, which acts as an acoustical cavity.

A sketch of the cavity and the mode of oscillation excited by knock is shown in Figure 1.6. The frequency of acoustic oscillation varies as the piston moves along the cylinder wall, but it is generally from 4 to 8 kHz. An accelerometer or force sensor is tuned to pick up vibrations in this frequency range and thereby discriminate against the general background noise of the engine. When knock is detected, the ignition timing is delayed so that the knocking dies out.

1.11 DECISION ANALYSIS AND CLASSIFICATION SYSTEMS

A major component of a diagnostics software system, which some would lump into the signal processing system and others would keep separate, is the decision-making algorithm. If certain failure models are available for the machine or system under test, these models can be incorporated into the software, and the parameters of the failure model can be evaluated based on the data. When these parameters get close

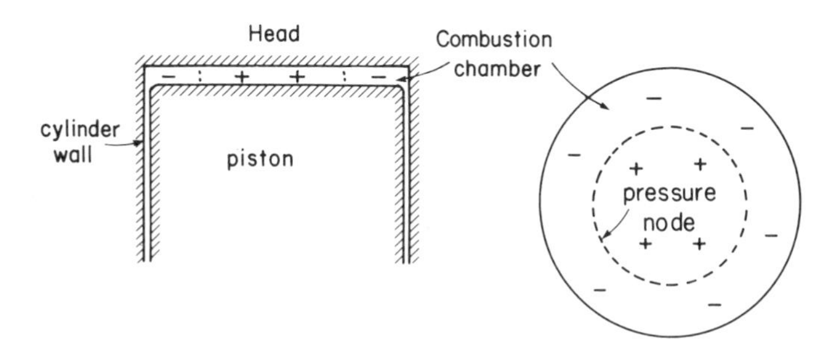

Figure 1.6 Acoustical mode of combustion chamber excited by detonation. The resonance frequency is usually in the range of 4–8 kHz and is used to detect "knock."

to a failure condition, the system can respond appropriately. The maximum likelihood procedure based on Bayes' estimation theory is often incorporated into these decision models.

The maximum likelihood procedure considers distributions of the output of a system, depending on whether it is in a normal or a failure mode. Any particular measurement of system output is evaluated in terms of the likelihood that the data belong to the normal or a failure population. The assignment of the output to one category or the other is weighted not only by the likelihood of its being in that category but also by the cost of possible errors (discussed in Section 1.2) termed Type I and Type II. A Type I error occurs if we say the system has failed when it has not. In this case, we incur the cost of an unnecessary stoppage and teardown. On the other hand, a Type II error occurs when we say the system is okay when it is not. In this case, we suffer the possibility of a system failure with the costs incurred when the process breaks down. These costs will influence the decision of assigning the reading to a failure or to a normal population.

Most fault decision systems are multidimensional; that is, several measures on the signature will be made to determine the machine condition. To illustrate decision analysis, however, we consider a one-dimensional case in which only a single metric x, a "diagnostic signal," determines the fault condition. This might be energy in a frequency band, or the amplitude of an impact event.

As noted at the beginning of this section, we expect that a machine without a fault (safe) will produce a distribution of the metric $p(x \mid s)$, shown in Figure 1.7. When machines are in a faulty condition, the metric will change, at least on average, and the probability distribution for the metric is $p(x \mid f)$, also shown in Figure 1.7. The basic problem is to select a decision level x_d so that when the metric exceeds this value, a faulty condition is announced, but when it is less than x_d, a safe condition is announced.

No matter what decision level is set, wrong decisions will be made. The area of $p(x \mid s)$ in the region $x > x_d$ defines the probability of a Type I error P_I. The area of $p(x \mid f)$ in the region $x < x_d$ defines the probability of a Type II error P_{II}. But the cost of an error is also important. Suppose the cost of a Type I error is K_I and that of a

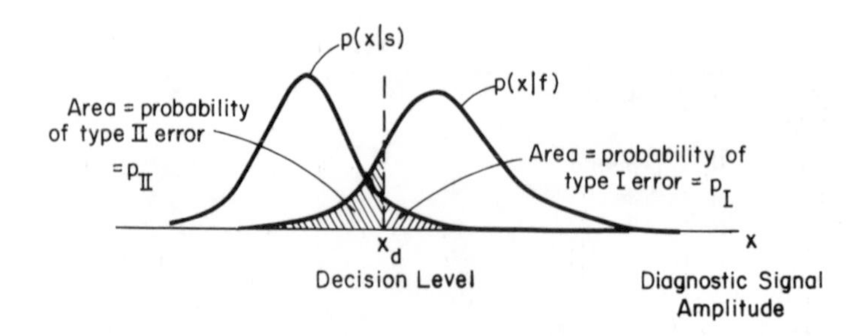

Figure 1.7 Probability densities of "good" and "bad" machines in terms of the diagnostic signal. The probabilities of Type I and II errors can be changed by adjustment of the decision level x_d.

Type II error is K_{II}. To make the expected cost for both errors the same, we require

$$P_{\mathrm{I}}K_{\mathrm{I}} = P_{\mathrm{II}}K_{\mathrm{II}} \qquad \text{or} \qquad \frac{P_{\mathrm{I}}}{P_{\mathrm{II}}} = \frac{K_{\mathrm{II}}}{K_{\mathrm{I}}}. \tag{1.1}$$

We then select x_{d} so that the ratio of error probabilities is the inverse of the ratio of the costs.

A diagnostic system designer must select the best diagnostic signal or metric, the probability distributions (for each metric) for faulty and safe conditions, and the costs for each error type. The probability densities $p(x \mid s)$ and $p(x \mid f)$ are particular design problems. When the product is relatively inexpensive, these distributions can be obtained by taking measurements on a large number of machines. For more expensive items, such statistics are difficult to obtain. In this case, faults may be artificially introduced, or, if a sufficiently complete model is available, an analytical development of this distribution may be used.

CHAPTER 2

Sources of Vibration

2.1 VIBRATION GENERATORS AS SOURCES

Machine operation involves the generation of forces and motions that produce vibrations. These generating events are called *sources*. This chapter describes several important sources in machinery noise and diagnostics. We may be interested ultimately in their frequency content and/or their temporal waveforms for either application.

Sources of interest include machinery imbalance, impacts, fluctuating forces in gear meshes and cams due to small variations in machining or local contact stiffness, electromagnetic forces in rotating machinery, combustion pressures, and fluctuating forces due to air and liquid flows.

2.2 EXCITATION OF VIBRATIONS IN A MACHINE DUE TO IMBALANCE

Figure 2.1 shows an operating machine and how internal imbalance forces produce vibration. The problem can be thought of as the sum of two other problems. In the first, the internal mechanisms are operating, and an externally applied force l_B is sufficient to cause the machine to stop vibrating. In addition, the negative of l_B, in the absence of internal motions of the machine, produces the observed vibration. The required blocking force for rotating imbalance is $m\varepsilon\omega^2$, but for other types of imbalance, the formula for this force is more complicated.

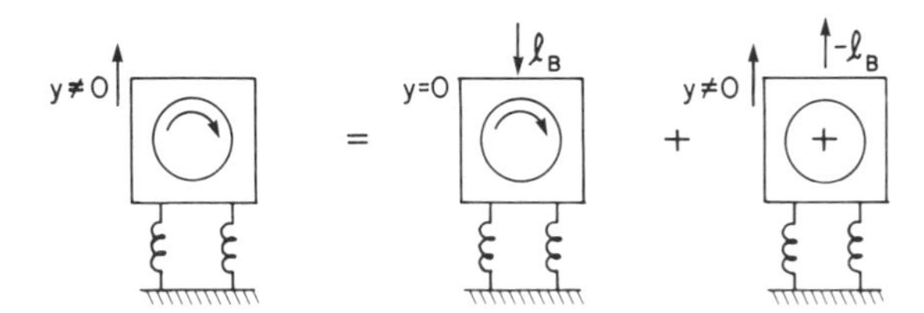

Figure 2.1 Decomposition of vibration generated by action of internal mechanisms to the force required to hold the machine housing at rest plus the negative of this force with mechanisms at rest.

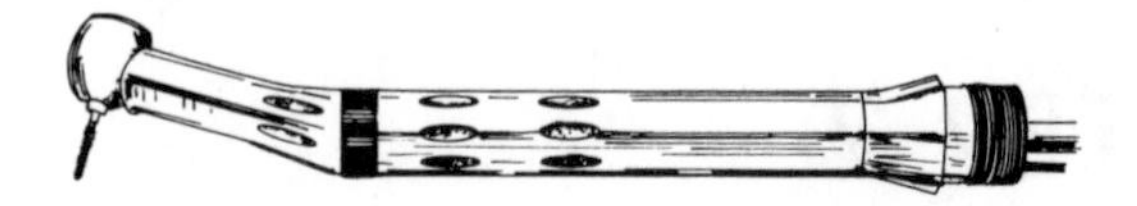

Figure 2.2 Modern dental drill operates at speeds in the range from 5 to 7 kHz. A small degree of imbalance can lead to annoying levels of sound radiated by such drills.

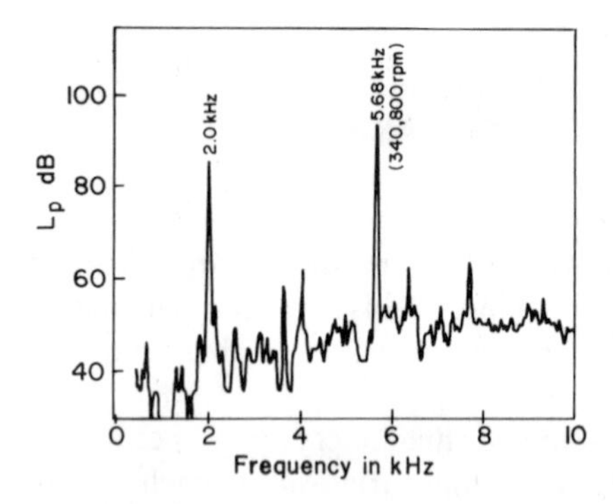

Figure 2.3 Radiated sound produced by a high-speed dental drill like that shown in Figure 2.2. The sound produced at 5.68 kHz is due to rotating imbalance. The spectral line at 2 kHz is due to imbalance in the rotating ball bearing cages that support the drill shaft.

An example of rotating imbalance that produces noise is the high-speed dental drill, shown in Figure 2.2. The rotating imbalance indicated in Figure 2.1 has relatively few harmonics, so the principal excitation due to imbalance is at the frequency of rotation, as shown in Figure 2.3. Most rotating machines turn at much slower rates, and the principal effect of imbalance is feelable vibration, not radiated sound, unless the imbalance leads to rattling or other such effects.

2.3 RECIPROCATING IMBALANCE

Reciprocating imbalance occurs in a crank and slider mechanism, such as automobile pistons or a sewing machine reciprocating needle bar. A sewing machine with its rotating and reciprocating components is shown in Figure 2.4. Suppose that we are concerned about the imbalance in this machine, because when placed in a cabinet excessive vibration can occur. Clearly some imbalance exists because if we hold the machine tightly while it is running, we can feel the forces that are produced. The issue is whether the imbalance can be improved; that is, can the machine be modified so that it vibrates less when in a cabinet.

Figure 2.5 shows how a reciprocating mass m can be thought of as a combination of two counterrotating masses, a and b, each with mass $m/2$. Clearly a

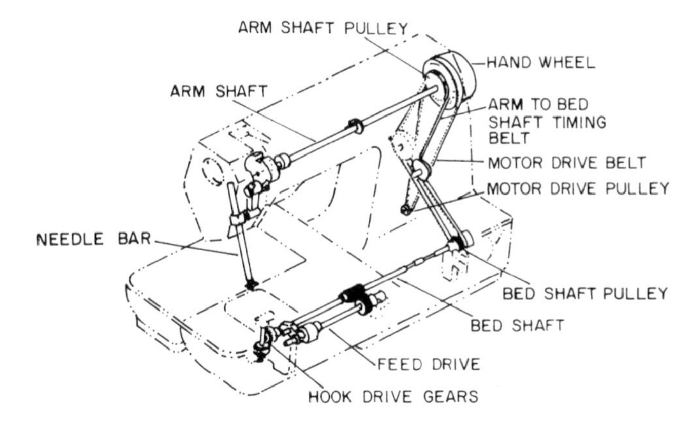

Figure 2.4 A simplified view of sewing machine mechanisms. The needle bar is a crank and slider mechanism that generates imbalance forces.

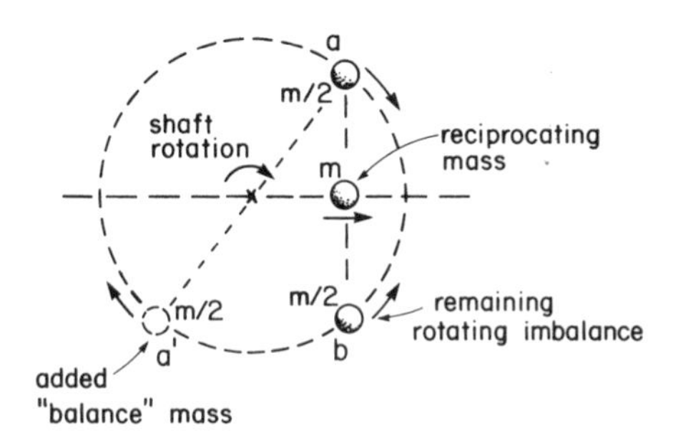

Figure 2.5 A reciprocating imbalance can be resolved into two counterrotating masses. If one of these is balanced out, a rotating imbalance remains.

rotating mass a' can balance out the effects of mass a. This is the balance procedure for simple rotating imbalance. But in this case the mass b produces a new rotating imbalance. Thus, adding rotating mass simply changes the nature of the imbalance; it does not significantly change its magnitude. Nevertheless, this may be useful in some cases if the particular imbalance, and its reaction force, is less effective in exciting the supporting structure.

We can use Figure 2.1 to suggest a general method of measuring imbalance forces. If the machine is held rigidly in place, the blocked force l_B can be directly measured. But this may be difficult to accomplish. If the frequency is low enough so that dynamically the machine acts as a simple mass, then its free vibration is simply

$$\text{free acceleration} = \frac{\text{blocked force (unknown)}}{\text{mass}}. \tag{2.1}$$

The blocked force is determined by measuring the free vibration and the inertia,

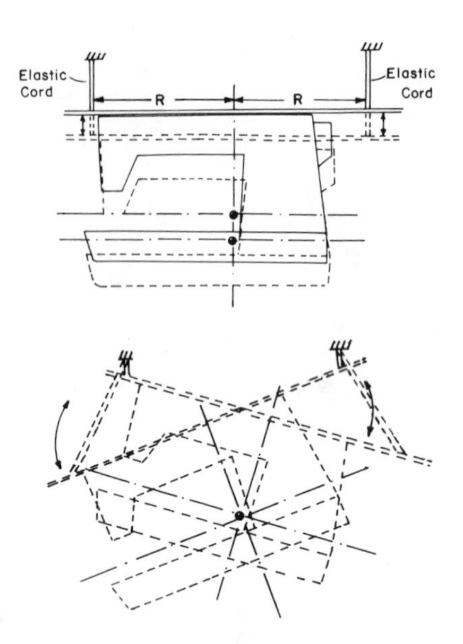

Figure 2.6 The radius of gyration can be determined by measuring the ratio of vertical bounce to rocking resonance frequencies. A simple analysis gives $\kappa/R = f_{vert}/f_{rock}$.

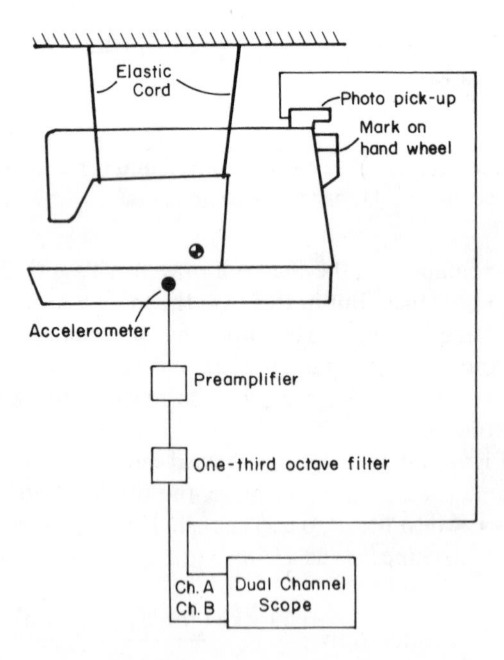

Figure 2.7 Setup for measuring the free vibration of a machine during operation. The total motion is resolved into translation of the center of gravity (CG) and rotation about the CG.

which for linear translation is simply the machine mass; however, to find the inertia for rotation involves finding the radius of gyration.

Measuring the moment of inertia for complex machine shapes may be rather difficult, but the experimental technique shown in Figure 2.6 can be useful. The machine's center of gravity has been determined, and the machine of mass M_m is supported by resilient cords tied to fixed locations on the machine at a distance R from the center of mass. We pull down the machine, let it vibrate vertically, and measure the frequency of vibration f_{vert} for this motion. We now rotate the machine about its center of gravity and release it without displacing the center of gravity. We let it vibrate freely, producing a rotation frequency f_{rot}. A simple analysis shows that the ratio of the rotation frequency to the vertical bounce frequency equals the ratio of the radius of the support distance R to the radius of gyration κ. The moment of inertia around this axis of the machine is $I_m = \kappa^2 M_m$. This is an effective way of measuring the moment of inertia about any axis of a machine when the location of its center of gravity is known.

We can find the free motion during the machine operation by using the technique in Figure 2.7. The machine is turned on and a photoelectric pickup is used to get a reference pulse for each cycle, typically when the needle bar is at top dead center (TDC). An accelerometer is placed on the desired location with a particular orientation, and its magnitude and phase of response relative to the photo pickup reference are displayed on a two-channel oscilloscope. A typical mounting of the accelerometer, in this case at the center of gravity in the machine, is shown in Figure 2.8. If at all possible, the basic suspension resonance in Figure 2.7 should be two octaves below the frequency at which the data are being taken. In this way, the response of the experimental system will be mass controlled in accordance with the theoretical assumption.

Figure 2.9 is a typical data display. The magnitude and phase of the vibration at the selected location in the selected direction is read relative to TDC. After a sequence of measurements of this kind is made, a computer program can then calculate and give the various imbalance forces and moments due to machine operation. The horizontal blocked force computed in this way from the measurements is shown in Figure 2.10.

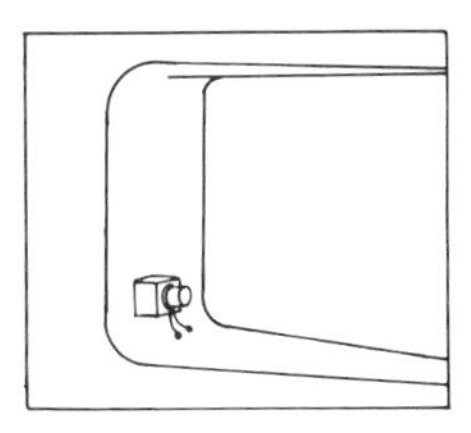

Figure 2.8 Accelerometer located at CG of machine to measure CG translation. Since the CG is not at the surface of the machine, a lightweight block is used to support the accelerometer at the proper position.

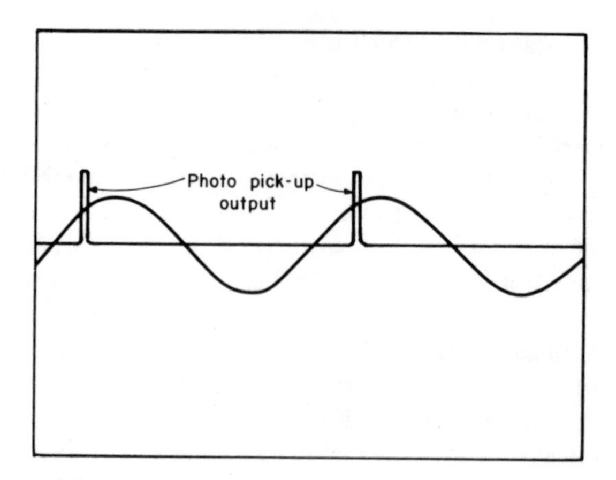

Figure 2.9 Sample waveform and timing pulse of system shown in Figure 2.7. This particular measurement shows nearly sinusoidal vibration.

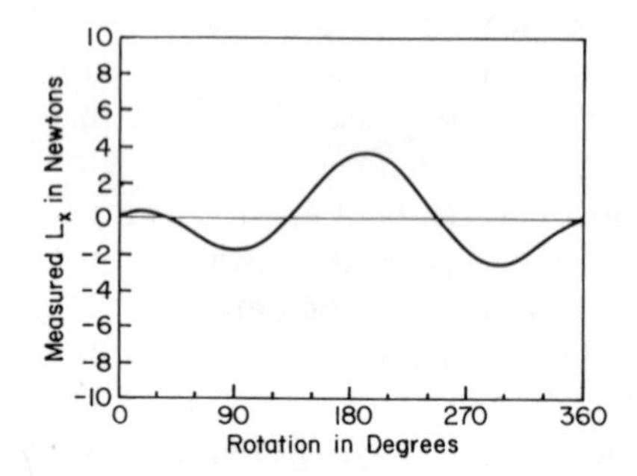

Figure 2.10 Horizontal component of imbalance force computed from free vibration measurements and machine mass. Note large component of second harmonic, typical of reciprocating imbalance.

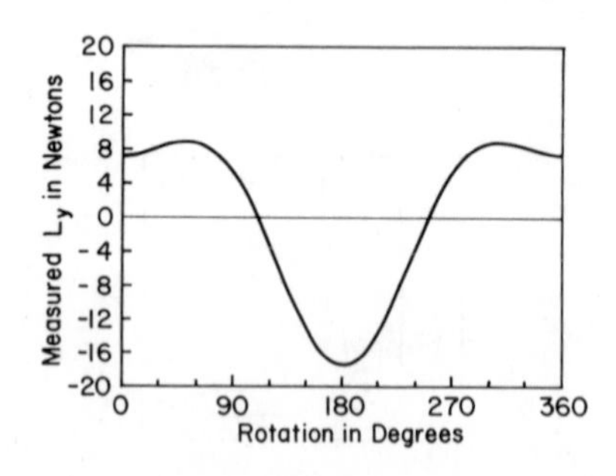

Figure 2.11 Vertical component of imbalance force computed from free vibration measurements and machine mass. Second harmonic is present but not as strong as for horizontal force in Figure 2.10.

The horizontal imbalance force in Figure 2.10 shows a rather large component of the second harmonic. The vertical force computed from the measurements (Figure 2.11) also shows a certain amount of second harmonic, but not quite as large relative to the fundamental as it is for the horizontal force. These horizontal and vertical forces can be combined to generate a resultant force vector, as shown in Figure 2.12. Mechanism analysis computer programs often generate force diagrams like Figure 2.12, and we can compare the experimental data directly with analytical results.

As another example, a purchaser of a larger number of electric motors has developed a QC procedure similar to that described earlier in this section, but which uses a blocked force measurement rather than a free vibration measurement. The test setup is shown in Figure 2.13. There are four force gauges supporting a mounting plate from a very rigid base, and the motors to be tested are mounted on the mounting plate. The magnitude and phase of these forces are measured, and then the location and magnitude of the motor imbalance are determined. We then decide whether the imbalance is acceptable in terms of motor bearing life and whether the motor should be rejected.

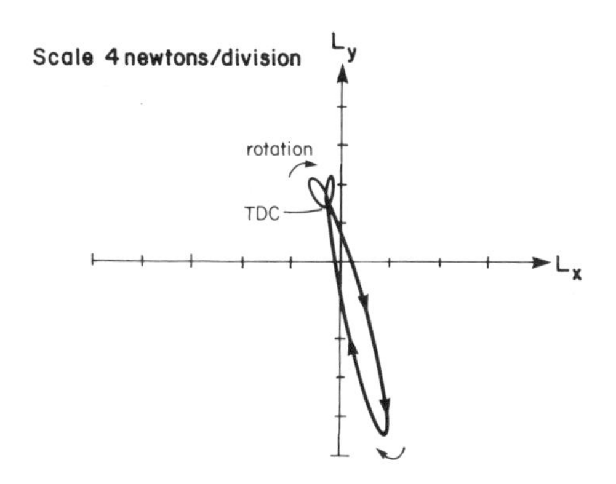

Figure 2.12 Resultant force for a machine cycle, computed from free vibration.

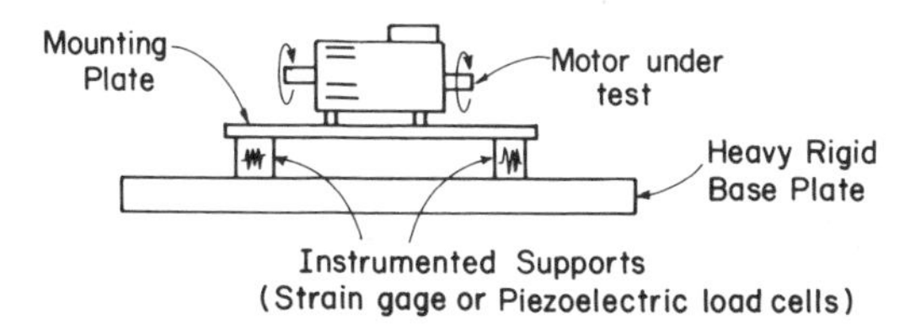

Figure 2.13 Measurement of imbalance by direct measurement of blocked forces. Although equivalent to a vibration measurement, this procedure may have some advantages for heavy machines.

2.4 IMPACT AS A SOURCE OF VIBRATION

Another mechanism-related source is impact. Like imbalance, the forces tend to be periodic at the operating rate of the machine, but since they are very abrupt the frequency spectrum contains a much larger range of harmonics. Thus we often concentrate more on the overall shape (or envelope) of the spectrum rather than on the strength of particular lines. We shall first study a rather simple model of impact.

We consider an elastic body of mass m and velocity v_0 that strikes a rigid surface, as shown in Figure 2.14. At time $t = 0$, this body comes in contact with the surface. Although the surface is rigid, the body, because of its elasticity, has a local strain at the contact point that we model by a contact stiffness of value K. To some degree, this may be considered a model for the impact hammer used in vibration testing.

We are interested in the force $l(t)$ applied to the surface by the collision or impact. This force in turn excites the structure and causes it to vibrate and radiate sound. Figure 2.15(a) is a plot of the velocity, which is a constant v_0 for $t < 0$.

Once the spring–mass system is in contact, it will begin to oscillate with a period determined from the resonance frequency:

$$f_{res} = \frac{\omega_{res}}{2\pi} = \frac{1}{2\pi}\sqrt{\frac{K}{m}}. \tag{2.2}$$

Since the resonance period is the reciprocal of the resonance frequency, we have

$$T_{res} = \frac{1}{f_{res}} = 2\pi\sqrt{\frac{m}{K}} \tag{2.3}$$

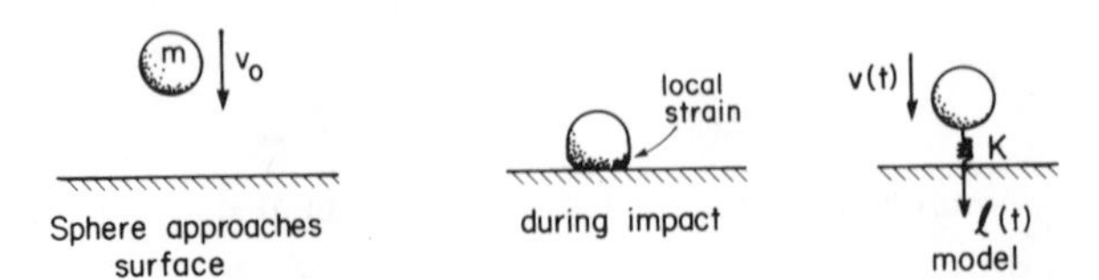

Figure 2.14 Impact of sphere against rigid surface showing derivation of model.

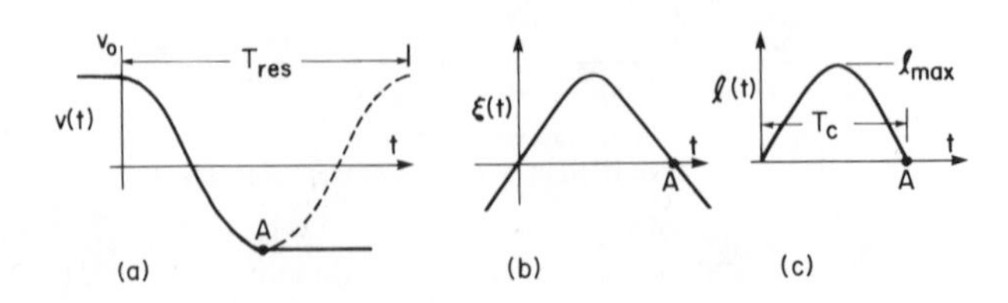

Figure 2.15 Velocity, displacement, and force for body impacting a rigid surface. Body leaves surface at A.

The displacement, which is the integral of the velocity, will be sinusoidal during the half-period in which the body stays in contact with the surface, as shown in Figure 2.15(b).

Since the displacement multiplied by the contact stiffness equals applied force, we obtain the force–time curve in Figure 2.15(c). The body stays in contact with the surface for half the resonance period, which we may call the contact time, T_c, given by

$$T_c = \frac{1}{2} T_{res} = \pi \sqrt{\frac{m}{K}} \tag{2.4}$$

Equation 2.4 shows that a body of stiffer material will stay in contact with the surface for a shorter time. During this contact period, the relationship between the various quantities involved is

$$\begin{aligned} v &= v_0 \cos \omega_{res} t, \qquad \xi = \frac{v_0}{\omega_{res}} \sin \omega_{res} t, \\ \xi_{max} &= \frac{v_0}{2\pi f_{res}}, \qquad l_{max} = K\xi_{max} = v_0\sqrt{Km}. \end{aligned} \tag{2.5}$$

This last relation shows the effect of body stiffness and mass on peak force. The spectrum of energy produced by the fluctuating force pulse, shown in Figure 2.15(c), is determined from the Fourier transform (see Appendix A):

$$\begin{aligned} \mathscr{L}(\omega) &= \int_0^{T_c} l_{max} \sin\left(\frac{\pi t}{T_c}\right) e^{j\omega t}\, dt = \frac{4 l_{max} T_c}{j\pi(\Omega^2 - 1)} e^{j\Omega/2} \cos\frac{\pi\Omega}{2} \\ &= -\frac{2mv_0}{1 - \Omega^2} e^{j\Omega/2} \cos\frac{\pi\Omega}{2}. \end{aligned} \tag{2.6}$$

where $\Omega = \omega/\omega_{res}$. Note that $\mathscr{L}(0) = \int l(t)\, dt = 2mv_0$ is the momentum or impulse imported by the body to the surface. Expressing the energy spectrum of the force pulse in decibels versus log frequency, we obtain the graph in Figure 2.16.

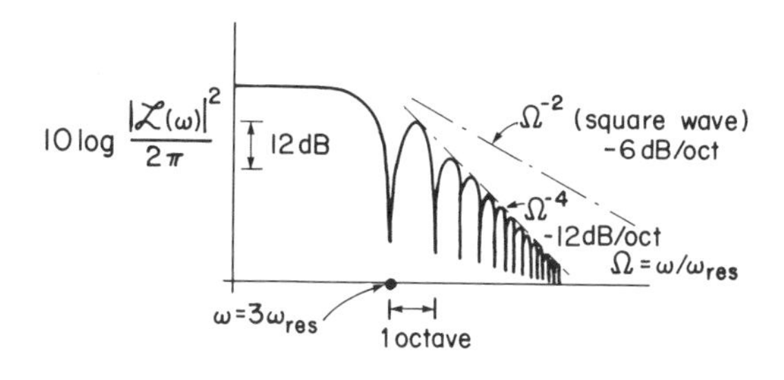

Figure 2.16 Energy spectrum of impact of sphere on a rigid surface. Parameters are defined in Figure 2.15.

We note from Figure 2.16 that at high frequencies the energy spectrum will decay at a rate of $1/\omega^4$ or -12 dB/octave. We can contrast this with the square-wave spectrum discussed in Appendix A, which varies as $1/\omega^2$ or -6 dB/octave. The reason for the higher rate of dropoff at higher frequencies in this case is that $l(t)$ starts from zero and increases linearly with time. This produces less high-frequency energy than does a wave that has a finite step in the force at $t = 0$. The general rule is that if the force onset is very abrupt, then the high-frequency behavior will drop off at 6 dB/octave, whereas if the force is applied in a linearly increasing fashion, the high-frequency spectral character will be -12 dB/octave.

The preceding discussion assumed that the impacting body hit a rigid surface, but often the surface is flexible, and this affects the force–time input to the structure. Examples might be steel parts hitting a thin metal chute used for guidance, or plastic pellets hitting cardboard sheets during the filling of boxes in packaging. We can model this situation as in Figure 2.17, where the particle is represented by its mass and the sheet of thickness h by its surface mass density, longitudinal wave speed, and thickness, or, more fundamentally, the radius of gyration of its cross section. A large, thin, flat structure of this type has a drive point mobility, which is a ratio of velocity to force given by (we derive this relation in Chapter 3)

$$Y_{\text{dp}} = \frac{\text{vel}}{\text{force}} = \frac{1}{8\rho_s \kappa c_l}. \tag{2.7}$$

Equation 2.7 is the mobility of a pure dashpot containing no stiffness or mass reactance component. The loss in the dashpot represents the energy carried away from the drive point by bending waves in the structure. This energy is eventually transmitted to other structures or dissipated within the sheet.

We model the system as in Figure 2.18. The mass with initial velocity v_0 comes in contact with the dashpot having a mechanical resistance R, the reciprocal of the structural point drive mobility. Analysis of this model results in the velocity versus time shown in Figure 2.19. Since the force in the dashpot is the resistance times this velocity, the force–time curve has the same waveform as the velocity does for $t > 0$, but it has a different waveform from that of the mass impacting the rigid surface. We now seek its spectrum, so we first calculate the Fourier transform of the force:

$$\mathscr{L}(\omega) = \int_0^\infty R v_0 e^{-Rt/m} \cdot e^{-j\omega t}\, dt = \frac{R v_0}{R/m + j\omega} \xrightarrow[\omega \to 0]{} m v_0. \tag{2.8}$$

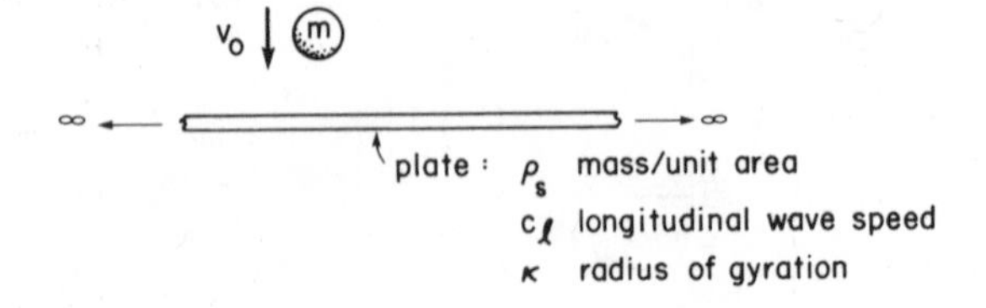

Figure 2.17 Impact of sphere against thin elastic sheet. Dynamics of the plate are discussed in Chapter 3.

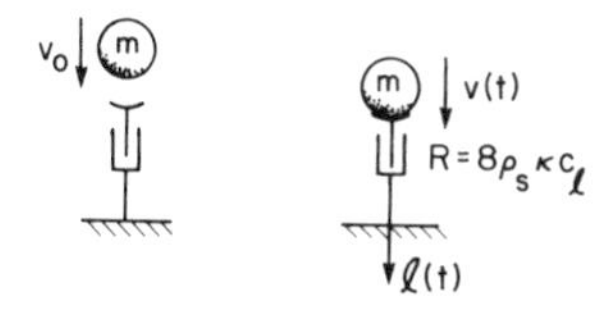

Figure 2.18 Mass-dashpot model for sphere impacting a large thin elastic sheet. Dissipation in the dashpot represents the vibrational wave energy carried away by the sheet.

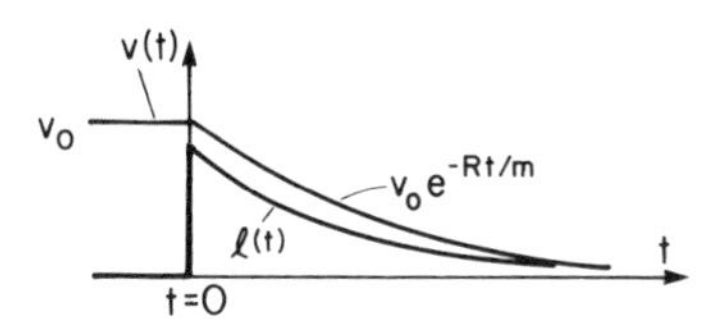

Figure 2.19 Velocity of the mass during impact with a thin elastic sheet. The force is the velocity v times the resistance R for $t > 0$ only.

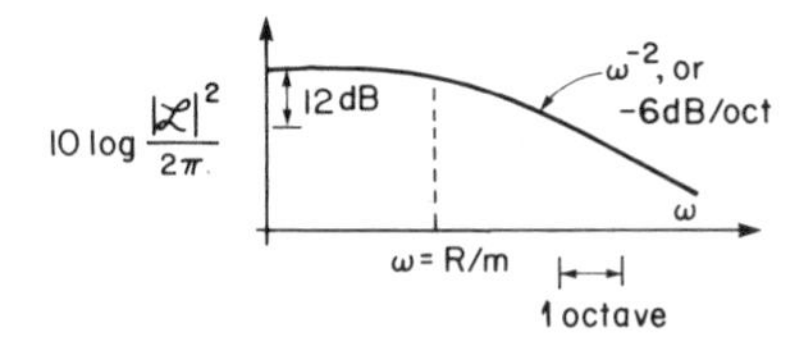

Figure 2.20 Energy spectrum of force produced by sphere impacting on a thin elastic sheet. Note difference in high-frequency spectrum between this curve and Figure 2.16.

Equation 2.8 again confirms that $\mathscr{L}(0)$ is the total impulse imparted to the structure.

We can now obtain the energy spectrum

$$E(\omega) = \frac{1}{2\pi}|\mathscr{L}|^2 = \frac{1}{2\pi}\frac{(Rv_0)^2}{\omega^2 + (R/m)^2}, \tag{2.9}$$

which is graphed in Figure 2.20. It has a constant value at low frequencies. At the (radian) frequency R/m, the characteristic frequency of the system of Figure 2.18, the force spectrum begins to roll off at 6 dB/octave. We have noted before that the 6-dB/octave dropoff is related to a finite step of force at time $t = 0$, in contrast to the linear increase of force produced when the sphere hits a rigid surface.

We have already noted that at $\omega = 0$ the value of $\mathscr{L}$ is equal to the integral of force over the time interval, which is the impulse or momentum supplied by the impacting object to the surface. For an elastic rebound, the initial momentum is $+mv_0$ and the final momentum is $-mv_0$, leading to a total momentum change of $2mv_0$. On the other hand, the ball striking the thin sheet has an initial momentum

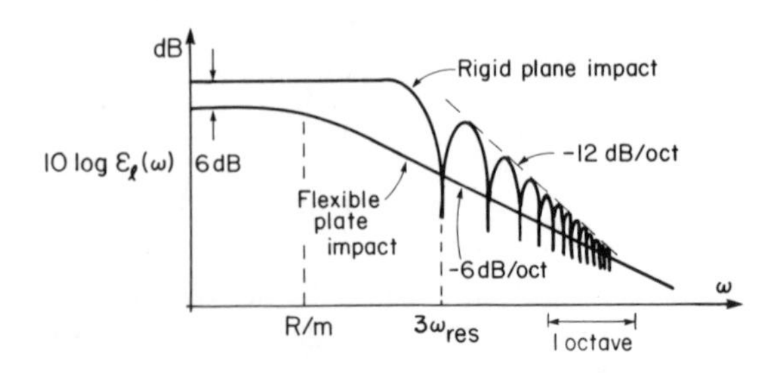

Figure 2.21 Comparison of force–energy spectrum, assuming that the characteristic frequency ω_{res} is greater than R/m.

mv_0 and a final momentum of 0, because the ball has stopped. Consequently, the Fourier transform amplitude in the second case will be one half that of the elastic rebound, giving a difference of 6 dB is the low-frequency energy. The ball striking an elastic sheet drops off at 6 dB/octave at high frequencies, compared to 12 dB/octave for the impactor against a rigid surface. The comparison is shown in Figure 2.21.

In summary then, the low-frequency behavior is dominated by the momentum given by the impactor to the surface, and the high-frequency shape of the spectrum is determined by the rate of onset of force: -6 dB/octave for a step in force, -12 dB/octave for a linear increase in force. If the force onset in the impactor is quadratic (square law), then the high-frequency behavior is -18 dB/octave. The frequency at which we enter the transition from the low-frequency to this high-frequency behavior is determined by the time constant of the interaction, which we can usually judge from the impacting waveform or from the parameters of the system. Thus, we can often make very good estimates of the excitation spectrum by simply observing the impacting mechanisms.

2.5 MECHANICAL DIAGRAMS

In electrical circuits with inductors, capacitors, and resistors, an electrical element has two roles: it is a picture of the physical component and may be used to construct the circuit; and it also represents a mathematical relationship between the flow variable (current) and the drop variable (voltage), as shown in Figure 2.22. For example, the inductor symbol represents a derivative relationship between the flow and the drop, the capacitor element an integral relationship between these quantities, and the resistance element a multiplicative relationship.

There are similar relationships between the force and velocity for various mechanical elements. For example, a mass has a relationship $l = m\,dv/dt$ (see the figure) and may be represented by the inductance element if the velocity is the flow and the force is the drop. On the other hand, a simple spring represents an integral relationship between the velocity and force, and a mechanical damper a multiplicative relationship.

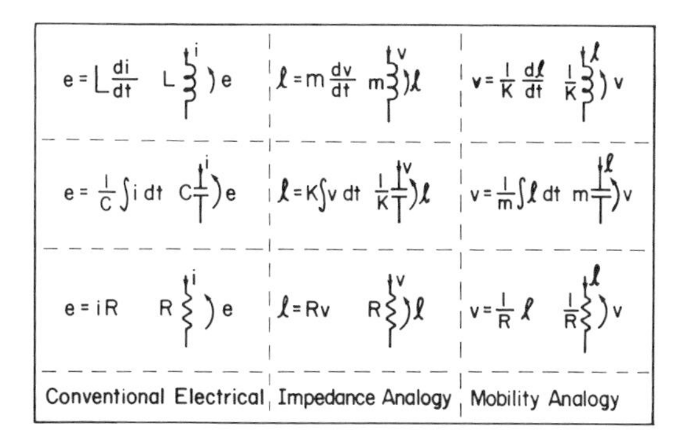

Figure 2.22 Diagram elements used in electrical and mechanical systems to represent connectivity and dynamics.

This association between flow and drop and the element types is not unique. If we reverse the roles of force and velocity and treat force as flow and velocity as drop, we can generate the relationships shown in the third column of Figure 2.22. Then the inductance element becomes a spring, the capacitance element becomes a mass, and the resistance element becomes mechanical conductance.

As noted, the electrical diagram of current as flow and voltage as drop not only meets the mathematical requirements for analyzing the circuit, but it also is topologically (i.e., appears) the same as the actual physical electrical circuit. It turns out that in mechanical systems we get a closer topological relationship between the diagram and the physical device being modeled when the mobility analogy, the rightmost column of the figure, is used. For this reason, the mobility analogy or diagram is more commonly used to represent the dynamics of mechanical systems.

As an example, consider a resilient mass that falls on a thin plate, a slightly more elaborate system than that described earlier. The situation is modeled in Figure 2.23(a), in which the impactor of mass m and contact stiffness K is attached to

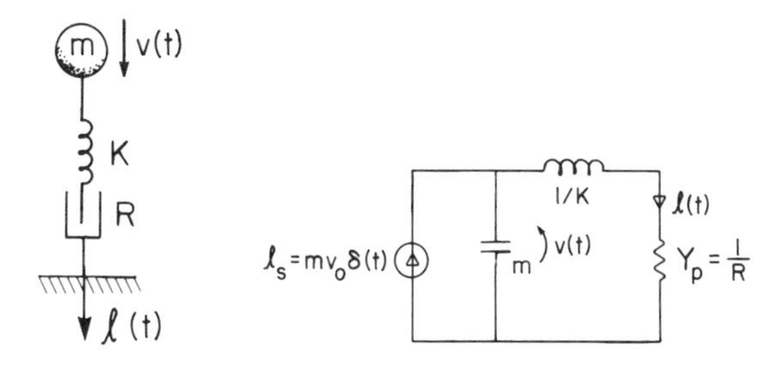

Figure 2.23 Dynamical model of sphere impacting a flexible sheet, including contact stiffness of sphere, and equivalent mobility analogy diagram showing the force $l(t)$ transmitted to the sheet.

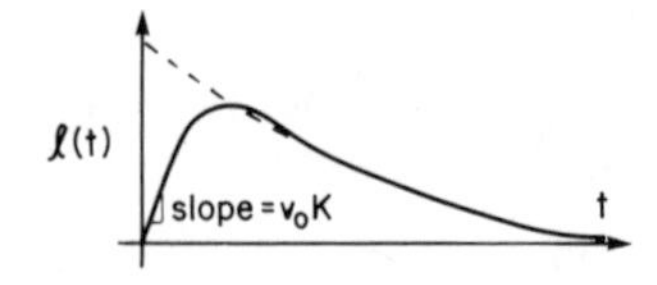

Figure 2.24 Force–time waveform for compliant sphere impacting elastic sheet. Compare with Figure 2.19 and note finite initial rate of rise due to compliance of the impactor.

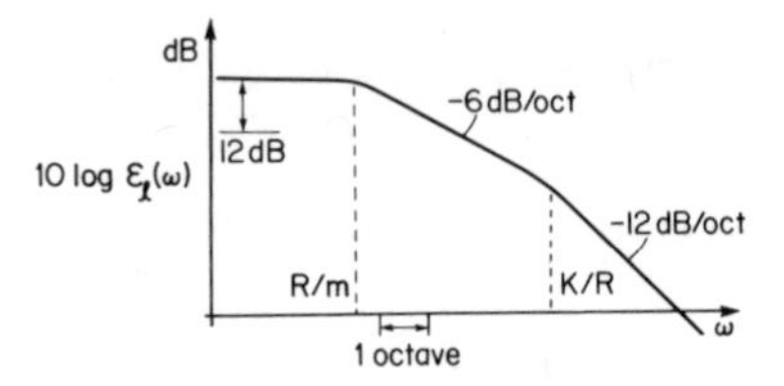

Figure 2.25 Force spectrum of compliant sphere striking an elastic sheet.

damper R, representing the thin-plate drive point impedance. The initial velocity v_0 is attained by allowing a source excitation force to be an impulse that acts at $t = 0$, where the impulse has initial momentum mv_0. Since neither the stiffness nor the damper have mass, the force flowing through them must be the same and is, therefore, the force applied by the mass to the thin plate. Consequently, in a mobility diagram the stiffness and damper are in series, as shown in Figure 2.23(b).

The initial impulse is due to a source of prescribed flow l_s that gives momentum mv_0 into mass m. This gives the mass an initial velocity v_0, just as an impulsive source of current would apply an initial charge to a capacitance. The temporal response to the impulse excitation l_s in Figure 2.23(b) is sketched in Figure 2.24, indicating an initial slope of force versus time due to the impeding effect of the spring or inductance-type element and then a gradual decay of the force due to the relaxation of momentum stored in the mass element.

The spectrum of this force excitation in Figure 2.25 has a finite value at zero frequency proportional to the initial momentum, as discussed earlier—a region of −6-dB/octave drop-off due to the rather abrupt increase in force, but then at very high frequencies, a −12-dB/octave drop-off, indicating that there is a finite slope of the initial force curve due to the contact stiffness. The diagram, therefore, gives us a way to use the techniques or circuit analysis or, indeed, to make an electrical circuit model of the system for display or computation of the applied force and its spectrum.

2.6 PISTON SLAP

Piston slap is a form of impact excitation within diesel engines that results from the clearance between the piston and cylinder wall and the high pressures that push on the top of the piston during combustion. As Figure 2.26 shows, the changing crank

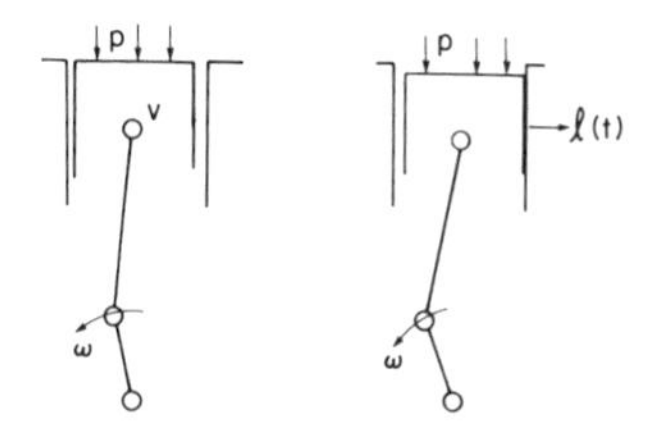

Figure 2.26 Impact of piston against cylinder wall due to cylinder pressure and clearance between piston and wall. The noise resulting from this impact is termed piston slap.

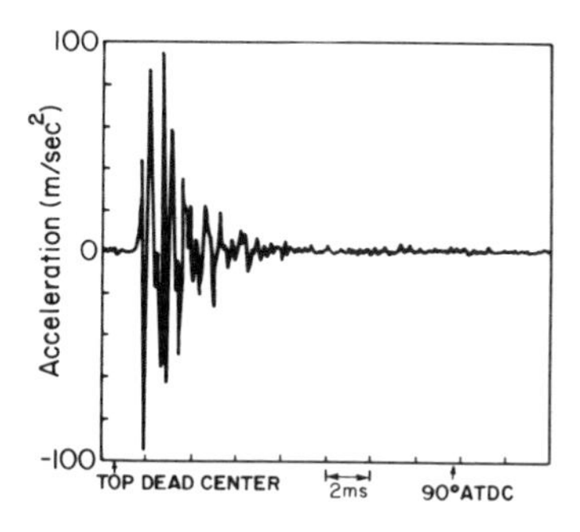

Figure 2.27 Acceleration measured on cylinder wall due to piston slap impact.

angle and these forces result in the acceleration of the piston across the clearance, causing the piston to impact the cylinder wall on the thrust side immediately after TDC.

An experimental rig was developed to illustrate and analyze this effect without the complication of other noise sources in diesel engines. This rig is an engine in which all of the mechanical elements have been removed, except for the components associated with the number 1 cylinder. The engine is motored by an electric motor driving it through a reduction gear. An accelerometer is mounted to the cylinder wall, access to which has been gained by removing a freeze plug.

The vibration of the cylinder wall due to a single piston slap is shown in Figure 2.27. The period of vibration of the cylinder wall due to this impact occurs over about 40° of engine rotation and lasts about 7 msec. As expected, the vibration begins shortly after TDC.

The spectrum of this vibration is shown as the lower curve in Figure 2.28. There are major peaks in the spectrum at about 1.3, 2, and 3.15 kHz. We have seen previously that the spectrum of motion due to an impact is determined both by the impactor and the impacted structure. We need to determine the mechanical behavior of these two parts of the system.

A measurement of the drive point mobility of the cylinder wall, shown in Figure 2.29, indicates that over the frequency range of interest, the cylinder wall is

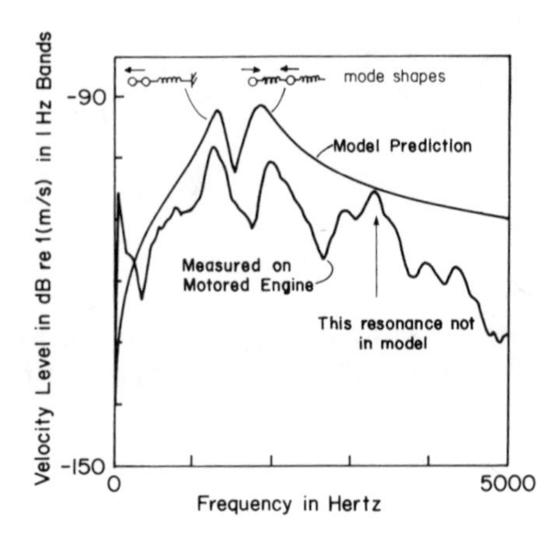

Figure 2.28 Comparison of predicted and measured spectra of cylinder wall velocity. Note difference in number of peaks.

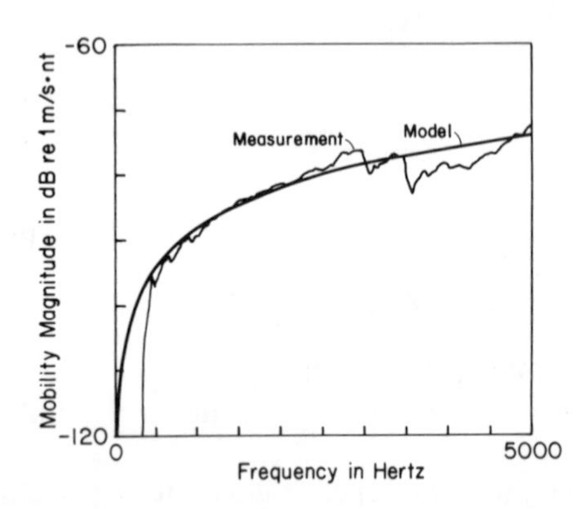

Figure 2.29 Measured drive point mobility of cylinder wall and comparison with mobility of an elastic foundation.

behaving very nearly as a simple elastic foundation. Clearly, the resonant characteristics shown in the preceding figure are not explainable from the behavior of the cylinder wall alone.

The drive point mobility of the piston can be measured if the piston remains on the crankshaft and is dropped below the crankcase as shown in Figure 2.30. An impedance head used to measure the drive point mobility is placed against the side of the piston skirt, and the data measured are also given in the figure. The data show a free resonance at about 1.7 kHz and antiresonances at about 1.4 and 2.8 kHz.

These results can be interpreted in terms of a mechanical model for the piston–cylinder wall combination, as shown in Figure 2.31. The cylinder wall is a simple

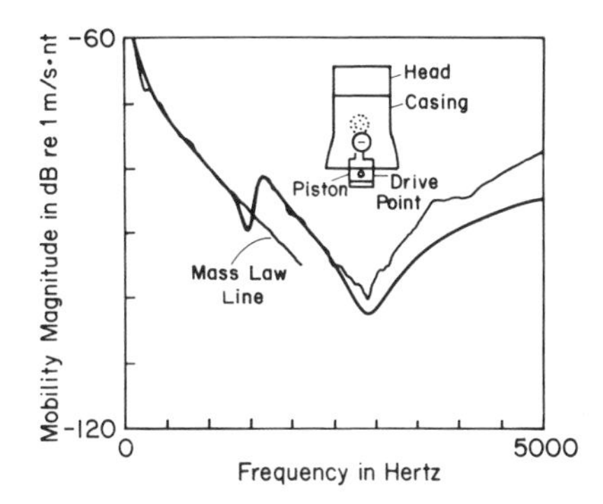

Figure 2.30 Measured magnitude of mobility of piston assembly and comparison with model.

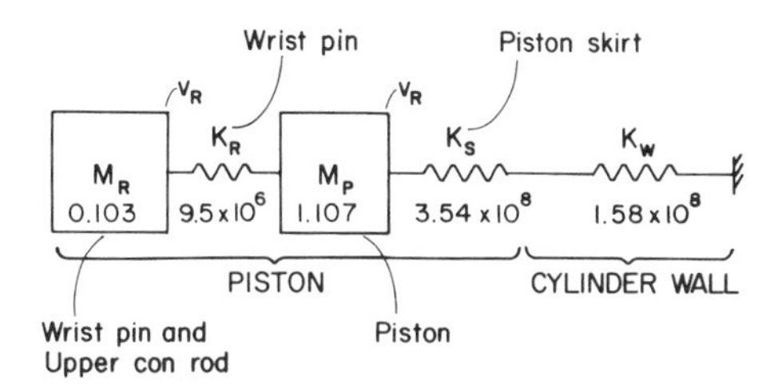

Figure 2.31 Two-degree-of-freedom model for cylinder wall and piston assembly combination.

stiffness, and the piston is a combination of two masses and two stiffnesses. One mass is that of the wrist pin and the upper part of the connecting rod, M_r, and another is the mass of the piston, M_p. The stiffness K_r is that of the wrist pin connecting the piston to the upper end of the connecting rod, and the stiffness K_s is that of the skirt of the piston, which is in contact with the cylinder wall. Values for these parameters are shown in the figure in SI units.

An analysis of this system, if we assume an initial velocity for the piston resulting from the gas pressure in the cylinder and the available clearance, predicts a vibration spectrum for the cylinder wall shown in the upper curve of Figure 2.28. The two modes of vibration associated with each of the two peaks are also sketched in the figure. Clearly, the vibration levels are somewhat overestimated, probably because any retarding effects due to piston rings or oil film on the translation of the piston across the free gap have been ignored. The mode of vibration at 3.5 kHz is not explained. It is very likely that this vibration is due to a rocking motion of the piston against the cylinder wall, but the analysis presented here cannot confirm this conjecture.

2.7 CRANK–SLIDER VIBRATION

A piston sliding within a cylinder is one example of a crank–slider mechanism, as noted in Section 2.3. Another is the needle bar in a sewing machine, which also has impacts associated with its operation. A simple experiment to study the vibrations produced by crank–slider action is shown in Figure 2.32. The slider is a slot that rides on a fixed pin and is driven from an eccentric on a rotating shaft. A free-body diagram of this system may be drawn as in Figure 2.33. From this diagram, we can determine the force between the slider and the fixed pin, L_{yp}:

$$L_{yp} = (g\cos A) + r\omega^2(\cos A)(\cos B) - r\omega^2(\sin A)(\sin B) - L\frac{d^2B}{dt^2}\frac{\frac{1}{3} + \cos^2 B - \sin^2 B}{P/mL}. \tag{2.10}$$

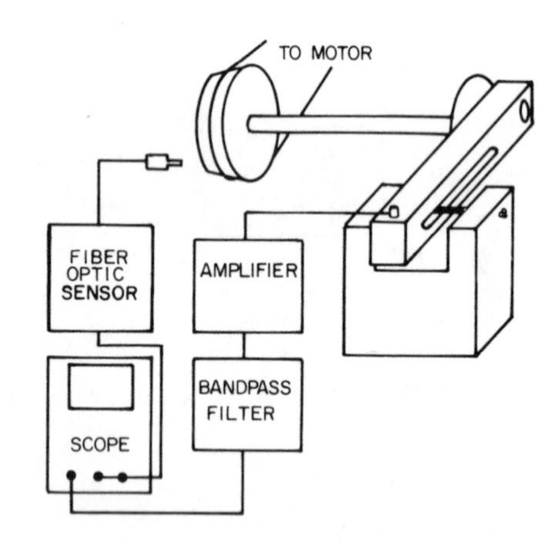

Figure 2.32 Experimental arrangement for getting impact data on a crank–slider mechanism.

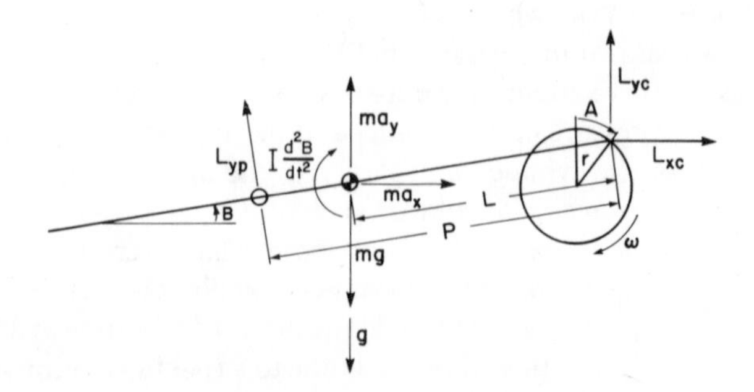

Figure 2.33 Free-body diagram of crank–slider showing force between pin slider and slot.

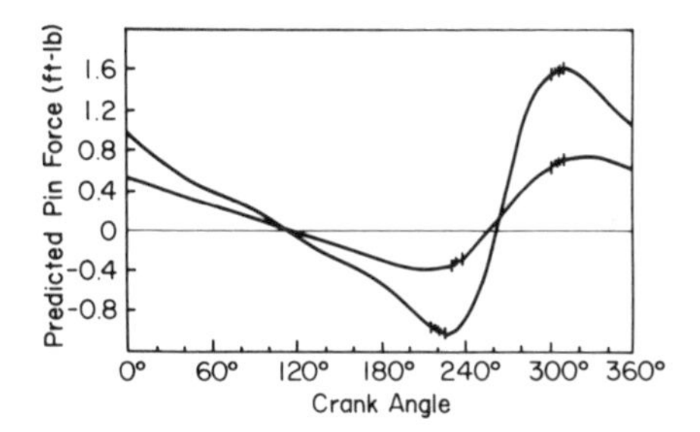

Figure 2.34 Theoretical pin force as a function of crank angle for two pin distances. Vertical bars are placed on the plots corresponding to periods in which impacts occur. Note coincidence of impacts and high pin forces.

This force, computed for two different values of the separation between the fixed pin and the center of rotation of the crank eccentric, is graphed in Figure 2.34. We would normally expect that the impacts or regions of high vibration will occur (as for piston slap) at a place where the force between the pin and the slider vanishes so that the slider can move freely with respect to the pin and cause an impact when it again contacts the pin. We shall indeed see evidence of such impacts in Chapter 6. The data, however, show that the impacts occur during the hatched intervals in Figure 2.34. These occur closer to the maxima in the force between the pin and the slider rather than near the regions of minimum force. This suggests some sort of frictional or slipstick vibration. The pin and slider were not well lubricated, so such a frictional chatter is a reasonable presumption to be drawn regarding the source of this vibration.

2.8 DISPLACEMENT SOURCES: EXCITATION BY DIMENSIONAL AND OTHER ERRORS IN GEARS, CAMS, AND SIMILAR MACHINE COMPONENTS

We shall illustrate the types of sources of machinery noise produced by errors in parts that are intended to mate and move together by considering the case of meshing gears. Consider the meshing of two gears of radii a_1 and a_2, as shown in Figure 2.35. The driving wheel is assumed to rotate at constant angular velocity Ω_1, and the driven wheel has an average angular velocity Ω_2 and an angular displacement

$$\theta_2 = \Omega_2 t + \delta\theta, \tag{2.11}$$

where $\Omega_2 = a_1\Omega_1/a_2$. The transmission error is $\delta\theta$. That is, if the output gear were to rotate at an absolutely constant angular velocity, $\delta\theta$ would vanish. Since $\delta\theta$ is time dependent, we consider a single frequency component of transmission error $\Theta e^{j\omega t}$, which produces a mesh force component. The interaction force due to this error will

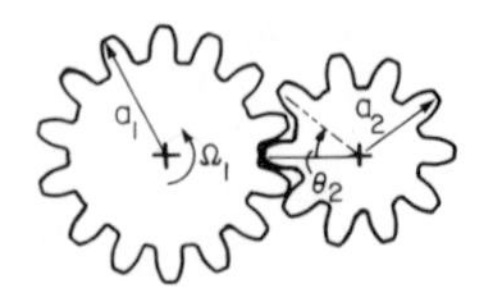

Figure 2.35 Gears in mesh with gear 1 at constant angular velocity. Transmission error produces fluctuations in θ_2.

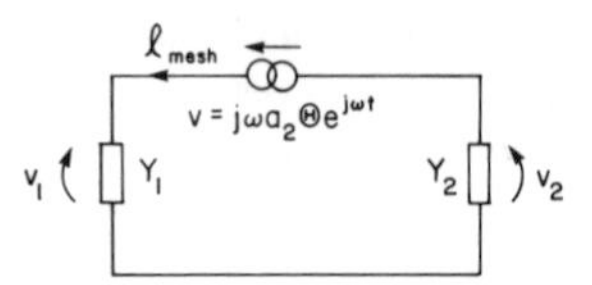

Figure 2.36 Mobility diagram for response of gear mesh due to transmission error excitation of gear mobilities Y_1 and Y_2.

be equal on the two gears, and their lineal displacements will differ by the amount $a_2\Theta e^{j\omega t}$. From the mobility diagram for this system (Figure 2.36), we see that the mesh force resulting from this error is

$$l = \frac{v}{Y_1 + Y_2} = \frac{j\omega a_2\Theta e^{j\omega t}}{Y_1 + Y_2}. \tag{2.12}$$

The input mobilities to the two gears may be very complicated, particularly at higher frequencies. It is frequently useful, however, to consider them as simple compliances, because the local deformation of the teeth is often large compared to the motion induced into the gear wheel itself.

For most systems the error may be expressed as a time function over the operating cycle of the machine. The actual values of the transmission error depend on such things as tooth deflections, inaccurate grinding of profiles, and variation of the pitch between gear teeth and the lead, which is the imaginary contact line between gear pairs. This line is parallel to the axis of the gear for a spur gear, but is a helical spiral for a helical gear.

An example of a double reduction gear set used in ship propulsion is shown in Figure 2.37. This is a double helical, two-stage reduction gear set. The input driving turbine's shaft speed is about 3600 rpm and the output shaft speed is 120 rpm. The intermediate shaft is at about 720 rpm. Figure 2.38 shows meshing gears in more detail. The interaction force between the gears l_{mesh} is inclined but perpendicular to the helical face. A lead error occurs when the contact line between the meshing gears departs from a true helical spiral along the interaction region.

It is well known that if the contact between the meshing gear teeth is a pure involute, then there is no transmission error; but if the tooth deflects under load, or if

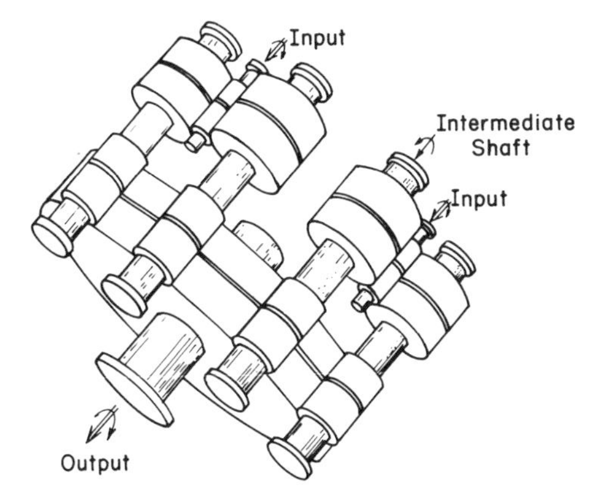

Figure 2.37 Double reduction and double input locked train gear. Each mesh is double helical with high contact ratio.

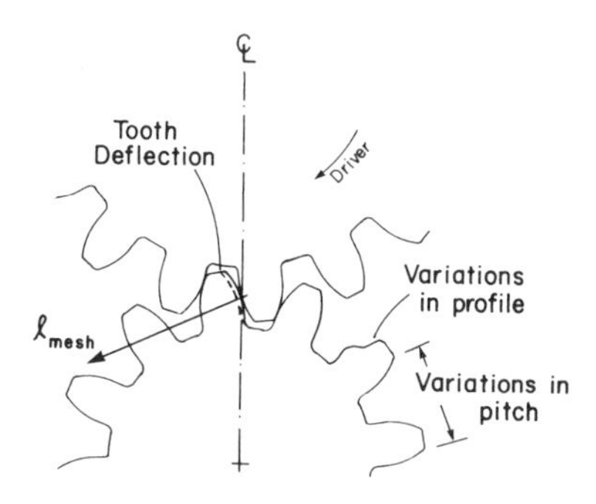

Figure 2.38 A pair of meshing gears showing important parameters. A true involute under load will have no transmission error, but the gear will then have a different shape at other loads.

there is local contact distortion due to Hertzian stresses, then profile errors (i.e., departures from a true involute shape) will occur. In addition, if gear teeth are not manufactured consistently, there will be variations in these quantities as meshing proceeds from tooth to tooth.

All of these types of errors are diminished if several teeth are in contact simultaneously, because this produces an averaging effect. The number of teeth in the circumferential direction in contact simultaneously is called the *circumferential contact ratio.* If the gear is helical, there will also be a number of teeth in contact in the axial direction, called the *axial contact ratio.* The total contact ratio is the product of the circumferential and axial contact ratios. Contact ratios are typically from 1 to 2 for spur cut gears and 3 to 10 for helical gears.

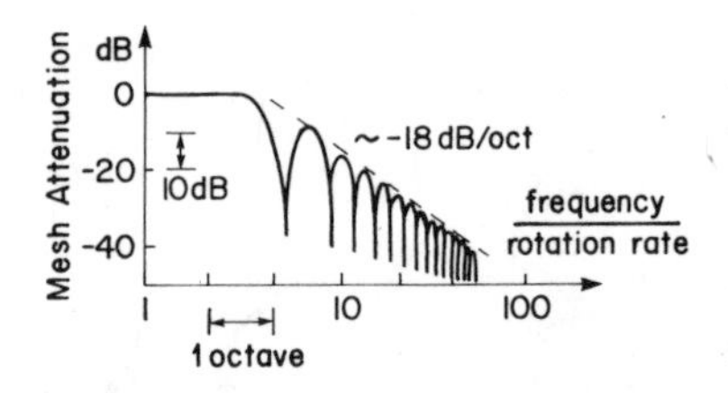

Figure 2.39 Sample of mesh attenuation function for spacing errors between teeth. Exact form of mesh attenuation depends on type of error and the contact ratio.

The reduction of transmission error due to the individual tooth error by the contact ratio is called the *mesh attenuation function*, and its general form is sketched in Figure 2.39. In gears with a large contact ratio, this attenuation may range from 6 to 12 dB. Typical values of transmission error range from 1 to 50 μ, depending on the quality of the gears that are cut. In the later discussion on diagnostics, we will show how we can use vibration data to infer transmission errors in gears. The direct measurement of such transmission errors is very complicated: bending of the gear tooth may be an important contributor to transmission error.

2.9 NOISE GENERATION BY FLUCTUATING MAGNETIC FORCES

Fluctuating forces in motors and generators can occur because of the variation in magnetic flux as the armature rotates. This is illustrated in Figure 2.40. For example, at one time there may be three slots in the magnetic circuit gap, but later there may be four. This produces a fluctuating attraction between the armature and the pole pieces. Since the pole pieces are attached to some kind of support structure, these fluctuating forces will make the motor vibrate. One way to reduce these forces is to skew the armature so that one segment slides into the gap as another one is leaving, much like trying to reduce the transmission error in a gear by making the gear face helical and increasing the contact ratio. As Figure 2.41 shows, the force fluctuation

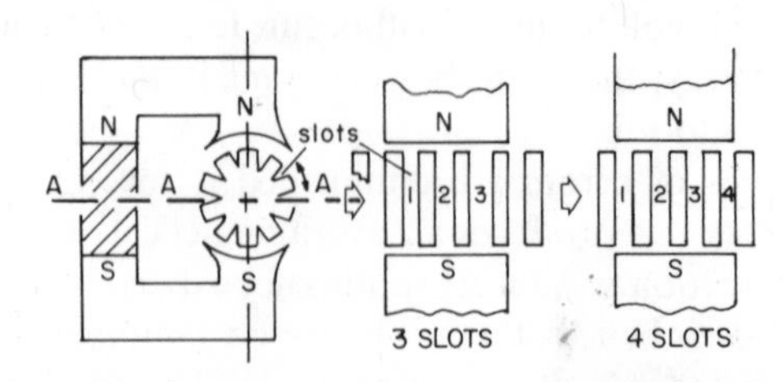

Figure 2.40 Variation of the number of poles in the magnetic gap of a motor as it rotates, which generates a varying flux and force on the magnetic structure.

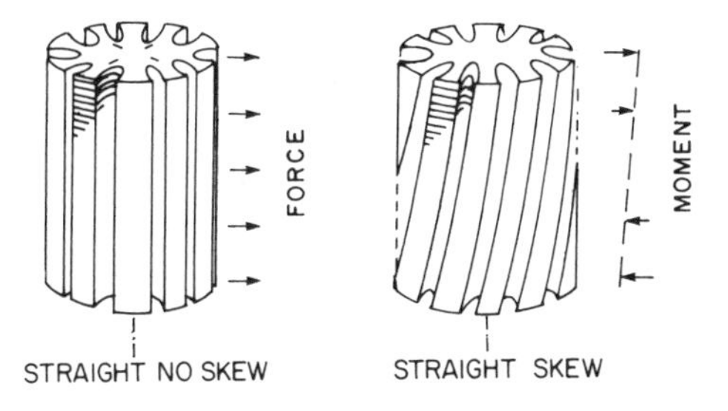

Figure 2.41 Straight and skewed armatures. Skewing reduces the fluctuating force on the armature but may increase moments.

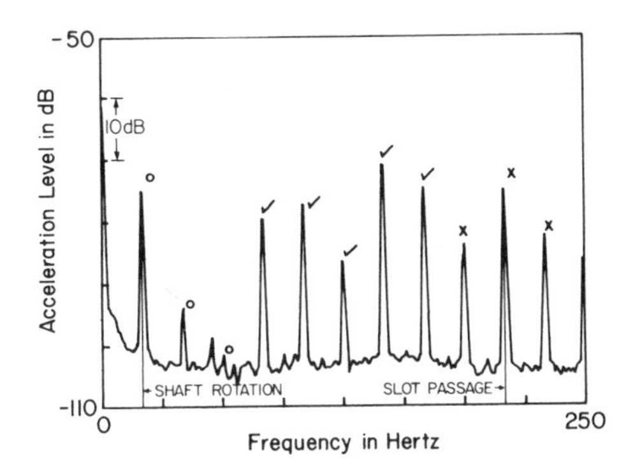

Figure 2.42 Acceleration spectrum of externally driven motor with deliberate imbalance and misalignment showing lines due to imbalance (○), misalignment and looseness (√), and magnetic forces (×).

in a straight, unskewed armature is uniform along the axis of the rotor, but skewing the armature produces a moment perpendicular to the armature axis. Depending on the structure supporting the armature, the moment might actually lead to higher vibration levels than the force does.

Figure 2.42 shows the spectrum of vibration of the motor case of an externally driven rig containing magnets and an armature. Deliberate misalignment of the bearings and imbalance in the armature were introduced. The three ○ lines are the vibrations due to imbalance. There is a rapid drop-off of amplitude with frequency for imbalance vibrations, as noted in Section 2.2. The lines indicated with √ are due to misalignment and looseness and tend to occur at a higher frequency range than those due to imbalance. The three × lines are due to magnetic fluctuations and tend to cluster around the slot frequency (rpm times the number of armature slots).

In the motor described here, the vibration at the slot frequency, or 10 times shaft speed, coincided with a mode of vibration of the motor case assembly having the shape in Figure 2.43. The solid lines are one extreme of vibration, and the dashed

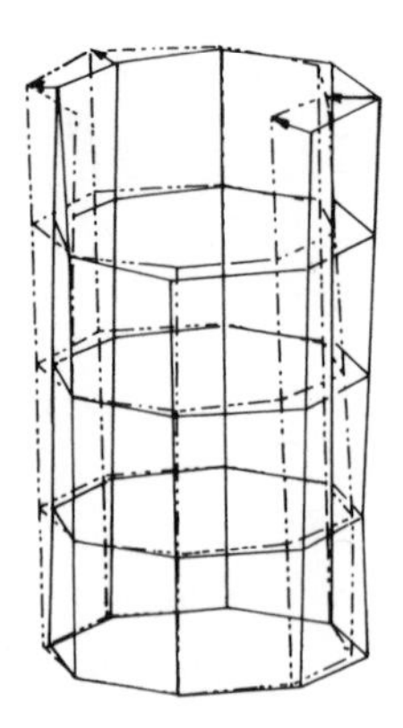

Figure 2.43 Mode shape for motor housing determined by modal analysis methods described in Chapter 3.

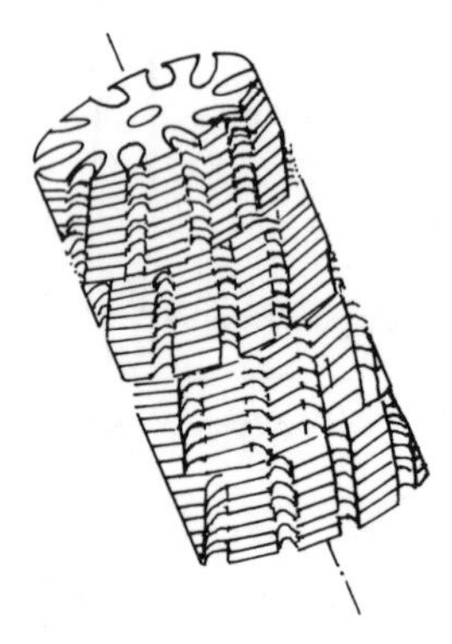

Figure 2.44 Modified slot shape for armature, termed "z-skew," designed to reduce both forces and moments on motor structure.

lines are another for this particular motion. The mode is essentially a cantilever or side-to-side motion of the motor case. Such a mode is very responsive to the moment applied by the magnetic forces; therefore, skewing of the armature would have only a modest effect in reducing vibration levels.

One resolution of this problem is to use a z-skew for the rotor slots, as shown in Figure 2.44. Such a slot shape produces substantial averaging of the forces and reduces the net fluctuating force and moment at the free end of the motor case. Winding of such an armature is done by having a straight slot, with shaped laminations. After the motor is wound, the protruding ends of the laminations are bent down into the armature to form the z-skew. Figure 2.45 shows the reduction in the vibration spectrum at the slot passage frequency in a motor with a z-skew armature. Also notice that in this case the imbalance and misalignment were removed prior to the change from straight skew to z-skew, so the only significant frequency of vibration is at the slot passage frequency.

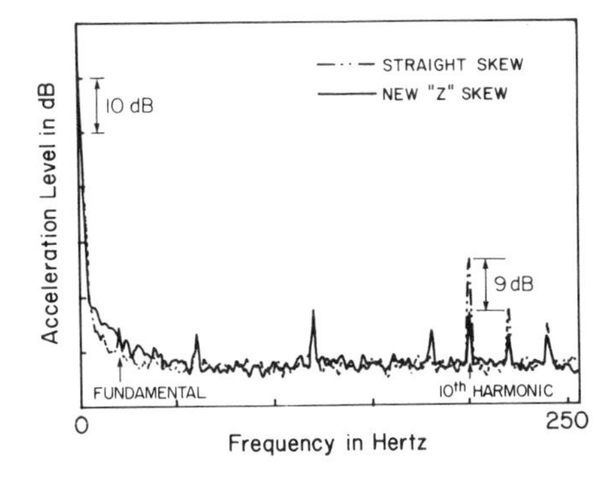

Figure 2.45 Motor case acceleration spectra comparing regular and *z*-skew armatures. Motor is externally driven, well aligned, and balanced.

2.10 DIESEL ENGINE COMBUSTION PRESSURE AS A SOURCE

The pressure in the combustion chamber of a diesel engine has been much studied. This work has mostly been concerned with engine operating performance, but in recent years there has also been interest because of combustion-induced, high-frequency vibration and noise in diesel engines.

A typical pressure–crank angle curve for cylinder pressure in a diesel engine is shown in Figure 2.46. If combustion does not occur, then the pressure rises and falls as shown by the motoring curve. If fuel is injected slightly before TDC, then, after a small delay period, there is rapid combustion during which the pressure rises quickly

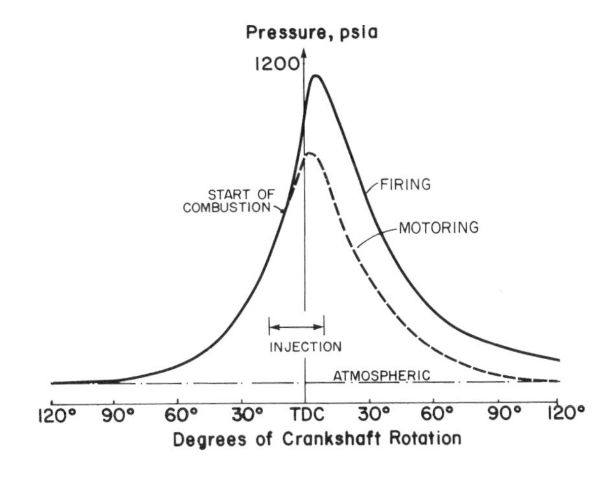

Figure 2.46 Pressure–crank angle diagram for a naturally aspirated diesel engine, compression ratio of 13.5:1, showing rapid pressure rise during combustion. Turbocharged diesels and spark ignition engines do not show such a rapid pressure rise.

and then falls off with a waveform similar to that of motoring but at a higher pressure. The extra area under the firing curve is the work done by the cylinder during the combustion stroke.

The rapid rise in pressure contains the high-frequency components of the excitation force, as we can see from Figure 2.47. This figure shows the dramatic increase in frequency components of excitation above 250 Hz when combustion takes place. Clearly, any information about the combustion process, particularly its onset and rate of rise, is determined by these high-frequency components, which are many times the basic rotation frequency of the engine.

The spectrum for motoring pressure drops off rapidly, about 40 dB per octave, and contains little high-frequency energy. Also shown are cases of the engine firing with no load and operating under full load. Since the shape of the combustion pressure curve is similar to that of the motoring curve except for the brief period of rapid pressure rise, we can infer that the increase of high-frequency energy when combustion takes place above 250 Hz is due to the rapid increase in pressure when the piston is approaching TDC.

The rate of dropoff of signal energy in Figure 2.47 for the octave from 1000 to 2000 Hz is about 14 dB. Recalling our discussion of how the high-frequency behavior of excitation spectra is related to waveform, we know that a step in force will produce a 6-dB/octave dropoff at high frequencies, whereas a linear increase in force will produce a 12-dB/octave dropoff. It is clear that this simple rule does not exactly predict the spectral shape of Figure 2.47. Since the *A*-weighted sound of a diesel engine is dominated by the frequency interval from 500 to 3000 Hz, the noise heard from the casing of a diesel engine is due to this more rapid rise in pressure when combustion occurs.

The data in Figure 2.47 are on a constant-bandwidth basis, but noise data are often expressed in constant percentage bands such as third-octave bands. The center frequencies of these bands and their band limits increase by $\sqrt[3]{2}$ from a lower band to

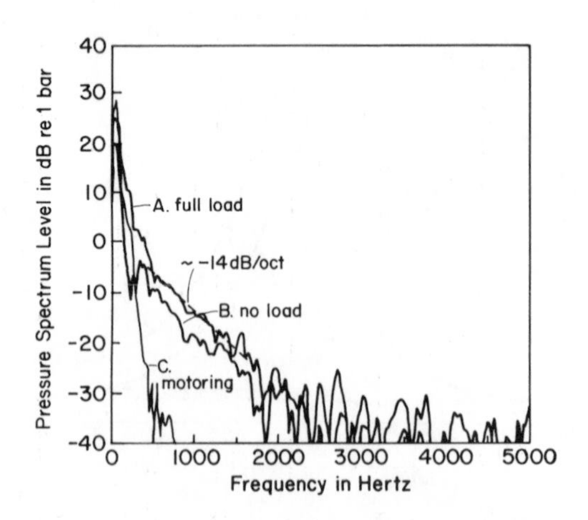

Figure 2.47 Spectrum of cylinder pressure in four-cylinder diesel showing rise at high frequencies due to combustion.

the next higher band. Therefore every three third-octaves is an octave, corresponding to a frequency ratio of 2:1. Every 10 third-octaves is very close to a decade because $(\sqrt[3]{2})^{10}$ is very close to 10. The band center frequencies are adjusted slightly to make this happen.

Since the bandwidth of the constant-percentage bands increases linearly with frequency, then, if a spectrum has a certain dependence on a constant-bandwidth basis, converting to constant-percentage bands will increase the levels by 3 dB/octave. Therefore, a −12-dB/octave dependence is converted to a −9-dB/octave or an f^{-3} dependence for energy or power when presented on a third-octave or octave-band basis. Based on this, the force level for combustion pressure is

$$\begin{aligned} L_{\text{eff}}\ (\text{dB re } 1\ \text{N}) = {} & 20\log(\text{max pressure in N/m}^2) \\ & + 10\log(\text{piston area in m}^2) \\ & + 10\log(\text{number of cylinders}) \\ & + 10\log(\text{rpm}) - 30\log f\ (\text{Hz}) + 90. \end{aligned} \tag{2.13}$$

In deriving this formula, we assume that the excitation is scaled by the maximum pressure, which means that the combustion pressure waveform is essentially independent of the size of the engine or the number of cylinders. The force level is then proportional to the piston area; then assuming that the firing of the cylinder produces incoherent vibrational energy at high frequencies, we add 10 log(number of cylinders). If each cycle is independent, we also add 10 log(rpm). The frequency dependence is $-30\log f$ or −9 dB/octave, which assumes a linear pressure rise at combustion onset. The 90 dB is a scaling factor due to the various constants.

2.11 TURBULENT FLOW AS A SOURCE OF VIBRATION

Machine noise and vibration generated by air or liquid flow occur primarily as a result of impingement of turbulence onto solid surfaces or the shedding of turbulence by obstacles in the flow stream. Although it is possible to have single-frequency oscillations under special flow conditions, the excitation spectrum is generally broadband with a haystack shape like that in Figure 2.48. We must scale

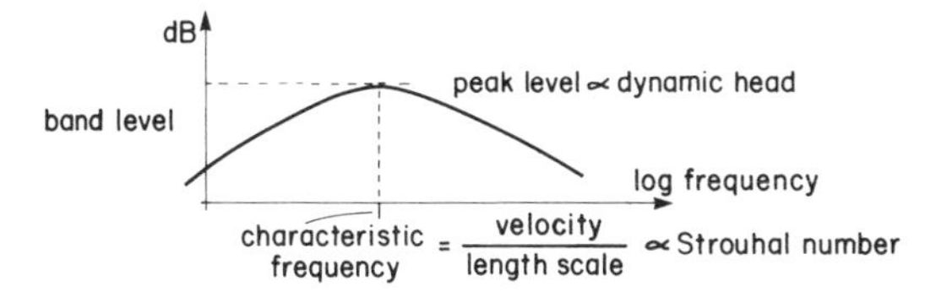

Figure 2.48 Typical broadband aerodynamic noise "haystack" spectrum indicating parameter dependence on flow velocity and length scales.

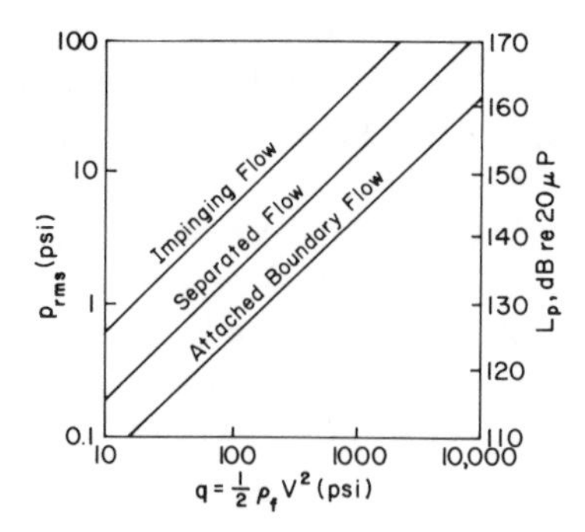

Figure 2.49 Relation between the dynamic head and overall pressure fluctuation for various types of flow.

this curve along the frequency axis and also along the vertical or overall pressure axis. The frequency axis scales according to the Strouhal number, which is the product of the frequency and a typical dimension divided by the velocity: $S = fD/v$. The peak in the spectrum is typically when this parameter equals 0.2. Thus a thicker flow with larger scale turbulence will have a lower-frequency spectrum than will a very thin flow with small scale turbulence, if the typical velocities in the two flows are similar.

The relationship between overall rms pressure and the dynamic head of the flow is shown in Figure 2.49 for an impinging jet flow, a separated flow, and an attached, well-behaved, turbulent boundary layer. Notice that the overall fluctuating pressure is essentially proportional to dynamic head. When the dynamic head increases by a factor of 10, the fluctuating pressure also increases by a factor of 10, or by 20 dB. Also note, however, that for the same dynamic head, there is about a 10-dB increase in pressure in going from an attached, well-behaved, turbulent boundary layer flow to separated flow, and approximately another 10-dB increase in going to an impinging flow. Thus the nature of the flow has an important effect on the overall pressure fluctuations. It also affects the frequency spectrum, because the scale of the flow for these different cases is likely to be different.

Impinging flows are frequently encountered with machines, such as punch presses, in which air jets are used to move parts away from the work area. The impingement of the flow on the part or on the surface of the machine produces fluctuating forces that cause some sound to be directly radiated and some vibration to be produced. In a punch press with an open configuration and made of heavy components, the directly radiated sound tends to be more important than the vibration induced, but in other machines where the flow impingement area may be isolated from the outside environment, the impingement force may be more important as a source of vibration.

We now consider the spectrum of pressure fluctuation produced in these random flows. The third-octave pressure spectrum plotted against Strouhal number is shown in Figure 2.50. Note that the spectrum is 11 dB down from the overall

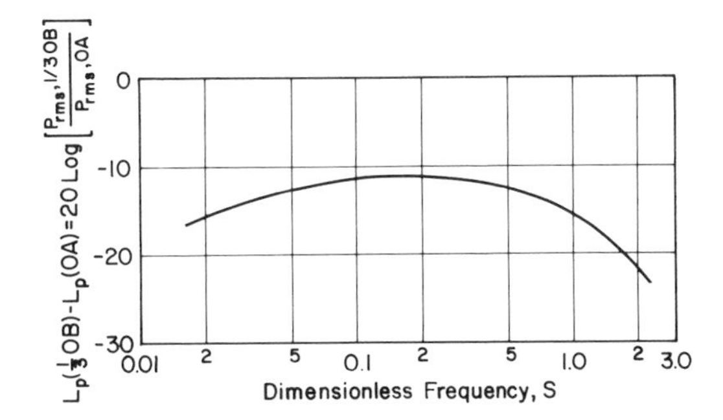

Figure 2.50 Third-octave band spectrum for aerodynamic noise pressures, normalized to overall pressure fluctuation.

pressure fluctuation at $S = 0.2$. Scaling by dimension and velocity will determine the frequency dependence of the turbulent flow, and using the relationship between overall pressure and dynamic head (Figure 2.49) lets us fix the magnitude of the spectrum. Combining these two procedures lets us make a reasonably good first estimate of the pressure spectrum produced by a flow of interest.

An example of this procedure is the situation in Figure 2.51, comparing the pressure fluctuations at the boundaries of a channel flow that in turn generates an impinging flow on a nearby surface. Since the flow has the same dynamic head at the two locations, the spectrum of impingement is scaled vertically upward by 20 dB to account for its difference in overall level, as shown in Figure 2.49. It is scaled down in frequency, however, to account for the larger characteristic dimension of the impinging flow.

If we are concerned with frequencies corresponding to a Strouhal frequency much less than 0.2, then the pressure fluctuations associated with such measurements are due to fairly slow phenomena as far as flow is concerned and are

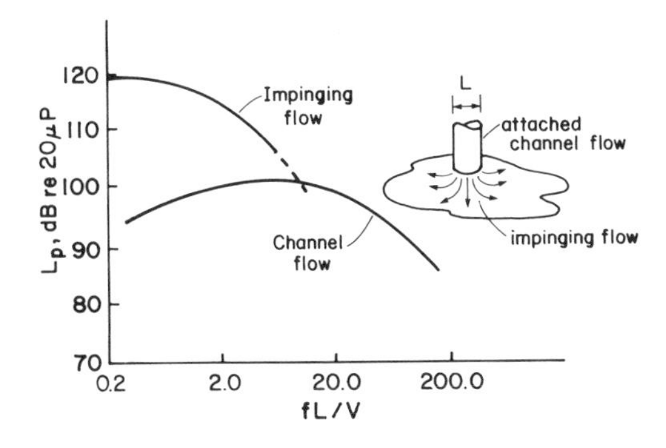

Figure 2.51 Comparison of pressure spectra for turbulent boundary layer and impinging wall jet with same dynamic head.

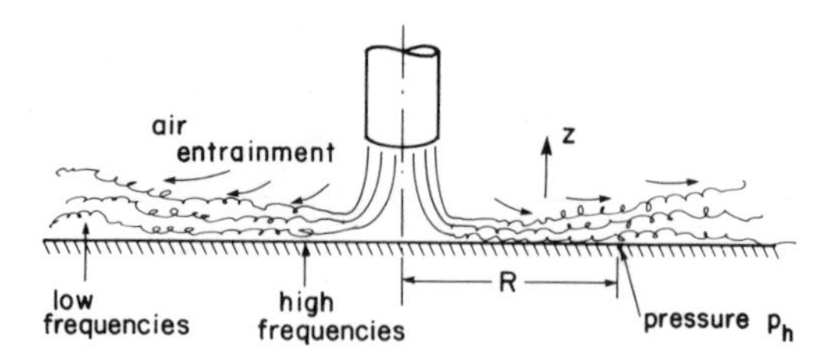

Figure 2.52 Sketch of a wall jet showing regions of high- and low-frequency excitation and the growth in thickness due to entrainment of the surrounding air.

dominated by large-scale motions of the fluid. On the other hand, if we make measurements so that the Strouhal frequency is several times that of the peak at 0.2, then we are dealing with very short time or high-frequency phenomena as far as the flow is concerned.

Another nondimensional number of great importance for flow phenomena is the Reynolds number Re:

$$\mathrm{Re} = \frac{VD}{\nu} = \frac{\text{inertial forces}}{\text{viscous forces}}, \tag{2.14}$$

where ν = (shear viscosity)/(density of fluid). The Reynolds number is a ratio of the inertial forces that operate in the fluid compared to the viscous forces. The flow of honey out of a jar, for example, is dominated by viscous stresses and is a low Reynolds number flow, whereas the flow of air out of the exhaust of a jet engine is dominated by inertial forces and is a high Reynolds number flow.

Let us look at impingement a little more closely. Imagine a jet of air, as shown in Figure 2.52, in which the jet flows from a tube perpendicular to the surface and then turns and flows radially away from the axis of the tube. As the flow spreads out along the plate, it continues to entrain the surrounding air into the flow stream, so the total mass of fluid within the moving region continues to increase. This entrainment causes the thickness of the flow to increase as it moves out radially; if there were no entrainment, the stream would have to get thinner because of the constantly increasing cross section of the flow.

If we measure the velocity profile at any radius, we find that very close to the boundary the flow increases rapidly, much like an ordinary boundary layer. The flow reaches a peak value V_{max} and then trails off gradually as the measurement point continues to move away from the surface, as shown in Figure 2.53. This flow is referred to as a *wall jet*.

Figure 2.53 shows that there are two length scales involved: an effective boundary layer thickness, and a total thickness of the jet. The jet thickness is defined to be $z_{1/2}$, the distance from the wall at which the flow velocity drops off to one-half of its maximum value. Our dimensional scaling parameter is $z_{1/2}$, and the velocity scale is V_{max}.

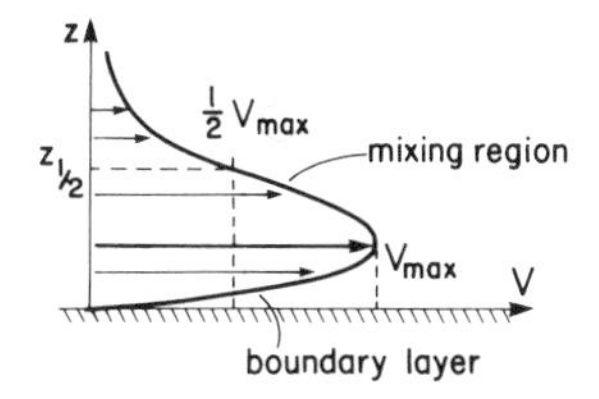

Figure 2.53 Velocity profile for wall jet and definition of the "thickness" of the jet to be $z_{1/2}$.

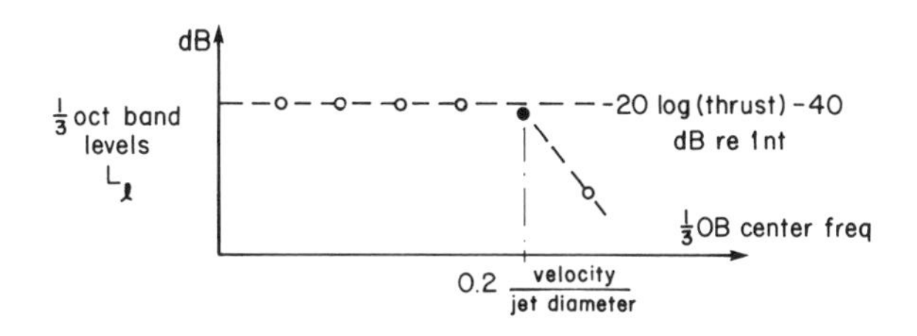

Figure 2.54 Effective wall jet force spectrum in third-octave bands, calculated assuming excitation of bending waves in a thin plate.

Experiments indicate that $z_{1/2}$ grows almost linearly with the radius or distance R from the axis of the jet, and that the peak velocity V_{max} varies as $1/R$. This means that the Strouhal frequency varies as $1/R^2$. Therefore, the high-frequency excitation of the structure is produced close to the jet axis, and the low-frequency excitation is produced farther from the jet axis.

It is possible to carry out a very detailed analysis of the excitation of a structural surface by a wall jet. The analysis results in a very simple expression for the effective rms force in any third-octave band:

$$l_{\text{rms, 1/3 OB}} = 10^{-2}\, l_{\text{thrust}}. \tag{2.15}$$

The total effective rms force in any third-octave band is approximately 1% of the thrust of the jet. This spectrum is frequency independent at low and intermediate frequencies. Within a small area near the jet axis, the flow is fairly smooth and has not developed its fully turbulent characteristics. Therefore, the spectrum will drop off at high frequencies, as sketched in Figure 2.54, at a Strouhal frequency based on the jet diameter at the point of impingement. A force spectrum of this kind is called "pink" noise; on a constant-bandwidth basis, it would drop off as 1/frequency, but on a constant-percentage bandwidth basis it is independent of frequency.

CHAPTER 3

Structural Response to Excitation

3.1 WAVE MOTION IN STRUCTURAL RESPONSE

At very low frequencies, structures can respond as a simple mass if "freely suspended" (see Section 2.3) or as elastic compliance if supported by a rigid base. At the higher frequencies associated with noise-related vibrations, and for some diagnostic signals, the structure supports wave motion. Later we can be more definite about the frequency range where this begins to happen, but for now we want to discuss the kinds of wave motion that can contribute to structural vibration.

All wave motion is a balance between the kinetic and potential energies of vibration. The kinetic energy of a particle of mass Δm with velocity v is $(1/2)\Delta m v^2$. Thus, if we can measure the velocity, we can determine the kinetic energy. The difference between various types of waves is the form in which the potential energy is stored. In compressional waves (an important component of structural response in machines), the potential energy is stored in the longitudinal strain of the material, since the motions are parallel to the direction of wave propagation. For thin structures, the direction of propagation will be parallel to the free surface, as shown in Figure 3.1. The compressional wave is sometimes referred to as a "P" wave in geophysical literature or as a "sound wave" in the material.

In-plane shearing motions, sketched in Figure 3.2, and longitudinal motions are both important in vibration storage and transmission, particularly at higher frequencies. These two wave types have high impedances. Thus such waves are difficult to damp because the damping material used must also have a high

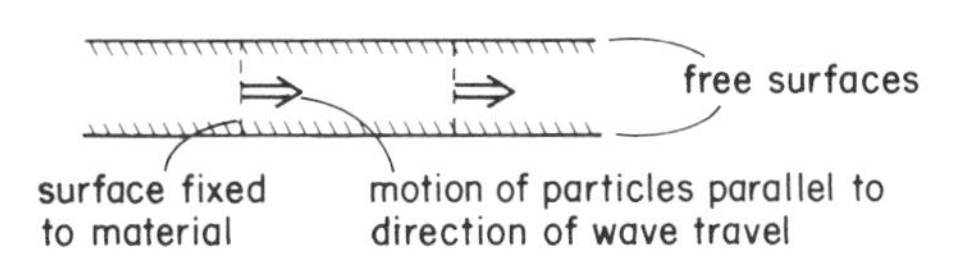

Figure 3.1 Sketch of longitudinal wave disturbance in thin rod or sheet. Motions are parallel to free surfaces.

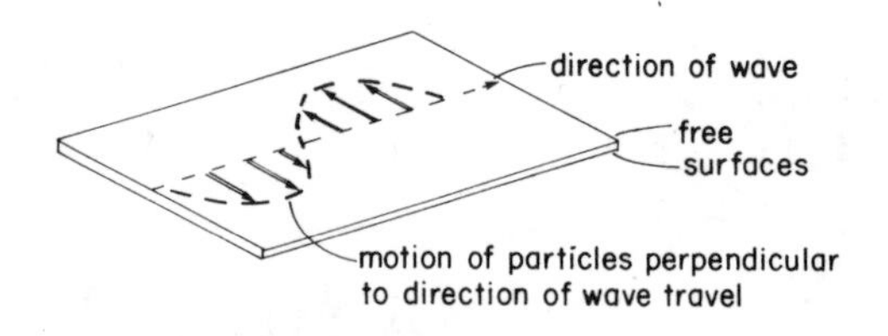

Figure 3.2 In-plane shear wave motion in a transverse wave. These waves, along with longitudinal waves, are the high-impedance degrees of freedom for most structures.

impedance. The impedance level of these waves can sometimes be a benefit because a low-impedance junction will reflect the energy and not allow it to be transmitted. However, it is often not possible to allow such a junction because the junctions must carry high static loads. In such cases, in-plane vibrations become efficient at vibrational energy transmission.

Another important wave type is bending, in which the potential energy is stored by flexure of the structure. See Figure 3.3. This motion is transverse to the direction of propagation, as for in-plane shear. Hence, bending waves are also transverse. Bending or flexural waves are important because they couple very well with the sound field. Most sound radiation from machine structures is due to flexural vibrations.

At high frequencies and for thick structures, another form of wave can occasionally be important—the surface or Rayleigh wave. Rayleigh waves are similar to ocean waves because there is a surface wave disturbance and this disturbance decreases in amplitude beneath the surface, as shown in Figure 3.4. The speed of Rayleigh waves is about 95% of the shear speed, about 3000 m/sec in steel. Thus, for example, at 10 kHz the wavelength is .3 m, so Rayleigh waves would only be expected to occur for relatively thick structures, such as the shafts in marine propulsion units. They are also high-impedance waves, and the previous comments about damping and transmission through high-impedance junctions apply to Rayleigh waves as well as in-plane waves.

As an illustration of the excitation of these various wave types, consider the marine gear in Figure 2.36. If we sketch the force interaction between one of the

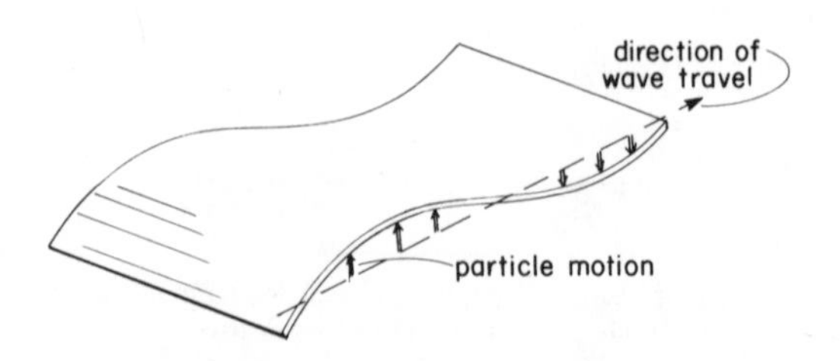

Figure 3.3 The bending or flexural wave is also a transverse wave, but in this case the motion is perpendicular to both the direction of wave travel and the free surfaces.

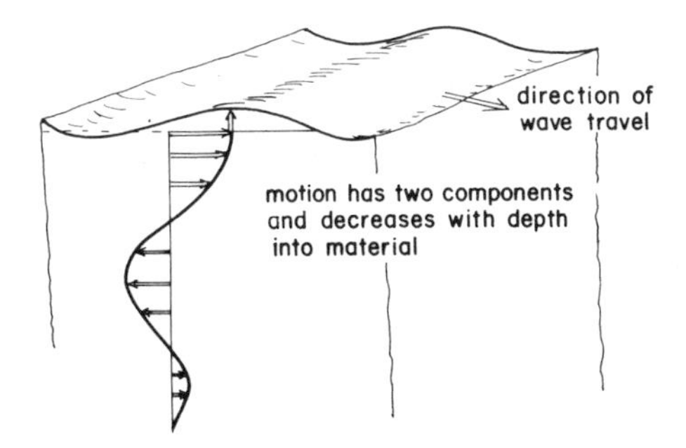

Figure 3.4 Rayleigh or surface wave has a complex pattern of motion and travels at 95% of shear speed. They are usually important only at very high frequencies.

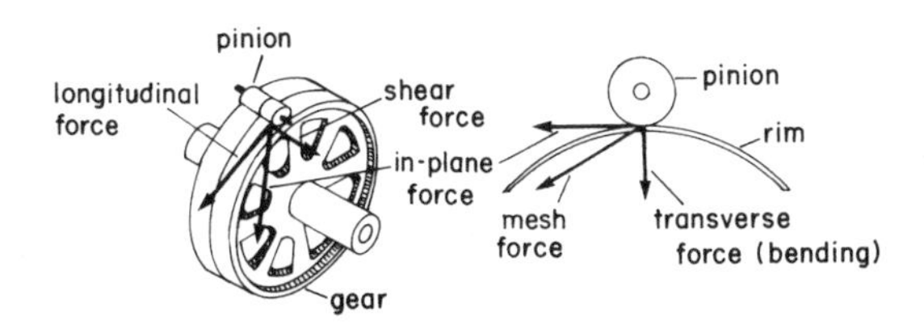

Figure 3.5 Excitation of various wave types in a helical gear mesh.

pinions and the output gear, shown in Figure 3.5, we can see that the intermesh forces tend to generate flexural vibrations, due to the force component perpendicular to the rim of the gear. Because of the helical nature of the gear, both in-plane shear and longitudinal vibrations in the rim of the gear structure are excited. In addition, when the vibration is transmitted through the gear into the large-diameter shaft of the output gear, Rayleigh waves may be generated, because the shaft structure is quite thick.

3.2 LONGITUDINAL WAVE MOTION

We now examine longitudinal waves in more detail. Suppose we have a beam or a plate as sketched in Figure 3.6, and we label surfaces of the material located at x and $x + \Delta x$ as shown. As the structure vibrates, the surface at x will move a distance u_x, and the surface at $x + \Delta x$ will move $u_{x+\Delta x}$. Thus, the strain produced by this vibration is

$$\frac{\{u_{x+\Delta x} + x + \Delta x - (u_x + x)\} - \Delta x}{\Delta x} = \frac{\partial u}{\partial x} \tag{3.1}$$

as $\Delta x \rightarrow 0$.

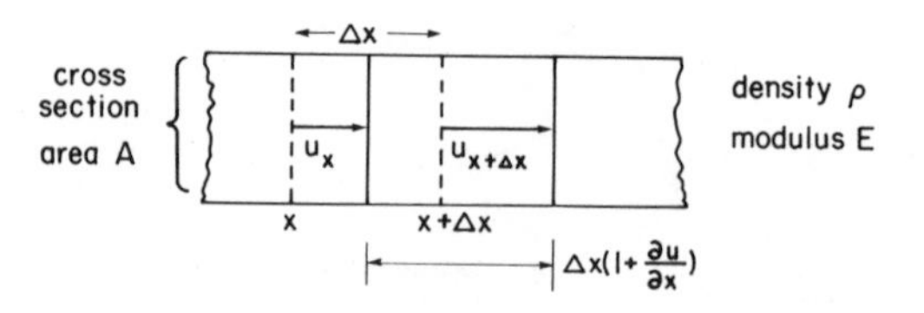

Figure 3.6 Thin rod or plate undergoing longitudinal in-plane vibration. Equation of motion is a balance between elastic restoring force and inertial reaction.

If Young's modulus of the material is E, then the stress is

$$\text{stress} = E \cdot \text{strain} = E\frac{\partial u}{\partial x} = \sigma. \tag{3.2}$$

From this definition, the stress is positive when the strain is positive, which means a pulling or stretching type of stress is assumed positive. The force in the structure is the stress times the cross-sectional area, and the difference in force at the two ends of the element results in a net force that accelerates the element:

$$l_{x+\Delta x} - l_x = \frac{\partial l}{\partial x}\Delta x = AE\frac{\partial^2 u}{\partial x^2}\Delta x = \rho l \Delta x \frac{\partial^2 u}{\partial t^2}, \tag{3.3}$$

or

$$\frac{\partial^2 u}{\partial x^2} - \frac{1}{c^2}\frac{\partial^2 u}{\partial t^2} = 0. \tag{3.4}$$

Equation 3.4 for the displacement is called the wave equation, and $c = \sqrt{E/\rho}$ is the wave speed.

The wave equation is solved by any function of the combination of coordinates $x \pm ct$ or by the sum of such functions. The significance of this can be seen from Figure 3.7. Suppose we plot a solution $f(x - ct)$ as a function of x for time

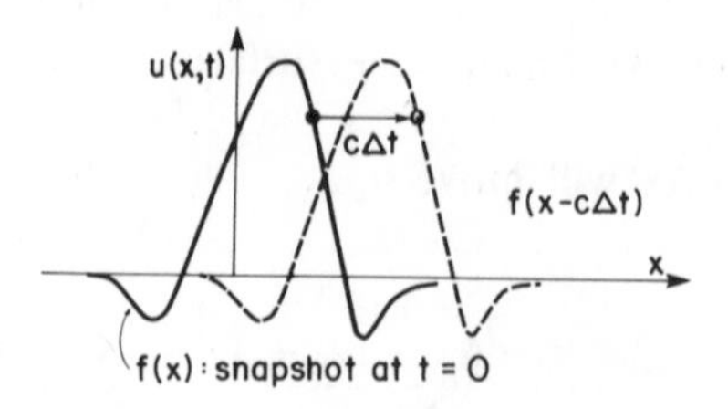

Figure 3.7 Translation of waveform $f(x - ct)$ to the right by $c\Delta t$ in time Δt; interpreted as a wave that travels at speed c.

$t = 0$. Now, we allow the time to increase by an amount Δt. The functional form is now $f(x - c\Delta t)$. A functional form means that if we put the same number into the argument of the function, we should always get the same result. To apply the same reasoning when the argument is $x - c\Delta t$, we must increase x by $c\Delta t$ to get the same value of the function. That is, the functional amplitude is reproduced if we now shift all positions x to the right by $c\Delta t$, as shown in the figure. The original waveform is therefore reproduced at this new spatial location. Clearly, the wave has moved to the right an amount $c\Delta t$ in time Δt, and therefore we infer that c is the speed of the wave, or wave speed. A function $g(x + ct)$ represents in a corresponding way a wave traveling at speed c in the $-x$ direction.

Another important concept is wave impedance. If we have a single wave with displacement $u = f(x - ct)$, then we can calculate the ratio of force to velocity at any position along the wave:

$$\left.\frac{l}{\partial u/\partial t}\right|_{\text{at } x} = \frac{-\sigma A}{\partial u/\partial t} = \frac{AEf'}{cf'} = \rho c A, \tag{3.5}$$

where ρc is called the characteristic impedance of the wave and, when multiplied by the cross-sectional area, gives us the wave impedance. Wave impedance has the same dimensions as mechanical resistance, although for other wave types it may, in general, be complex (i.e., have an imaginary part). For steel, $\rho c = 4 \times 10^7$ (mks); for air it is 4×10^2 (mks). Thus, the range of wave impedance for various wave types and materials can be quite large. The importance of the speed of the various wave types for the same structure (same density) is evident from Equation 3.5 insofar as impedance level is concerned.

Let us now look at the case where the wave encounters a change of impedance due to a change in cross-sectional area of the material. This case is illustrated in Figure 3.8, where we denote the structure to the left of the junction as "1" and the structure to the right of the junction as "2." For the two pieces of material to stay together at the junction, the displacements must be equal: that is, $u_1 = u_2$. If we imagine control surfaces within the material on each side of this junction and move them toward the junction, then as they get very close to the junction the amount of mass enclosed by the surfaces goes to zero, so the net force on the two surfaces must

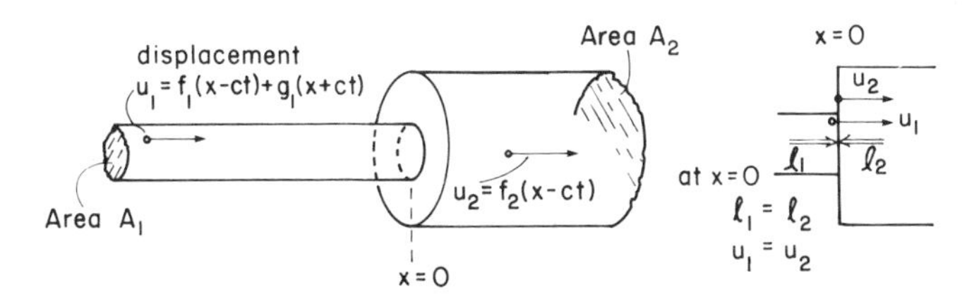

Figure 3.8 Junction between two rods of different cross-sectional area and conditions for longitudinal motion and forces.

be the same, since only mass can absorb force. Therefore, $l_1 = l_2$ at the junction. Thus

$$l_1 = l_2 \rightarrow A_1\sigma_1 = A_2\sigma_2 \rightarrow A_1 E\frac{\partial u_1}{\partial x} = A_2 E\frac{\partial u_2}{\partial x} \tag{3.6}$$

The displacement wave field in system 1 consists of an incident wave f_1 and a reflected wave g_1, whereas the wave on the second side of the junction consists of only a transmitted wave $f_2(x - ct)$ moving in the $+x$ direction. We assume that f_1 is known. Applying the conditions of equality of force and displacement at the junction gives us a pair of simultaneous equations for the reflected and transmitted waves that are readily solved.

$$f_2 = \frac{2f_1}{1 + A_2/A_1} \qquad \text{(wave on second side)} \tag{3.7}$$

$$g_1 = -f_1\frac{1 - A_2/A_1}{1 + A_2/A_1} \qquad \text{(reflected wave).} \tag{3.8}$$

Notice that waveforms f_2 and g_1 have the same shape as the incident wave f_1. Also notice that if there is no discontinuity in area (i.e., if $A_2 = A_1$), then $f_2 = f_1$ and g_1 vanishes, as we would expect. Also if A_2 vanishes (i.e., we have a free end of a beam), then $g_1 = -f_1$ and the wave is reflected with its sign reversed. If $A_2 \gg A_1$, then $g_1 = f_1$, and the stress amplitude is doubled at $x = 0$ in system 1.

In this discussion, we have assumed the same material on both sides of the junction; therefore, the characteristic impedance for the waves is the same. If we had allowed different materials on the two sides of the junctions, then, instead of the ratio A_2/A_1, we would have had $\rho_2 c_2 A_2/\rho_1 c_1 A_1$, a ratio of wave impedances rather than of areas. More generally, it is the impedance ratio that determines the magnitude of the reflected and transmitted waves; but when the materials are the same, the impedance ratio becomes the area ratio.

To see how these results can be used to determine the wave pattern for a finite structure, consider beams of different diameters connected together as in Figure 3.9. We plot the x coordinate along the axis of the beam, and time downward. Then the line $x - ct =$ const., called a characteristic line, will slope downward to the right, and the line $x + ct =$ const. will slant downward to the left. Any value of the waveform $f(x - ct)$ will then have a constant value along the characteristic $x - ct =$ const., and the reflected wave $g(x + ct)$ will have a constant value along the characteristic $x + ct =$ const.

If we select a particular location x and draw a line downward, then the displacement of the wave at this location can be plotted as a function of time along this line. Consider a displacement pattern, perhaps produced by an impact on the left end of this system, and follow it as a function of time. For a while, the disturbance follows the characteristic $x - ct$ until it comes to the junction between

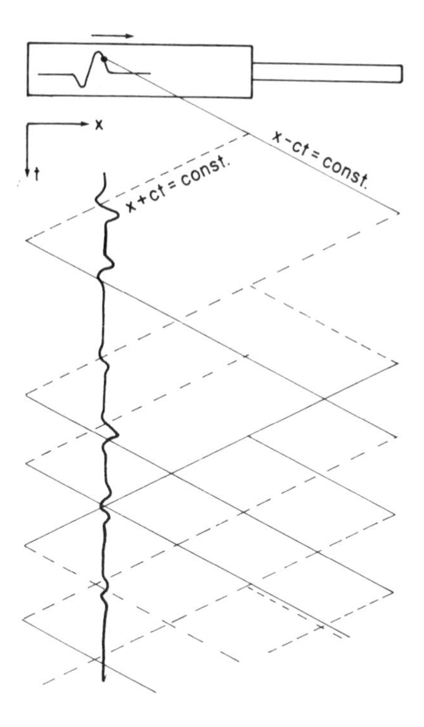

Figure 3.9 Demonstration of the use of characteristics in a one-dimensional system to build up the reverberation time pattern at a particular location.

the two beams, at which time a portion of the waveform, governed by Equation 3.8, will be reflected. Since the wave impedance is reduced, there will be a change in sign of this waveform, indicated by the dashed characteristic $x + ct = \text{const.}$

When the characteristic intersects our downward line $x = \text{const.}$, a wave signal will be produced. That characteristic continues to the left until the free end is reached, at which time it reflects again with a change in sign, so the characteristic is shown solid.

In the meantime, the original characteristic continued to the far right free end, reflected with a change in sign, became dashed, and traveled back again to the observer's location, producing an additional pulse at that time. As we follow these characteristics and their reflections at various junctions, we see that as time goes on the pattern gets more and more complicated, but, in principle, we can construct the entire sequence of arrivals of pulses at the observer's location.

This sequence of pulses due to the reflections at the ends and junctions is a form of reverberation, and a simple time pulse is stretched out into a series of waveforms due to reverberation. Reverberation results in a spreading of the energy in time. If there is also dissipation, then the amplitudes of the waves are also decreased because of the losses suffered during propagation. Reverberation, as such, is important in noise problems because it increases the vibration levels. Reverberation is also important for diagnostics, particularly if we are interested in the waveform, because the waveform is changed by reverberation.

3.3 MOBILITY OF A FINITE ROD

Let us now specialize the discussion to a rod of finite length and uniform diameter that is excited at one end by a sinusoidal force $l = L_s e^{j\omega t}$, as shown in Figure 3.10. Since the time dependence is $e^{j\omega t}$, we can write the two waveforms as

$$u = (Ce^{-jkx} + Be^{+jkx})e^{j\omega t}, \tag{3.9}$$

where $k = \omega/c_1 = 2\pi f/c_1 = 2\pi/\lambda$, and we now denote the longitudinal wave speed $c_1 = \sqrt{E/\rho}$.

The first term is the wave $f(x - c_l t)$, and the second term is $g(x + c_l t)$. The unknown amplitudes B and C are determined by setting the force equal to zero at $x = L$ and equal to the applied load at $x = 0$. Therefore:

$$\begin{aligned} l = 0 \rightarrow \frac{\partial u}{\partial x} &= 0 \qquad \text{at } x = L, \\ l = -EA\frac{\partial u}{\partial x} &= L_s e^{j\omega t} \qquad \text{at } x = 0. \end{aligned} \tag{3.10}$$

We can now find the drive point mobility or ratio of velocity to force at the input point $x = 0$:

$$Y_{\text{dp}} = \left.\frac{u}{l}\right|_{x=0} = \frac{-j}{\rho c_l A}\cot\frac{2\pi L}{\lambda} = \frac{-j}{\rho c A}\frac{\cos(\omega L/c_1)}{\sin(\omega L/c_1)} \tag{3.11}$$

where $\lambda = c_1/f$ is the longitudinal wavelength.

It is very easy to show that if the length of the beam is very small compared to a wavelength, then the input mobility is just the reciprocal of the mass reactance of the beam moving as a rigid body:

$$Y_{\text{dp}} = \frac{1}{j\omega\rho AL}. \tag{3.12}$$

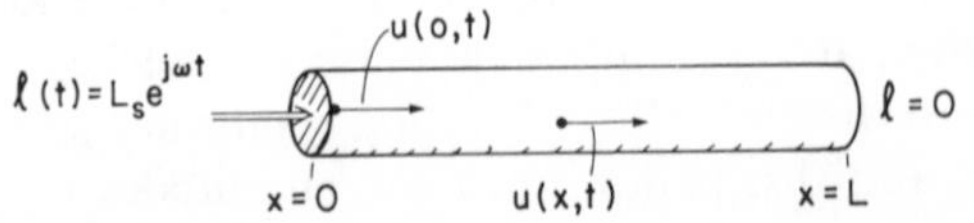

Figure 3.10 Excitation of a free-free rod into longitudinal vibration by a force at one end.

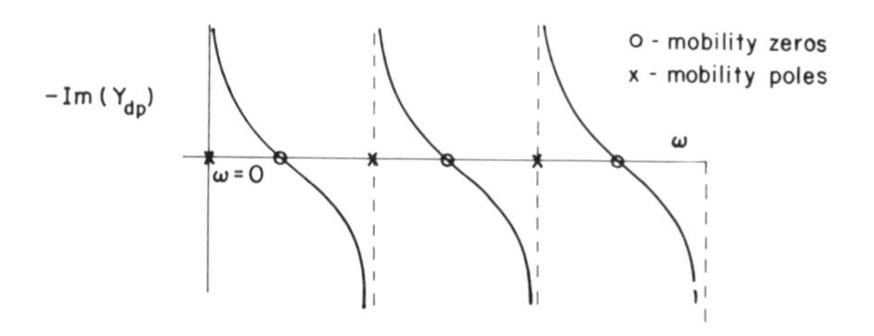

Figure 3.11 Drive point mobility for a free-free longitudinal rod. Note alternation of poles and zeros.

We note that the mobility in Equation 3.11 is purely imaginary, and if we plot the negative of its imaginary part, we obtain the plot in Figure 3.11. The figure shows a series of maxima or peaks in mobility corresponding to resonances, and a set of zeros in mobility corresponding to antiresonances of the system. At resonances, there can be motion for zero input force, whereas at antiresonances an arbitrary amount of force will not produce any motion at the driving point. Resonance occurs when kL is an integral multiple of π, which means that the beam length is an integral number of half wavelengths of compressional waves, as sketched in Figure 3.12.

The zeros in mobility, the antiresonances, correspond to frequencies at which the length of the rod is an odd number of quarter wavelengths, as illustrated in Figure 3.13. At an antiresonance, the rod undergoes a resonant vibration of sufficient amplitude to keep the applied force from moving the rod at the drive location. This condition is similar in concept to a dynamic absorber—a resonator that is placed on a structure and tuned to the operating frequency of the machine

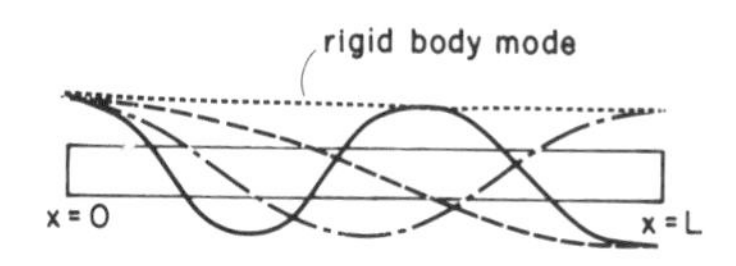

Figure 3.12 Lowest four resonant mode shapes of free-free rod in longitudinal vibration.

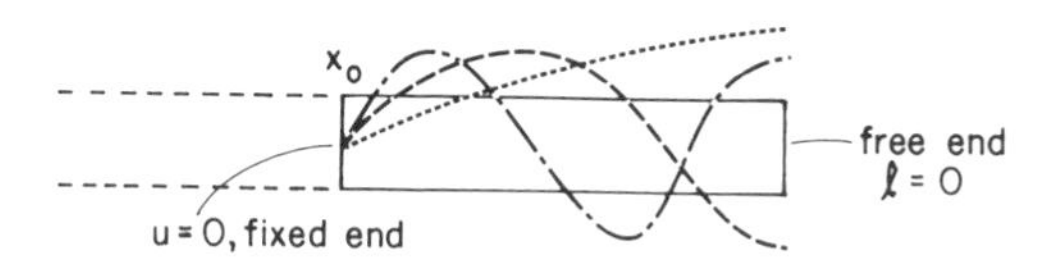

Figure 3.13 Zeros in mobility correspond to resonance of a fixed free rod since an applied force of any magnitude can be countered by the stress due to the resonance. Such resonance also causes zeros in transfer functions.

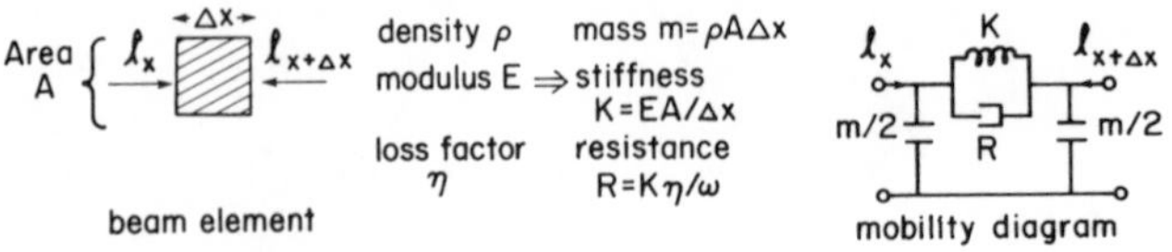

Figure 3.14 Mobility diagram for a lossy beam element.

with the idea that the resonator will vibrate and produce whatever force is required to hold the machine at rest at the attachment location.

If we drive the structure at an interior point, then whenever the distance from that driving point to one of the free ends becomes a quarter wavelength, or any odd multiple of a quarter wavelength, that segment will vibrate enough to hold the driving point fixed. For this reason, the zeros in the mobility function depend on the location of the drive point, whereas the resonances or the poles of the mobility depend on the overall system properties and not on the drive point location.

3.4 STRUCTURAL DAMPING

We can introduce the notion of damping by thinking of each small segment of the beam as a spring that, if lossy, has a small dashpot connected in parallel with it, as shown in Figure 3.14. The quantity $\omega R/K$ is the loss factor η. If we assume a time dependence $e^{+j\omega t}$, then this loss can be incorporated into the regular equations by replacing Young's modulus with a complex modulus $E(1 + j\eta)$. If we substitute this into the formula for the wave number $k = 2\pi/\lambda$, we now find that the wave number is complex

$$k \rightarrow k' = k\left(1 - \frac{j\eta}{2}\right). \tag{3.13}$$

The factor of e^{-jkx} represented a wave in the $+x$ direction; that factor now becomes $e^{-jkx - k\eta x/2}$. The real part of the exponential represents an amplitude decay by the factor e or 1 neper in a distance $2/k\eta = 2c/\omega\eta$. The time corresponding to this decay is equal to the distance traveled divided by the speed of propagation, which gives $2/\omega\eta$. Since 1 neper = 8.7 dB, a decay of 8.7 dB in a time $2/\omega\eta$ gives a decay rate

$$\mathrm{DR} = \frac{8.7}{2/\omega\eta} = 27.3 f\eta \ \mathrm{dB/sec}. \tag{3.14}$$

The decay rate may also be expressed as the time required for the vibration amplitude to decay by 60 dB, which is the reverberation time T_{R}:

$$\mathrm{DR} = \frac{60\ \mathrm{dB}}{T_{\mathrm{R}}\ \mathrm{sec}}, \tag{3.15}$$

so

$$T_R = \frac{2.2}{f\eta}. \tag{3.16}$$

A technique for measuring structural damping is illustrated in Figure 3.15. The structure is impact excited by a small hammer, and an accelerometer senses the resulting vibration. The accelerometer signal is passed through a filter set, typically a one-third or a full octave band, into a processor that converts the signal into levels in decibels against time, which is then presented on the oscilloscope. We then match a straight line to the decay rate. Knowing the center frequency of the bandpass filter, we can determine the loss factor from Equation 3.14. The processor is, in effect, a very fast sound-level meter. Its operation is illustrated in Figure 3.15(b), which shows the output from the filter, a decaying ringing sound that is then squared and short-time averaged to produce an envelope. Then a logarithmic detector converts this exponential decay into a straight-line decay, which is shown on the scope. An alternative procedure is to use the Hilbert transform, discussed in Appendix A, to generate the vibration envelope. The processes of averaging and taking the

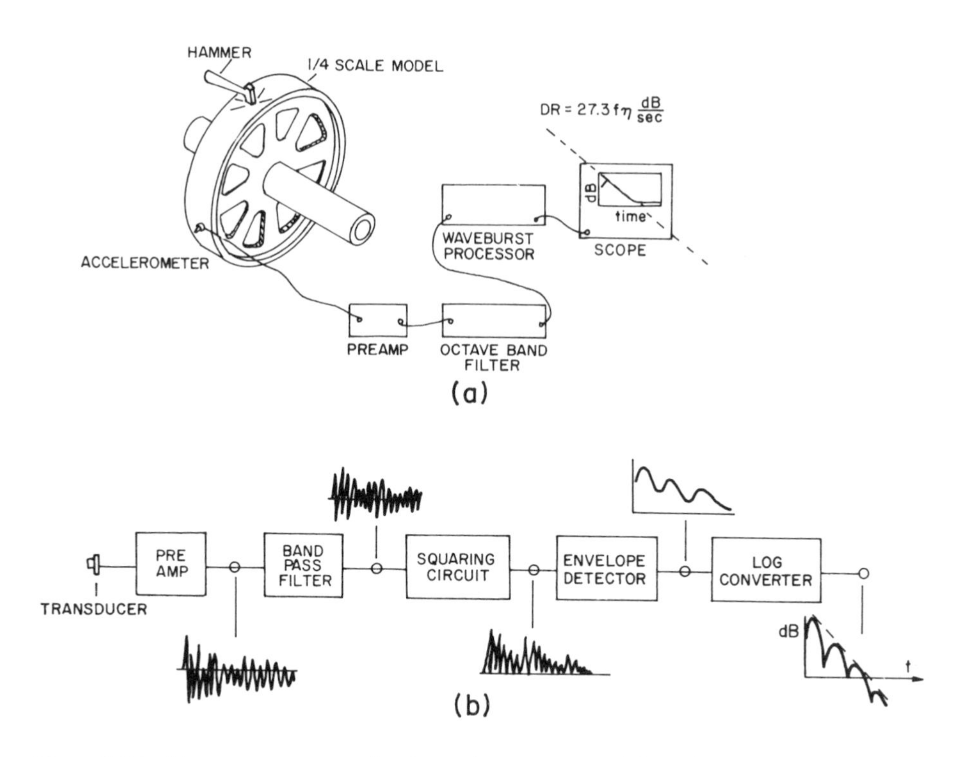

Figure 3.15 Measurement setup for determining structural damping by decay rate. When modal beating produces a fluctuating envelope as shown, other methods of measuring decay rate may be necessary.

logarithm are unaffected by this alternative except that they are likely to be done by digital rather than by analog instrumentation.

It often happens that only one or two modes may be decaying in the frequency band selected by the filter set. With very few modes, the modal vibrations at the observation point may go in and out of phase, leading to "beating" or large fluctuations in the decay curve. Although each individual mode is decaying exponentially, they interfere with each other at the pickup location and the output of the processor fluctuates wildly, making it difficult to measure the decay rate. If this happens, the "integrated impulse" technique can be most helpful. The technique generally involves either special instrumentation or digital storage of the decay trace and a numerical integration of the decay trace by a computer.

The various relationships between loss factor, decay rate, reverberation time, and so on, although developed in the context of longitudinal waves, are independent of the wave type. If we measure a certain decay rate, the loss factor of the material and the structure are determined regardless of whether the actual vibration is composed of longitudinal waves, in-plane shear, bending waves, or whatever. We note also, however, that the effectiveness of a given damping *treatment* may be different for the wave types.

3.5 BENDING WAVES

To analyze the dynamics of a beam or plate undergoing flexural or bending vibration, let us consider, as before, a small element of the material between the surfaces x and $x + \Delta x$. In bending motion, this segment is translated and bent as shown in Figure 3.16 so that the left end moves up an amount y_x and the right end moves up an amount $y_{x+\Delta x}$. This segment bends around a surface in the middle of a plate called the neutral plane; and if the curvature is negative as shown, the "fibers" of material above the neutral axis are stretched and those below it are compressed. The extension of any fiber is proportional to its original length, its distance z from the neutral plane, and the curvature. The extension is given by $-z\Delta x(\partial^2 y/\partial x^2)$. If this fiber has a Young's modulus E and cross-sectional area dA, then this stretching

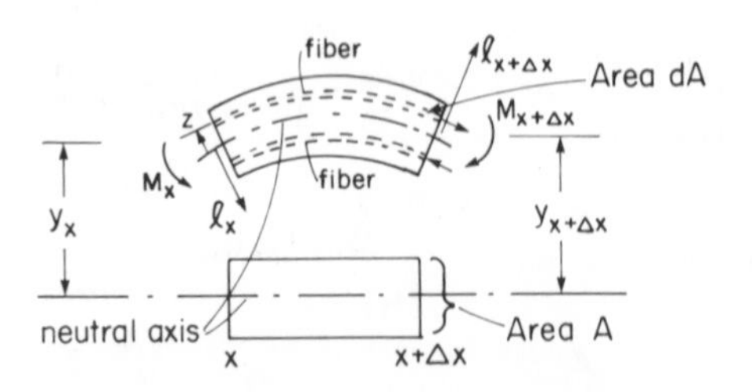

Figure 3.16 Forces and coordinates used in deriving dynamics of a beam element in flexural or bending motion.

requires an amount of force at the end of the fiber given by

$$dl = -z\Delta x \frac{\partial^2 y}{\partial x^2} E\,dA. \tag{3.17}$$

A fiber above the neutral axis requires a force outward from the face for its extension, and a fiber below the axis requires a force inward to the face for its compression. These positive and negative forces require that a moment be applied to the end of the face, the value of which is

$$M = \int_A z\,dl = -AE\frac{\partial^2 y}{\partial x^2}\kappa^2 \equiv -D\frac{\partial^2 y}{\partial x^2}, \tag{3.18}$$

where A is the total cross section of the beam and κ is the radius of gyration. The collection of terms $D = AE\kappa^2$ is often termed the *bending rigidity*.

Since all the quantities involved are continuous functions of the position x, then there will, in general, be a difference in the moment between the two ends of the beam element, which is given by

$$M_{x+\Delta x} - M_x = -\Delta x E A\kappa^2 \frac{\partial^3 y}{\partial x^2} = l_x \Delta x, \tag{3.19}$$

which must be balanced by a torque due to the shear forces at the end of the beam, $l_x\Delta x$. If there is a difference in these shear forces at the end of the beam element, there will be a net vertical unbalanced force on the element that causes it to accelerate in the y direction:

$$l_{x+\Delta x} - l_x = -\Delta x E A\kappa^2 \frac{\partial^4 y}{\partial x^4} = \rho A \Delta x \frac{\partial^2 y}{\partial t^2}$$

or

$$\kappa^2 c_1^2 \frac{\partial^4 y}{\partial x^4} = -\frac{\partial^2 y}{\partial t^2}. \tag{3.20}$$

We end up, therefore, with an equation for the displacement y in terms of derivatives with respect to position x and time t. The governing equation in this case, however, is different from the wave equation that we found before. Its solutions are quite complicated to express directly in the space and time variables. Consequently, we restrict our attention to solutions of the form $e^{j\omega t}$ and introduce the bending wave number $k_b = \sqrt{\omega/\kappa c_1}$. With this substitution, Equation 3.20 becomes

$$\frac{d^4}{dx^4} y = k_b^4 y, \tag{3.21}$$

which has a solution

$$y = \{Ae^{-jk_b x} + Be^{+jk_b x} + Ce^{-k_b x} + De^{+k_b x}\}e^{j\omega t}. \tag{3.22}$$

Equation 3.22 shows that there are two solutions that decay exponentially and do not represent wave motions. There are also two solutions that represent waves with a propagation speed $c_b = \omega/k_b = \sqrt{\omega\kappa c_l}$. The speed with which the bending wave travels depends on frequency, unlike the situation with compressional waves where all frequencies travel at the same speeed. Since we can always think of any waveform as a summation of frequency components, we see that different frequency components of this wave will travel at different speeds and, therefore, the shape of the wave will become distorted and change as the wave travels. This phenomenon is known as *dispersion.*

Another consequence of dispersion is that the energy carried by the wave does not travel at the same speed as the phase. To see this, consider two waves, corresponding to frequencies $\omega - \Delta\omega$ and $\omega + \Delta\omega$, and their corresponding wave numbers $k_b - \Delta k_b$ and $k_b + \Delta k_b$, as shown in Figure 3.17. Since these waves have slightly different wavelengths, there will be some places where they are in phase and other places where they are out of phase. This leads to a modulation of the waveform as shown. The individual wavelets move through this modulation at the phase speed, but the question is, "What is the speed of the modulation, which is the envelope of energy contained by the wave?" If we mathematically combine the two waves, we get

$$\begin{aligned} &\cos[(k_b - \Delta k_b)x - (\omega - \Delta\omega)t] + \cos[(k_b + \Delta k_b)x - (\omega + \Delta\omega)t] \\ &\quad = 2\cos(k_b x - \omega t)\cos 2(\Delta k_b x - \Delta\omega t). \end{aligned} \tag{3.23}$$

We see that the solution is a product of a wave of frequency ω and wave number k_b, the individual phase wave, and a modulation term with frequency $2\Delta\omega$ and wave number $2\Delta k_b$. The velocity of this wave, since it is a function of the form $f(x-ct)$,

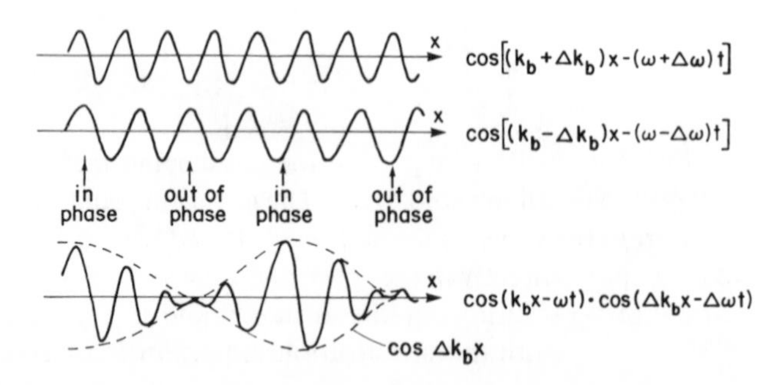

Figure 3.17 A modulation envelope is produced by the beating of two traveling waves of similar frequency. The speed of the modulation envelope is the group speed of the wave.

is the ratio of the coefficient of t to the coefficient of x; that is,

$$c_g = \frac{\Delta\omega}{\Delta k} \doteq \frac{d\omega}{dk}. \tag{3.24}$$

The quantity c_g is called the group or energy velocity and governs the speed of energy propagation through the medium. If ω and k are linearly related, as they are for longitudinal waves,

$$\frac{\omega}{c_l} = k_l, \quad \omega = k_l c_l, \quad c_g = \frac{d\omega}{dk} = c_l, \tag{3.25}$$

then the group speed equals the phase speed, whereas for bending waves,

$$c_g = \frac{d\omega}{dk} = 2k_b \kappa c_l = 2\sqrt{\omega\kappa c_l} = 2c_b, \tag{3.26}$$

the group speed is twice the phase speed. This case is opposite that of gravity waves on water, in which there is a difference in the phase and group speeds, but for water waves the group speed is half of the phase speed. Individual wavelets move from the back end of the group wave pattern toward the front end and disappear.

3.6 MEASUREMENT OF POWER FLOW IN BENDING WAVES

The vibrational energy density for a wave is half kinetic and half potential, or twice the kinetic energy density. The power flow in a traveling wave with velocity

$$v_t = V_t e^{-jk_b x + j\omega t} \tag{3.27}$$

is the energy density ρv_t^2 times the energy speed c_g:

$$\Pi = \rho\langle v_t^2\rangle c_g = \rho c_b |V_t|^2 = \frac{\omega\rho |V_t|^2}{k_b}. \tag{3.28}$$

A standing wave

$$v_s = V_s \cos(k_b x) e^{j\omega t} \tag{3.29}$$

has energy density $\rho\langle v_s^2\rangle$ but carries no power because it consists of equal-amplitude waves in the two opposite directions. In the general situation in Equation 3.22, both kinds of waves are possible, and it is desirable to be able to separate them experimentally.

The key is to notice that the slope dv/dx and the amplitude v are in phase for the standing wave v_s and $\pi/2$ out of phase for a traveling wave. Therefore, if for a general field $v = v_t + v_s$ we take the imaginary part of the product, we obtain

$$k_b|V_t|^2 = \mathrm{Im}\left(v - \frac{dv^*}{dx}\right) = \frac{1}{2\Delta x}\mathrm{Im}(v_{x+\Delta x} + v_x)(v^*_{x+\Delta x} - v^*_x)$$

$$= -\frac{1}{2\Delta x}\mathrm{Im}(v_{x+\Delta x}v^*_x - v_x v^*_{x+\Delta x}) \tag{3.30}$$

$$= -\frac{1}{\Delta x}\mathrm{Im}(v_{x+\Delta x}v^*_x)$$

Using Equation 3.28, we can express the power as

$$\Pi = -\frac{\rho\kappa c_l}{\Delta x}\mathrm{Im}(v_{x+\Delta x}v^*_x). \tag{3.31}$$

The power in a bending wave can therefore be found by making vibration measurements at two closely separated locations, 1 and 2, and measuring the cross spectrum. Since accelerometers are most likely to be used in this measurement, we can use the cross spectrum of acceleration $G_{a_1a_2}(\omega)$ to obtain

$$\Pi = \frac{\rho\kappa c_l}{2\omega^2\Delta x}\mathrm{Im}\{G^*_{a_1a_2}(\omega)\}. \tag{3.32}$$

In Chapter 5, we use a similar technique for measuring power flow in a sound wave.

3.7 RESONANCES OF A FINITE BENDING BEAM

We have noted that dispersion will modify source waveforms as they propagate. The change of waveform is not terribly important for noise problems because the ear tends to be more responsive to spectral energy rather than to details of the waveform. The greatest effect of dispersion in noise problems is on the distribution of the resonance frequencies of a system and the number of resonant modes available for storing vibrational energy.

Consider a beam simply supported at both ends, as shown in Figure 3.18. We use the general form of Equation 3.22 and apply the simply supported condition at

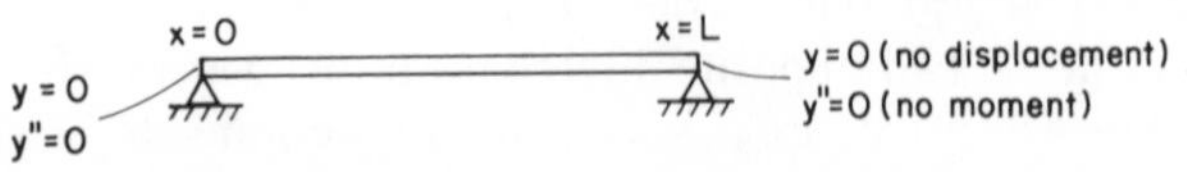

Figure 3.18 A simply supported beam has no moment or motion at its ends. The extra conditions are needed because the bending equation has fourth derivatives.

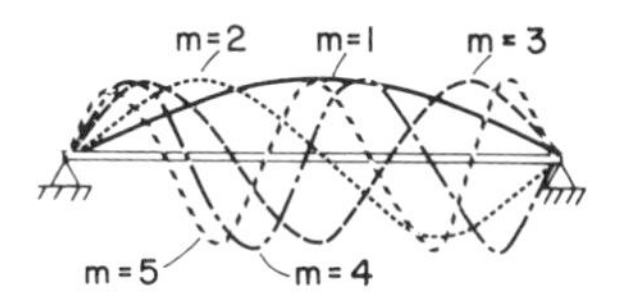

Figure 3.19 Sketch of first five modes of vibration in a simply supported beam. The mode shapes are the same as for a string, but the frequencies do not increase in an harmonic sequence.

the origin. This requires no displacement and no moment (second derivative equal to zero) at the origin, resulting in $C = D = 0$. In addition, requiring that the second derivative vanish at $x = L$ gives $B = -A$. Requiring that the displacement vanish at $x = L$ gives

$$\sin k_b L = 0 \qquad \text{or} \qquad k_m = \frac{m\pi}{L}. \quad (m = 1, 2, \ldots) \tag{3.33}$$

This means that only values of the bending wave number $k_b = m\pi/L$ are allowed, and since k_b is directly related to the frequency it means that only certain resonance frequencies are allowed for the structure. The resonance frequencies are $\omega_m = k_m^2 \kappa c_l$, and the allowed mode shapes are $\sin k_m x$, which are sketched in Figure 3.19.

Let us mark the allowed values of wave number $k = m\pi/L$ along the k axis as shown in Figure 3.20. Each $\times$ on this axis corresponds to a mode shape corresponding to a particular value of m. The $\times$'s are separated by a distance π/L. This diagram describes the resonances of either longitudinal or bending modes of a beam insofar as wave number is concerned. When we convert to frequency, we find that the uniform distribution of resonances in k causes the longitudinal resonance frequencies to be equally spaced. The resonances of a bending beam will get farther apart as we increase frequency, because for bending resonances frequency varies as the square of the wave number.

In either case, there is a monotonic relationship between wave number and frequency; that is, as one increases, the other increases. This allows us to calculate the total number of resonances up to any frequency in a very simple way. Knowing the number of resonances within any frequency interval is important in structural dynamics, because each resonance represents a way for the structure to store

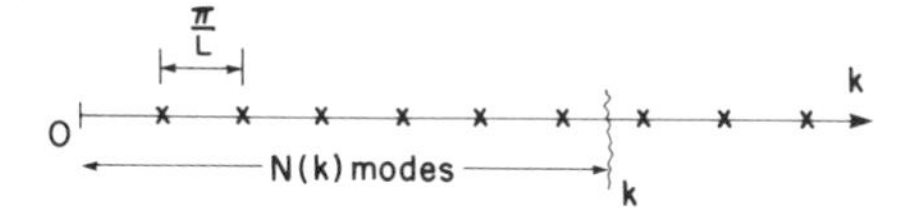

Figure 3.20 One-dimensional wave number lattice showing the principle of counting resonant modes in a wave number interval.

vibrational energy. Frequency bands with many resonances are bands with high structural mobility. The more modes that are able to accept vibrational energy from a source, the greater the response of the structure will be to that source. We will find that the modal density plays a very important role in determining the magnitude of structural vibration.

Referring to Figure 3.20, we have plotted the resonant wave numbers separated by distance π/L. Such a diagram could apply to a longitudinal or a bending one-dimensional structure. If we select a value of k, we can ask, "How many modes of vibration have been encountered as the wave number has increased to that value?" That quantity is the mode count

$$N(k) = \frac{k}{\pi/L} = \frac{kL}{\pi}. \tag{3.34}$$

The next step is to inquire how rapidly the number of modes changes as wave number changes; that is, we take the derivative of Equation 3.34 to determine the modal density in wave number:

$$n(k) = \frac{dN(k)}{dk} = \frac{L}{\pi}. \tag{3.35}$$

The modal density in wave number does not depend on the dynamics of the structure. It is the same for bending waves or longitudinal waves. The dynamics come in when we ask how $n(k)$ relates to a modal density in frequency. Since the number of modes, whether considered to be distributed in frequency or in wave number, should be the same for the corresponding frequency and wave number intervals, we have $n(\omega)\Delta\omega = n(k)\Delta k$, or

$$n(\omega) = n(k)\frac{dk}{d\omega} = \frac{n(k)}{c_g} = \frac{L}{\pi c_g}. \tag{3.36}$$

This is the modal density in radian frequency. It depends on the group velocity of the waves involved in the vibration, and there is now a difference between the modal densities for longitudinal and bending waves. An additional way of expressing modal density is in terms of cyclic frequency (or hertz) or by the reciprocal of modal density, which is the average frequency separation between modes:

$$n(f) = n(\omega)\frac{d\omega}{df} = \frac{2L}{c_g}, \qquad \frac{1}{n(f)} = \overline{\delta f} = \frac{c_g}{2L}. \tag{3.37}$$

As an example, for longitudinal waves the average frequency separation is $\delta f_l = c_l/2L$, so for a steel rod 1 m long, the average frequency separation for longitudinal modes is about 2500 Hz. On the other hand, the average frequency separation for bending wave modes is

$$\overline{\delta f_{\mathrm{b}}} = \frac{c_{\mathrm{b}}}{L} = \frac{\sqrt{\omega \kappa c_{\mathrm{l}}}}{L} = 3\frac{\sqrt{f\,(\mathrm{Hz}) \times h\,(\mathrm{mm})}}{L\,(\mathrm{m})}, \tag{3.38}$$

where the simplified formula applies to steel or aluminum beams. If the thickness of the beam is 1 mm and its length is 1 m, then the average separation between bending modes at 100 Hz is 30 Hz, but at 10 kHz, the average separation between modal resonances has increased to about 300 Hz.

3.8 MODE COUNT IN TWO-DIMENSIONAL STRUCTURES

We now consider the modal density of the two-dimensional simply supported bending plate with area $A_{\mathrm{p}} = L_1 L_2$ sketched in Figure 3.21. The mode shapes for such a structure are

$$\psi_M = 2 \sin k_{m_1} x_1 \sin k_{m_2} x_2, \tag{3.39}$$

where $k_{m_1} = m_1 \pi / L_1$, $k_{m_2} = m_2 \pi / L_2$, and $m_1, m_2 = 1, 2, \ldots$. Since the basic relationship between frequency and wave number for bending waves is $\omega = k^2 \kappa c_{\mathrm{l}}$, the resonance frequencies are

$$\omega_M \equiv \omega_{m_1, m_2} = k^2 \kappa c_{\mathrm{l}} = \left\{ \left(\frac{m_1 \pi}{L_1} \right)^2 + \left(\frac{m_2 \pi}{L_2} \right)^2 \right\} \kappa c_{\mathrm{l}}. \tag{3.40}$$

We can represent the resonant modes of this structure by the two-dimensional grid or lattice shown in Figure 3.22. Allowed values of k_1 and k_2 are separated by the intervals π/L_1 and π/L_2, respectively. Each point on the lattice therefore represents an allowed pair of values k_1 and k_2 that correspond to a mode of vibration. Furthermore, the distance from the origin to the lattice point is proportional to the resonance frequency according to Equation 3.40. Thus, the sequence of lattice points in this diagram enclosed by an expanding circle determines the sequence of resonant modes encountered as we increase the frequency of excitation.

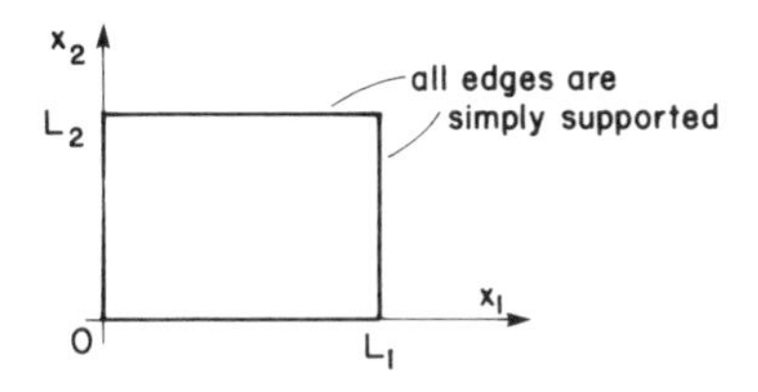

Figure 3.21 Geometry of simply supported thin plate used to determine modal density of two-dimensional structures.

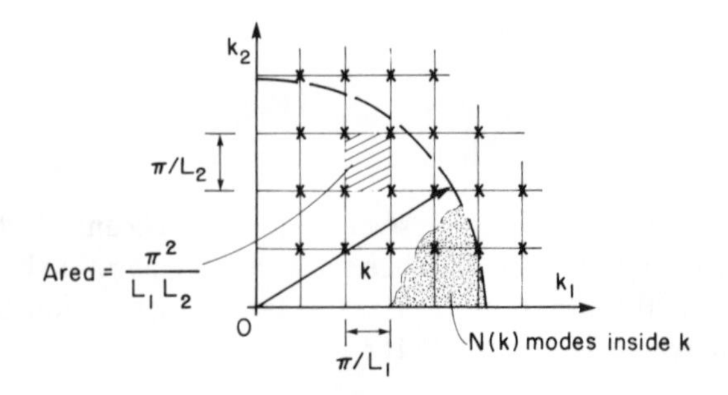

Figure 3.22 Two-dimensional wave number lattice for supported rectangular plate. The number of modes encompassed by the quarter circle of radius k is $N(k)$.

Let us now ask how many modes of vibration have been encountered up to a value of wave number k—or the corresponding frequency ω—when the circle of radius k has expanded. We note that each lattice point has an area associated with it that is a cell of dimensions π/L_1 by π/L_2, shown hatched in the diagram. Since this cell has an area of π^2/A_p and the area of the quarter-circle is $\pi k^2/4$, then the number of modes encountered up to k is approximately

$$\frac{\text{total area}}{\text{area per mode}} = \frac{k^2 A_p}{4\pi} = N(k). \tag{3.41}$$

After we find the number of modes up to wave number k, we determine modal density, wave number, and frequency as in the one-dimensional case. In two dimensions the modal density in wave number is

$$n(k) = \frac{dN}{dk} = \frac{kA_p}{2\pi}, \tag{3.42}$$

and the modal density in radian frequency is

$$n(\omega) = n(k)\frac{dk}{d\omega} = \frac{n(k)}{c_g} = \frac{kA_p}{2\pi c_g}. \tag{3.43}$$

Since $k_b = \omega/c_b$, we have

$$n(\omega) = \frac{\omega A_p}{2\pi c_b c_g} = \frac{\omega A_p}{4\pi c_b^2} = \frac{A_p}{4\pi \kappa c_l}. \tag{3.44}$$

Equation 3.44 shows how important the wave speed is in determining modal density. Low-speed waves, like bending, contribute more resonant modes than do high-speed waves, such as in-plane, either longitudinal or shear.

The modal density becomes independent of frequency, at least on average, for two-dimensional bending waves in a plate. Since the only geometric factor here is the total plate area, we can assume that the different plates with different shapes having the same total area will have similar modal densities. Converting to cyclic frequency gives

$$n(f) = n(\omega)\frac{d\omega}{df} = \frac{A_p}{2\kappa c_1}, \qquad \overline{\delta f} = \frac{2\kappa c_1}{A_p} = \frac{hc_1}{\sqrt{3}\,A_p} \tag{3.45}$$

where we have used the relation $\kappa = h/2\sqrt{3}$ between radius of gyration and thickness for a homogeneous plate.

As an example, consider an aluminum plate with a wall thickness of 3 mm and an area of 1 m^2. If we take the longitudinal velocity to be 5100 m/sec, we get an average frequency separation of 9 Hz. A one-third octave band centered at 500 Hz has a bandwidth of about 125 Hz. We would expect approximately 14 modes of vibration to occur in this band. If we go to the 100-Hz third-octave band, where the bandwidth is 25 Hz, then we would expect slightly fewer than three modes to resonate. In the 2.5-kHz third-octave band, we would expect about 70 resonant modes.

3.9 THE EFFECTS OF CURVATURE ON MODAL DENSITY

To examine the stiffening effects of curvature, let us take our plate of length L_2 and width L_1 and roll it to form a cylinder around the x_2 axis with radius $a = L_1/2\pi$. If the ends of the cylinder remain simply supported, then the mode shapes in the x_2 direction will be as they were before: $\sin m_2\pi x_2/L_2$. In the x_1 direction, however, we must have periodic dependence in the angle x_1/a subtended at the center of the cylinder. The functions can either be sines or cosines, since both have a periodicity of 2π in this angle. The mode shapes for the cylinder may then be written

$$\psi = \sin\frac{m_2\pi x_2}{L_2}\begin{pmatrix}\cos\\ \sin\end{pmatrix}\frac{m_1 x_1}{a}, \tag{3.46}$$

where now the "wave number" k_1 is the coefficient of x_1: $m_1/a = 2\pi m_1/L_1$.

Notice that the allowed mode numbers in the x_2 direction are still separated by π/L_2, but the allowed mode numbers in the x_1 direction are separated by $2\pi/L_1$, which is twice the separation that we had before. It looks, therefore, as though we have lost some modes by the rolling process. However, recall that both sine and cosine mode shapes are present. Therefore each lattice point in this new wave-number space corresponds to a pair of resonant modes which we can indicate as sketched in Figure 3.23.

The constant-resonance frequency loci for the cylinder, however, have a different shape from those of the flat plate. Waves traveling around the cylinder in

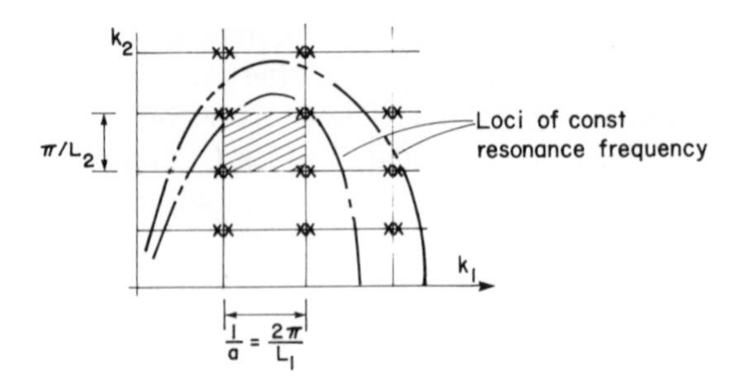

Figure 3.23 Modification of wave number lattice when the plate is rolled into a cylinder along its x_2 axis.

the x_1 direction are affected very little by curvature so that the form of the locus for wave numbers near the k_1 axis is essentially circular, as in Figure 3.22. However, the curvature of the cylinder causes the speed of waves in the x_2 direction to increase and, therefore, a smaller wave number will correspond to the same frequency of vibration. This has the effect of making constant-frequency loci appear as sketched in Figure 3.23.

One can see that the number of lattice points enclosed by a constant-frequency locus will be somewhat smaller than that for the flat plate. Therefore, at the lower frequencies, the modal density and total mode count of cylinders are less than those of a flat plate of the same area.

The stiffening effect of curvature is due to the buildup of in-plane membrane stresses in the cylinder when we try to bend the cylinder along its axis. These membrane stresses must propagate at the longitudinal wave speed in the material and can only be effective in stiffening the cylinder as long as the frequency of vibration is such that a longitudinal wavelength can encircle the cylinder. At very high frequencies, the membrane stresses do not have enough time to propagate to establish their influence, and the structure reverts to flat-plate dynamics. This transition occurs when the longitudinal wavelength equals the circumference of the cylinder at the "ring frequency":

$$f_{\text{ring}} = \frac{c_l}{2\pi a}. \tag{3.47}$$

Near the ring frequency, therefore, the frequency loci make a transition from the cylindrical type of locus shown in Figure 3.23 to the flat-plate locus shown in Figure 3.22, and when this occurs, a fairly large number of modes will resonate. This produces a small peak in the modal density near the ring frequency, as indicated in Figure 3.24. Above the ring frequency, the cylinder dynamically becomes a flat plate, whereas below the ring frequency the membrane stresses operate and affect the modal density and other aspects of structural behavior.

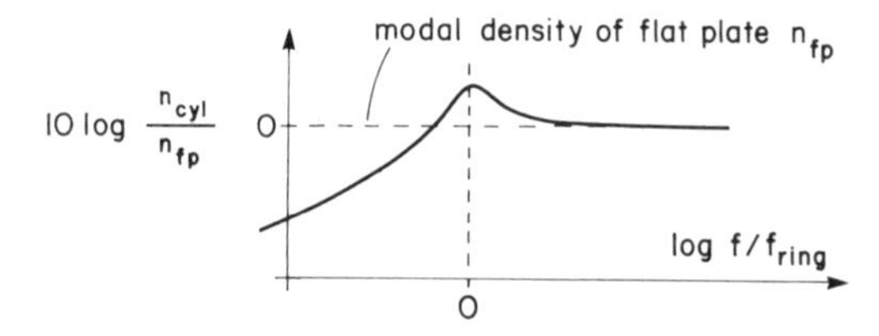

Figure 3.24 Sketch of modal density of cylinder showing the role of ring frequency in the relaxation of membrane effects.

3.10 MODAL DENSITIES OF SOME MACHINE STRUCTURES

Modal densities and mode counts are determined in practice both by measurement and calculation. The model gear structure in Figure 3.25 has been carefully studied. Experimental measurements of the mode count have been made of the shaft by a frequency sweep technique indicated in the figure. This mode count is graphed in Figure 3.26. At sufficiently high frequencies, the shaft can be thought of as a two-dimensional structure, a cylinder. The wave motions are as in-plane shear, speed c_s, and longitudinal modes, speed c_l, and, possibly, Rayleigh waves, speed c_R. Considering the unrolled shaft area and the speed of the shear and longitudinal modes, we can also draw a theoretical curve for the mode count versus frequency, using Equation 3.43 with $c_g = c_s$ and c_l, respectively. It is clear that the measurement is consistent with the theory at high frequencies, but there is an upward shift of the data, and an augmented number of modes at lower frequencies. This augmentation is probably due to bending waves and results in the permanent offset of mode count as a function of frequency at higher frequencies.

For engineering purposes, it is sometimes more useful to know the number of modes in a one-third octave band rather than to use a cumulative mode count. Third-octave mode counts, determined theoretically, for the rim and the hub of a

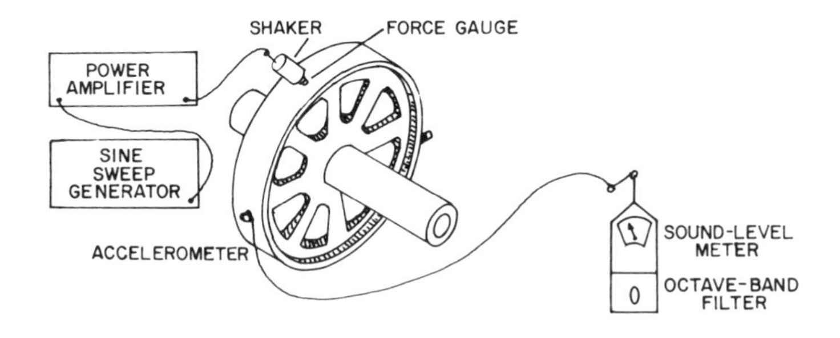

Figure 3.25 Measurement of the mode count of a gear using a frequency sweep method.

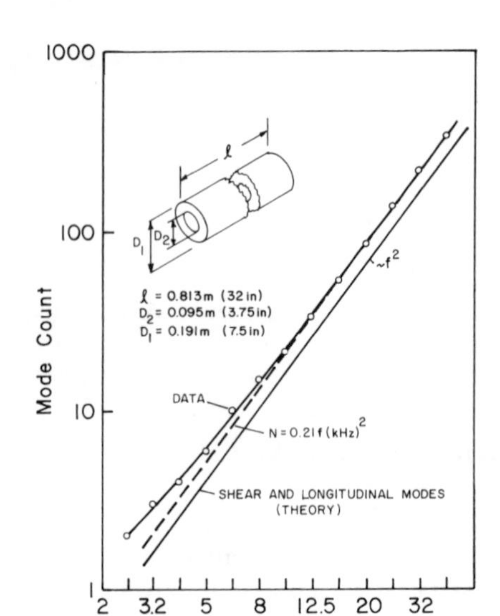

Figure 3.26 Mode count for shaft with circular hole. Above 8 kHz the shaft is a two-dimensional structure. Offset between measurement and estimate is due to the bending modes that dominate at low frequencies.

scale model gear are shown in Figure 3.27. Note that in the rim the bending and in-plane vibrations, longitudinal and shear, are important at all frequencies. In the shaft or hub section, one-dimensional shear and bending modes dominate at low frequencies, but there is a transition in behavior at about 16 kHz due to the occurrence of new circumferential mode types in which the hub begins to act as a

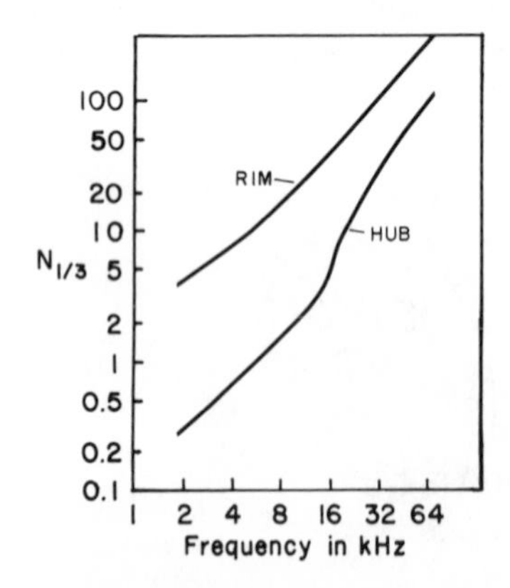

Figure 3.27 Mode counts in third-octave bands for two structural elements of a model gear. Note transition in hub mode counts when another class of modes begins to resonate.

two-dimensional structure. These new modes add to the ability of the hub structure to carry vibrational energy. Increased vibration transmission in certain frequency intervals will often occur because some component of the structure is suddenly able to respond and transmit more energy because it has more available modes.

3.11 RESPONSE FUNCTIONS AND RECIPROCITY

To describe the detailed response of a structure to forces is a highly complicated problem. Suppose we excite the structure at a point. Three components of force and three components of moment can be applied at that location, and the structure will respond with the corresponding translational and rotational velocities. These six forces and six velocities mean that there are, in principle, 36 mobilities, of which 21 are independent, that can be defined at each location.

Lowering our sights a bit, however, we consider the case in which a single force component is applied to a structure, and we are interested in a single component of response velocity at another position. It is fairly common to represent this system as a two-port network, as shown in Figure 3.28. Mobilities Y_{11} and Y_{22} are called *input* mobilities, because they relate velocity and force at the same port. Mobility Y_{12} is called a *transfer* mobility because it relates the velocity at one location to the force at another. It is possible to show that transfer mobility is reciprocal; that is, that $Y_{12} = Y_{21}$, so only three mobilities are needed to represent the system.

The principal of reciprocity will be used in this book, so it is worthwhile to discuss its applicability to systems like machines. The basic requirements for its applicability are threefold:

1. The system should be passive; that is, no sources other than those being considered should exist. Excitations in an operating machine other than the source being considered do occur, but if the frequencies are different or if the phase of the source is randomized, this requirement can usually be met.
2. The system should be linear; that is, the response should be at the same frequency as the excitation and proportional to it. Since oil films, contact stiffnesses, and other machine elements are often nonlinear, this requirement may not always be met, and it should always be considered as a possible cause of ambiguity in the data.

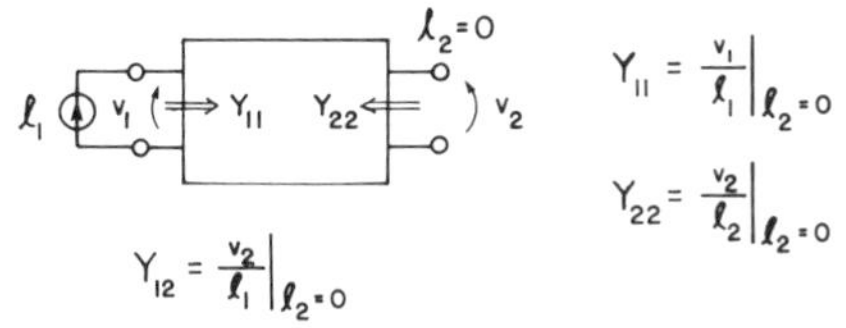

$$Y_{11} = \left.\frac{v_1}{f_1}\right|_{f_2=0} \qquad Y_{22} = \left.\frac{v_2}{f_2}\right|_{f_2=0} \qquad Y_{12} = \left.\frac{v_2}{f_1}\right|_{f_2=0}$$

Figure 3.28 Input and transfer mobilities for a two-port system. The transfer mobility Y_{21} is not shown because it equals Y_{12} by reciprocity.

3. The system should be bilateral; that is, when the phase or direction of the excitation is altered, the response should change in the same manner. Moving systems, such as rotating or reciprocating mechanisms, will probably not meet this criterion, at least not near their operating frequencies. But at higher frequencies, where the system is nearly stationary during a cycle of vibration, reciprocity could apply.

3.12 RESPONSE OF A SINGLE-DEGREE-OF-FREEDOM RESONATOR

We begin by considering the input mobility of a one-degree-of-freedom resonator. We shall later see that no matter how complicated the structure is, it can usually be thought of as a combination of resonators. Therefore, this simple system is a building block to be used as a way of modeling most structures. The resonator is sketched in Figure 3.29 along with its mobility diagram.

The velocity v is given by the applied force l_s times the total mobility of the system:

$$Y_{dp} = \frac{1}{1/Y_m + 1/Y_k + 1/Y_r} = Y_k \frac{1}{1 - \omega^2/\omega_0^2 + j(\omega/\omega_0)\eta}, \tag{3.48}$$

where the mobility of the stiffness is $Y_k = j\omega/K$, the dashpot is $Y_r = 1/R$, the mass is $Y_m = 1/j\omega m$, and the formulas for the resonance frequency $\omega_0 = \sqrt{K/m}$ and the loss factor $\eta = R/\omega m$ have been used.

In Figure 3.29, the power dissipated by the resistance element Π_{diss} is $(1/2)R\langle v^2 \rangle$, the kinetic energy T is $(1/2)M\langle v^2 \rangle$, and the potential energy V is $(1/2)K\langle x^2 \rangle = (1/2)K\langle v^2 \rangle/\omega^2$. The denominator of Equation 3.48 can therefore be represented by

$$\frac{1}{Y_{dp}} = R + j\omega M - \frac{K}{j\omega} = \frac{1}{\langle v^2 \rangle}\left\{\Pi_{diss} + \frac{1}{2j\omega}(T - V)\right\}. \tag{3.49}$$

Although derived for a special case, Equation 3.49 is quite general. When a structure is driven at some frequency, if the kinetic and potential energies balance out, the mobility at the drive point is real. The frequency at which this occurs is a resonance frequency of the system. At other frequencies, the drive point mobility is

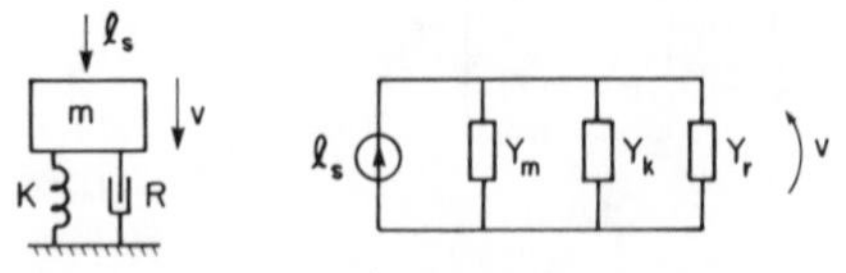

Figure 3.29 Single-degree-of-freedom resonator, a useful model for each resonant mode of a structure.

complex, and its sign indicates whether the potential energy is less than or greater than the kinetic energy stored by the structure.

At very low frequencies, Equation 3.48 shows that the drive point mobility becomes the mobility of the spring, and the resonator is said to be stiffness controlled. As ω becomes very large compared to the resonance frequency ω_0, the mobility of the mass dominates the equation and the drive point mobility equals Y_m. We say the system is mass controlled in this frequency range. Near $\omega = \omega_0$, the stiffness and mass mobilities cancel, and only the mobility of the damper is left. In the neighborhood of the resonance frequency, the system is said to be damping controlled.

The power dissipated in the damper is

$$\Pi_{\text{diss}} = \langle v^2 \rangle R_{\text{dp}} = \langle l^2 \rangle \omega_0 \eta M |Y_{\text{dp}}|^2, \tag{3.50}$$

where R_{dp} is the real part of Y_{dp}^{-1}. In Figure 3.30, we graph $|Y_{\text{dp}}|^2$ as a function of frequency. In a low-frequency or stiffness-controlled region, $|Y_{\text{dp}}|^2$ increases as the square of the frequency or +6 dB/octave. In the interval around the resonance frequency, the response is damping controlled. The most common way of representing the boundaries of this region is when $|Y_{\text{dp}}|^2$ equals half of its maximum value, the so-called half-power bandwidth of the resonator, since the dissipated power is proportional to $|Y_{\text{dp}}|^2$. The half-power bandwidth is the resonance frequency ω_0 times the loss factor η.

A more useful damping bandwidth for noise problems is to imagine that the resonator is a rectangular filter of unknown bandwidth. This filter has a resonant response within its bandwidth and is zero outside. Since the resonant response is that of a simple damper, this method is useful for estimating a response. It turns out that this equivalent noise bandwidth is $\pi/2$ times the half-power bandwidth, as shown in Figure 3.30. Thinking of each modal resonator as a simple rectangular filter is a very useful way of estimating the response of a complex structure with many modes of vibration.

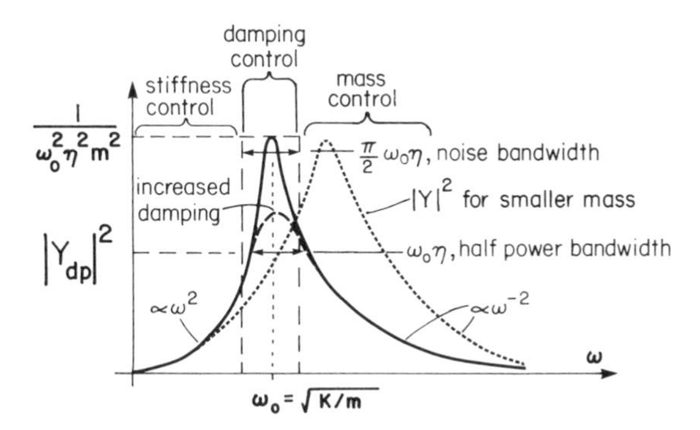

Figure 3.30 Mechanical drive point mobility of a resonator showing effects of changing mass and damping.

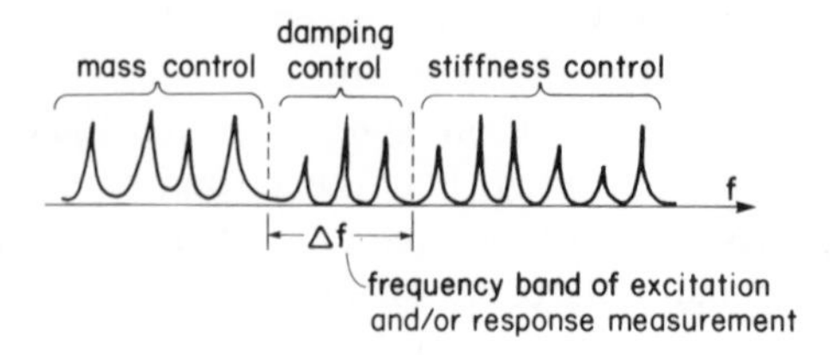

Figure 3.31 Illustration of how response in the frequency band Δf has contributions from modes that respond in stiffness, mass, and damping controlled motions.

The response in the mass-controlled region at high frequencies goes as $1/\omega^2$ or -6 db/octave. If we consider another resonator with a smaller mass, then the resonance frequency will shift up as shown in Figure 3.30 and the response in the mass-controlled region will increase for the smaller mass, but the response in the stiffness-controlled region will not change, because we have not changed the structural stiffness.

If we think of excitation as being concentrated in a particular frequency band Δf, as shown in Figure 3.31, then modes that resonate above that frequency interval will respond to the excitation as though they are stiffness controlled. On the other hand, modes that resonate below the excitation frequency interval will respond in a mass-controlled fashion. Modes that resonate within the excitation frequency band will respond according to their damping. If, for example, we are concerned with a mass-controlled phenomenon (often true of sound transmission through panels), then changing the damping or stiffness of the structure will not directly affect that behavior. Therefore, making a structure heavier or stiffer or increasing its damping might not change the response in the desired way unless the phenomenon depends on the parameter being changed.

3.13 STRUCTURAL RESPONSE TO IMBALANCE

Vibration problems often arise when rotating or reciprocating machines are supported in a structure with resonances excitable by the imbalance forces in Section 2.3. We shall take as our example a sewing machine sitting in its cabinet.

The machine and cabinet combination is shown in Figure 3.32. We are interested in the vibration of the overall system produced by the imbalance forces of the machine. The mode shape ψ of the combined system is important because it determines how well the imbalance forces are able to excite structural vibration. The first step is to measure the blocked force by measuring the vibration of the machine when it is suspended freely or by measuring the forces required to hold the machine fixed in place, as discussed in Section 2.3.

If the free velocity of the machine is V_F, then the blocked force l_B required to hold the machine in place is the acceleration $j\omega V_F$ times the machine mass M_m. The force could be a moment, in which case the appropriate "mass" is the moment of inertia of the machine. A mobility diagram of the system is shown in Figure 3.33.

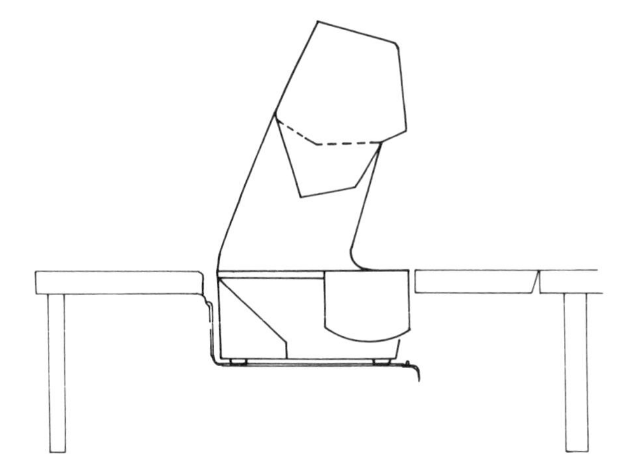

Figure 3.32 Sketch of a sewing machine sitting on a cabinet. The vibration of the machine and cabinet as a combined system is excited by imbalance forces discussed in Chapter 2.

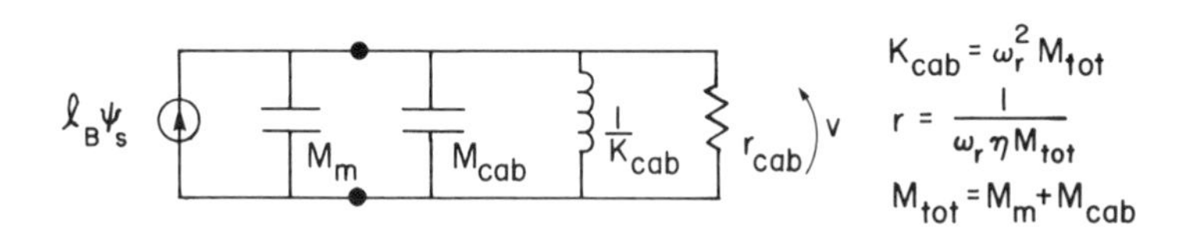

Figure 3.33 Equivalent diagram of a machine-cabinet combination. At the system resonance frequency ω_r, the machine can be regarded as a lumped mass and the stiffness K_{cab} is due to the cabinet structure.

The effective force on a system mode is the blocked force times the mode shape amplitude ψ evaluated at the machine location. The total effective modal mass M_{tot} is the mass of the machine M_m plus an effective cabinet mass M_{cab} determined by the kinetic energy of vibration of the cabinet referenced to the machine velocity.

The cabinet stiffness is determined by the resonance frequency and the combined mass of the machine and cabinet: $K = \omega_0^2 M_{tot}$. The effective mobility of the damper r_{cab} is determined by the cabinet loss factor and the total effective mass: $r_{cab} = (\omega M_{tot}\eta)^{-1}$. At resonance, the mass and stiffness cancel each other, so the modal velocity is determined by the system damping.

The ratio of velocity to the free velocity at resonance is therefore

$$\left|\frac{V(x_s)}{V_F}\right| = \psi^2(x_s)\frac{M_m}{M_{tot}}\frac{1}{\eta}. \tag{3.51}$$

The machine is placed in the cabinet and tested for vibration as shown in Figure 3.34. The machine speed is controlled with a variable transformer, because it

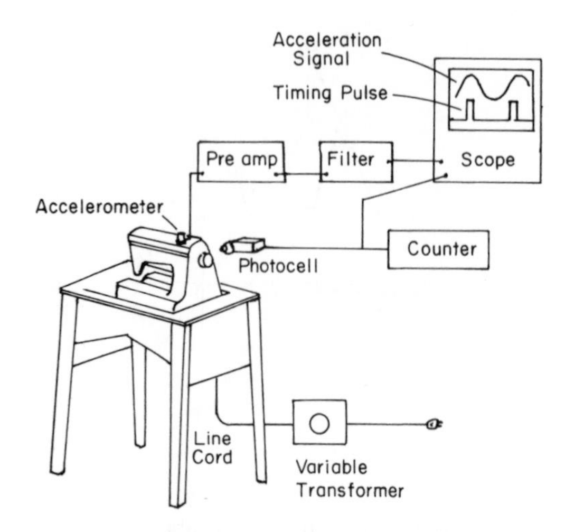

Figure 3.34 Setup for measuring mode shape of combined machine and cabinet system. Frequency of operation is adjusted to coincide with modal resonance frequency.

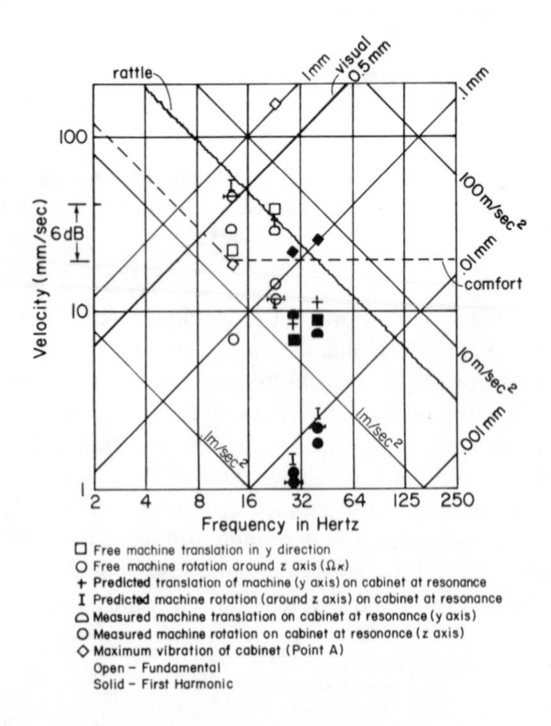

Figure 3.35 Comparison of measured machine and cabinet vibration amplitudes with criteria values established on basis of visual, tactile, and rattle limits.

is necessary to measure the vibration at a specific machine speed—the resonance frequency of the machine-cabinet mode.

A summary of some vibration measurements and a comparison with the vibration criteria are shown in Figure 3.35. Vibration criteria are discussed in Appendix B. The various data values are described in the legend, but, in general, the comparison between the vibration levels calculated according to Equation 3.51 and the measured data is quite good. Also, at some speeds there is considerable amplification of the measured vibration over the free vibration of the machine, showing the importance of the mode shape and damping in increasing the response amplitude under resonance vibration conditions.

We also note that a location on the cabinet is vibrating about 15 dB more than the machine vibrates. There are places, particularly on the cabinet, where the vibration exceeds the visual or rattle criteria, and there are places on the machine during operation where the comfort criteria is exceeded.

3.14 MODAL RESPONSE OF FINITE STRUCTURES

If we consider the general structure in Figure 3.36, driven by forces p with displacement response y, then the equation of motion is

$$\rho\ddot{y} + r\dot{y} + \Lambda y = p, \tag{3.52}$$

where ρ is mass density, r is a viscous damping coefficient, and Λ is a stiffness operator depending on the nature of the structural system. For one-dimensional bending, $\Lambda = \rho\kappa^2 c_1^2(d^4/dx^4)$. The spatial operator Λ has eigenfunctions ψ_m (mode shapes) and eigenvalues ω_m (resonance frequencies) determined by the equation

$$\frac{1}{\rho}\Lambda\psi_m = \omega_m^2\psi_m. \tag{3.53}$$

For the two-dimensional plate, we noted earlier that the mode shapes are

$$\psi_m = 2\sin k_{m_1}x_1 \sin k_{m_2}x_2, \tag{3.54}$$

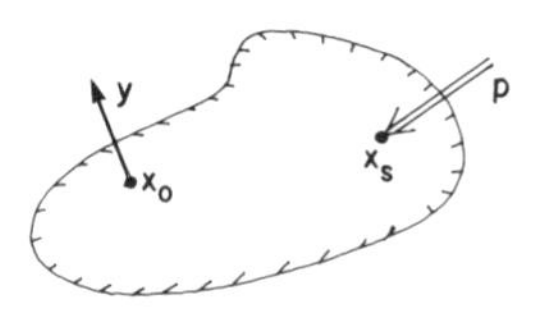

Figure 3.36 Excitation and response of linear continuous structure. The excitation may be distributed or localized at a point.

where the amplitude of the mode shape is chosen so that $\langle \psi_m^2 \rangle = 1$. These mode shape functions can be used to expand the spatial dependence of the displacement and the applied force to give

$$y = \sum_m Y_m(t)\psi_m(x) \qquad \frac{p}{\rho} = \sum_m \frac{L_m(t)\psi_m(x)}{M}, \tag{3.55}$$

where the Y_m's are modal displacements, the L_m's are modal forces, and M is the total mass of the structure.

If we substitute Equation 3.55 into Equation 3.52, we obtain the set of modal equations

$$\ddot{Y}_m(t) + \frac{r}{\rho}\dot{Y}_m(t) + \omega_m^2 Y_m = \frac{L_m(t)}{M}. \tag{3.56}$$

Each modal force produces a modal amplitude response according to a simple resonator equation, and every mode behaves as though it were a simple resonator independent of all the others responding to its own modal force. After computing the response of each modal resonator, we combine the responses according to Equation 3.55 into a total overall system response.

The resonator mass that appears in Equation 3.56 is the total mass of the structure. This is one way to represent modal mass; other textbooks express it differently. Our expression for the modal mass and its equality to total system mass is a result of our normalization of the eigenfunctions so that $\langle \psi_m^2 \rangle_\rho = 1$. If we have a point source of magnitude L_s at frequency ω, then we can form the ratio of the velocity at an observation point x_o to the force at the drive point x_s to get the transfer mobility

$$Y_{so} = \frac{j\omega y(x_o, t)}{L_s} = \frac{j\omega}{M}\sum_n \frac{\psi_n(x_s)\psi_n(x_o)}{\omega_n^2 - \omega^2 + j\omega\omega_n\eta}. \tag{3.57}$$

Each term in Equation 3.57 is the mobility of a simple resonator with an additional factor, the product of mode shape amplitudes at the source and observation points. The quantity $\psi_n(x_s)$ represents the effect of a particular source location on the generalized force for each mode and the second factor $\psi_n(x_o)$ gives the relative amplitude of response at the observation location. Suppose that we excite the system with a frequency $\omega = \omega_n$ and that the response amplitude at that frequency is dominated by the nth mode. Then the measured response is

$$V(x_o) = L_s \frac{\psi_n(x_s)\psi_n(x_o)}{\omega_n \eta M} \qquad (\omega = \omega_n) \tag{3.58}$$

By taking measurements at various observation and excitation positions, we can trace the mode shape to an acceptable degree of accuracy. If we put the source and observation positions at the same point (i.e., generate a drive point mobility) and

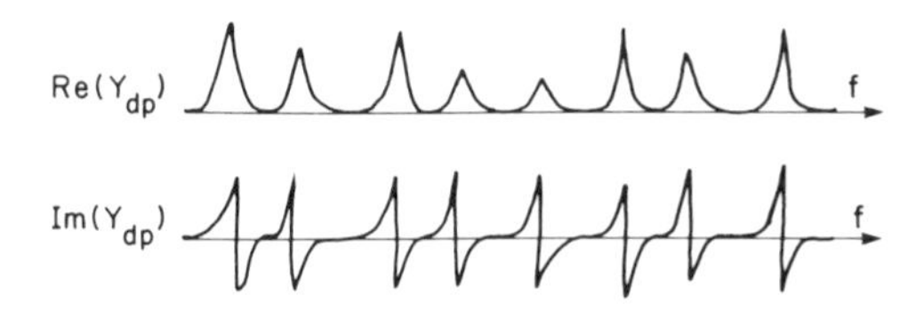

Figure 3.37 Real and imaginary parts of the drive point mobility of a resonant structure. Each modal resonance has a contribution to the mobility like that of a simple resonator.

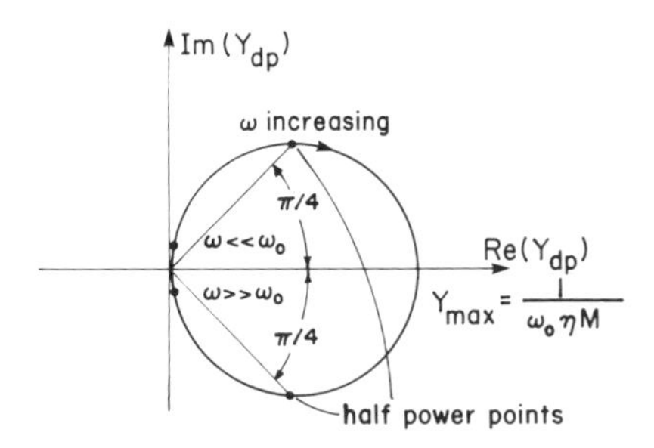

Figure 3.38 Nyquist diagram for the mobility of a single mode showing how damping may be obtained from the phase of the response.

graph the real part of Equation 3.57, we get the top graph in Figure 3.37. The imaginary part of the expression is the bottom graph. The real part is always positive, because for a drive point the power must be positive (i.e., from the source into the structure). The imaginary part relates to the reactance of the system and may either be masslike or stiffnesslike at any frequency and have positive or negative values. If we choose different locations for x_s and x_o, then the numerator in Equation 3.57 can be negative and the signs of some of the pulses in Figure 3.37 may change.

The *relative* signs of the real and imaginary parts do not change, however. If we plot real versus imaginary, we generate the Nyquist plot for one mode (Figure 3.38). The computer programs that develop mode shape and damping information using transfer functions have elaborated the ideas presented here. When there is overlap in modal frequency response or when the mode shape functions are complex (i.e., different parts of the structure move at relative phases to one another), such programs can still extract mode shapes.

3.15 AVERAGE DRIVE POINT MOBILITY

Let us now use these ideas to calculate the response of the structure to a source of noise in a frequency band Δf. We suppose that the source has a power spectral density of force, $W_1(f)$, that is flat over this frequency interval. Concentrating on one mode, $n = r$, we obtain the mean square force due to the noise spectrum for this mode:

$$\langle l^2 \rangle = W_l(f)\frac{\pi}{2}f\eta, \tag{3.59}$$

where $(\pi/2)f\eta$ is the equivalent noise bandwidth of the resonator.

Within this band, the transfer mobility for the resonator is $\psi_r(x_o)\psi_r(x_s)/\omega_r\eta M$, and the mean square velocity is

$$\langle v^2 \rangle = \langle l^2 \rangle |Y_{dp}|^2 = \frac{W_l(f)(\pi/2)\eta f_r}{4\pi^2(\eta M)^2 f_r^2}\langle\psi_m^2(x_o)\rangle\langle\psi_m^2(x_s)\rangle. \tag{3.60}$$

This gives the mean square velocity for a single mode, and if the modal density is $n(f)$ then the total mean square velocity due to the $n(f)\Delta f$ modes in the band is $(\langle\psi^2\rangle = 1)$

$$\Pi_{diss} = R\langle v^2 \rangle = 2\pi\eta f\langle v^2 \rangle M = \frac{W_l(f)n(f)\Delta f}{4M} = \frac{\langle l^2 \rangle n(f)}{4M}. \tag{3.61}$$

The left side is the power dissipated due to structural damping, which must equal the power injected into the structure by the point force. This power is equivalent to the mean square force times an average structure input point mobility:

$$\overline{G}_{dp} = \frac{n(f)}{4M} = \frac{1}{4M\overline{\delta f}} \tag{3.62}$$

The quantity $\overline{G}_{dp}$ represents the ability of the structure to absorb power from the noise source, and it depends on the number of modes available to absorb power. It is proportional to the modal density and inversely proportional to the mass of the structure. For a plate, $n(f) = A/2\kappa c_1$, and

$$\overline{G}_{dp} = \frac{1}{8\kappa c_1 \rho_s} \quad \text{(plate)}. \tag{3.63}$$

Figure 3.39 shows a drive point accelerance, the ratio of acceleration to force at the driving point on the rim of a model gear. The accelerance increases as the frequency increases. There is a mode at about 300 Hz and another one slightly over 1000 Hz, but above 2000 Hz the number of modes increases rapidly, accompanied by an increase in the average accelerance. We have seen that the modal density of the driven structure greatly affects the input mobility. In this regard, note the peak in the accelerance at about 20 kHz due to a shear wave cross-resonance on the face of the gear. At this frequency, there is an extra group of resonant modes that occurs and results in increased responsiveness of the structure in that frequency range.

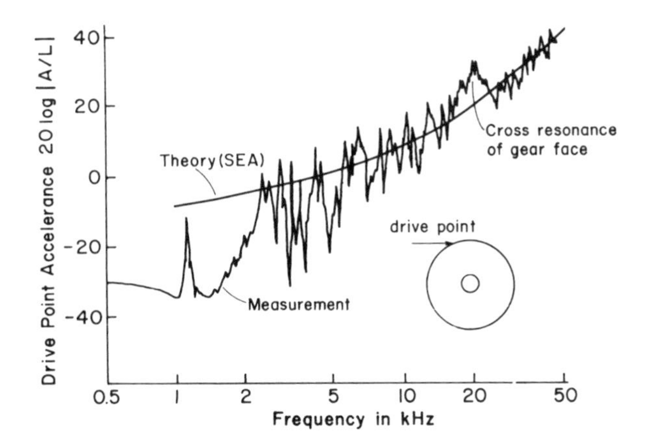

Figure 3.39 Drive point accelerance for model gear structure showing effect of increase in modal density as frequency increases.

3.16 TRANSFER FUNCTIONS

We now consider the transfer function for structures. The magnitude squared of the transfer function is by definition

$$|Y_{12}|^2 = \frac{\langle v_2^2 \rangle}{\langle l^2 \rangle}. \tag{3.64}$$

This relation is given by Equation 3.60, where v_2 is a typical reverberant velocity on the structure. The ratio of the mean square velocity to force is

$$\frac{\langle v_2^2 \rangle}{\langle l_1^2 \rangle} = \frac{\bar{G}}{\omega \eta M} = |Y_{12}|^2. \tag{3.65}$$

We can use these results to discuss a problem of considerable importance in the mounting of machinery. Figure 3.40 shows a machine sitting on a foundation of

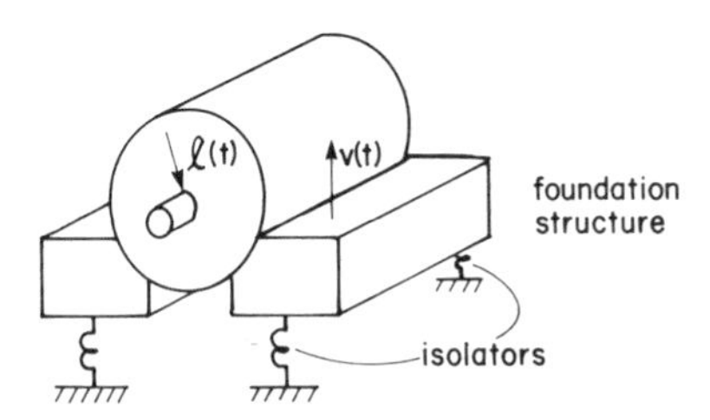

Figure 3.40 Machine is mounted on a foundation that, in turn, sits on isolators. The problem is to design the foundation to minimize the force transmitted by the isolators.

area A. The problem is to design the foundation so that it has the least motion possible at the mounting points where spring isolators are to be placed. The question is whether we should build as much mass or as much stiffness as possible into the foundation to reduce this vibration. The answer depends somewhat on whether we consider the mounting point to be at the same location where excitation forces are applied to the foundation (a "drive point") or at an average "reverberant" point. If we express the vibration in terms of bending rigidity and mass of the foundation, the mean square velocity at a general reverberant position is (D = bending rigidity)

$$\langle v^2 \rangle = \langle l^2 \rangle \frac{1}{8\omega\eta A \rho_s^{3/4} D^{1/2}}, \tag{3.66}$$

where we have used the relation

$$G = \frac{1}{8\sqrt{\rho_s D}}. \tag{3.67}$$

Assuming that the surface density of the foundation structure is proportional to the thickness, we find that the mean square velocity is inversely proportional to the cube of the thickness. At a drive point, however, the ratio of mean square velocity to force is

$$\frac{\langle v_{dp}^2 \rangle}{\langle l^2 \rangle} = \bar{G}^2 = \frac{1}{64 D \rho_s} \propto \frac{1}{h^4}. \tag{3.68}$$

At such a location, the response varies inversely with the fourth power of the thickness of the material. Also, we notice that the drive point velocity is inversely proportional to the bending rigidity, whereas the reverberant response is inversely proportional to the square root of the bending rigidity. Thus if the force is carried to the foundation at the mounting points for the spring isolators, we will reduce transmission equally by increasing the rigidity of the foundation structure or its mass. If supports are at an "average" or reverberant location, increased mass is slightly more effective than increased stiffness.

3.17 FORCE TRANSMISSION IN MACHINE STRUCTURES: AN APPLICATION OF RECIPROCITY

Mechanical vibration transmission is often expressed as a ratio of velocity to force, but very often we are interested in the transmission of force. When a force is applied at one place in a structure, our intuition tells us that it will tend to diminish as the vibrational energy spreads through the structure, but this is not always true.

Consider a machine sitting on a rigid foundation as shown in Figure 3.41. We model this situation as shown in part (b) of the figure, with an input force L_{in}

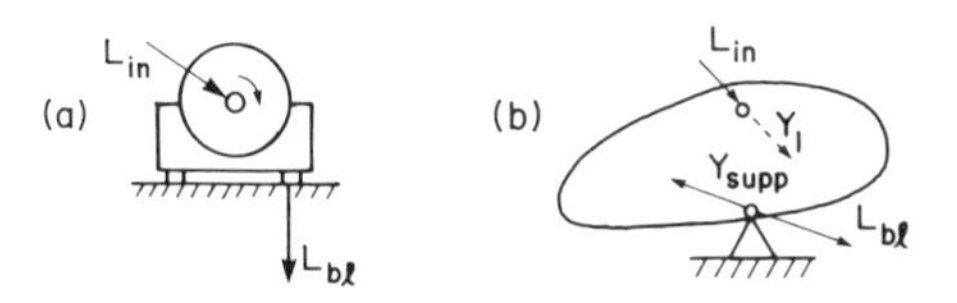

Figure 3.41 Structural transmission from an internal force in (a) to a support point is modeled as in (b).

representing the forces due to imbalance or whatever, driving the structure at a point where it has an input mobility Y_1. At the rigid support point, the blocked force is L_{bl}, and the point mobility looking into the structure from the supporting point is Y_{supp}. The power injected into this structure by the force is

$$\Pi_{in} = \langle L_{in}^2 \rangle G_1 = \frac{\langle L_{in}^2 \rangle \langle \psi_1^2 \rangle}{4 M_1 \delta f} N_1, \tag{3.69}$$

where we have used the previously derived formula, Equation 3.62, for the point conductance of a system that has N_1 modes and an applied excitation of bandwidth Δf. We have left in the mode shape factor to include the possibility that the drive point location might not be "a typical average" location.

The mean square velocity of the structure is determined by this input power and the damping and is

$$\langle v^2 \rangle = \frac{\Pi_{in}}{\omega \eta_1 M_1} = \frac{\langle L_{in}^2 \rangle G_1}{\omega \eta_1 M_1}, \tag{3.70}$$

resulting in the previously derived expression for the transfer function of this simple system:

$$\frac{\langle v^2 \rangle}{\langle L_{in}^2 \rangle} = |Y_{12}|^2 = \frac{G_1}{\omega \eta_1 M_1}. \tag{3.71}$$

If the system is linear, then the mean square blocked force will be proportional to the mean square velocity by a factor Γ or

$$\langle L_{bl}^2 \rangle = \Gamma \langle v^2 \rangle. \tag{3.72}$$

We now imagine a reciprocal experiment in which there is a prescribed velocity v'_{supp} at the support point. This prescribed velocity results in a mean square force $\langle L'^2_{supp} \rangle$ at the support point, determined from

$$\langle v'^2_{supp} \rangle = \langle L'^2_{supp} \rangle \cdot |Y_{supp}|^2, \tag{3.73}$$

and a power input to this structure at the support point, given by

$$\langle L'^2_{\text{supp}}\rangle G_{\text{supp}} = \Pi'_{\text{supp}} = \omega\eta_1 M_1 \langle v'^2\rangle. \tag{3.74}$$

The ratio of structural reverberant velocity to the support drive velocity is, therefore,

$$\frac{\langle v'^2\rangle}{\langle v'^2_{\text{supp}}\rangle} = \frac{G_{\text{supp}}}{\omega\eta_1 M_1 |Y_{\text{supp}}|^2} = \frac{\Gamma G_1}{\omega\eta_1 M_1}, \tag{3.75}$$

which by reciprocity determines Γ to be

$$\Gamma = \frac{G_{\text{supp}}}{|Y_{\text{supp}}|^2 G_1} = \frac{R_{\text{supp}}}{G_1}. \tag{3.76}$$

Suppose that we now locate an isolator at the place where the structure was previously supported by the rigid point. The complex mobility of the structure at this location is

$$Y_{\text{supp}} = G_{\text{supp}} + jB_{\text{supp}}, \tag{3.77}$$

where the real part is the conductance and the imaginary part is the susceptance. We now add the isolator as shown in Figure 3.42. We assume that we have chosen an isolator that has as small an amount of mass as possible; that is, it is a combination of damping and compliance with a mobility that is large compared to that of the mobility of the structure:

$$|Y_{\text{isol}}| \gg |Y_{\text{supp}}|. \tag{3.78}$$

The mobility looking into the system now from the rigid support point is $Y_{\text{comb}} = Y_{\text{isol}} + Y_{\text{supp}}$, which by our hypothesis is approximately equal to the isolator mobility alone.

If we apply a force at the support point for the isolator, then the power accepted by the system depends on G_{comb}, the real part of Y_{comb}. Since this force flows

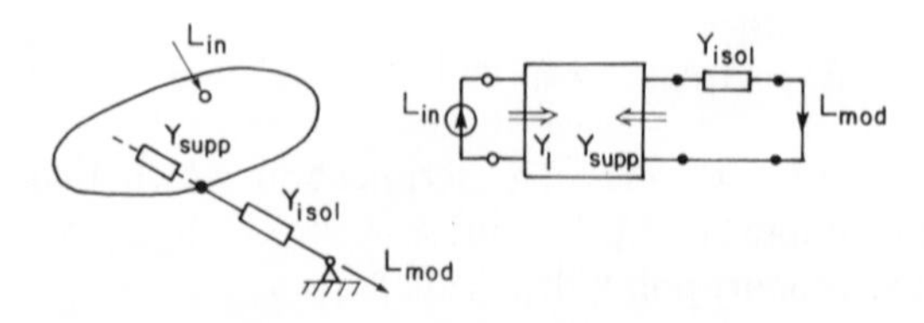

Figure 3.42 Modification of system diagram when an isolator is placed between the attachment point and the support.

through the isolator without attenuation (the isolator has no mass), we find that G_{comb} is approximately equal to G_{supp}, so part of Equation 3.75 is unchanged. We, therefore, have a new value of Γ, call it Γ_{mod}, given by

$$\Gamma_{mod} = \frac{\Gamma |Y_{supp}|^2}{|Y_{isol} + Y_{supp}|^2} \simeq \Gamma \left| \frac{Y_{supp}}{Y_{isol}} \right|^2, \tag{3.79}$$

which represents a reduced transmitted force because of the large value of Y_{isol} compared to Y_{supp}.

Equation 3.79 shows how the transmitted force is reduced by making the isolator softer. There is a limit as to how soft the isolator can be, since its static deflection due to the weight of the machine must be limited. Indeed, we are usually trying to reduce static deflection, because if the isolator is too soft there is a tendency for the machine to move or change its orientation in a way that is not always controllable or desirable.

3.18 ISOLATOR PERFORMANCE ON FLEXIBLE STRUCTURES

The analysis to this point represents standard isolator design practice. Often however, the structure that the isolator sits on is not rigid enough. This is particularly true in some modern buildings where a machine may be placed on a floor that is quite flexible; it may also be a problem in mounting machinery in very heavy ship structures, particularly if one tries to use a stiff isolator design so that static deflection is minimized. Then it may turn out that even an apparently stiff foundation may not be stiff enough to provide isolator performance as designed.

Again consider the machine to be excited by an internal force. Its mobility at the support point is Y_{supp}. The isolator that attaches it to the foundation has mobility Y_{isol}, and the foundation has a mobility Y_{found}, as shown in Figure 3.43. The mechanical diagram for this system is shown in Figure 3.44. Using this diagram, we

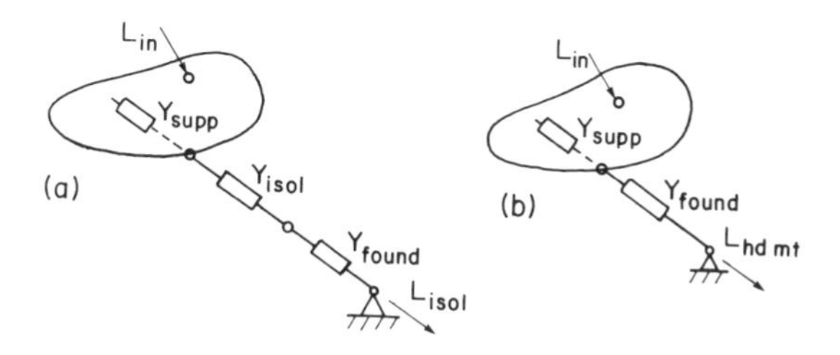

Figure 3.43 An isolated machine that sits on a foundation that has a high mobility may not have effective isolation unless $Y_{isol} \gg Y_{found}$.

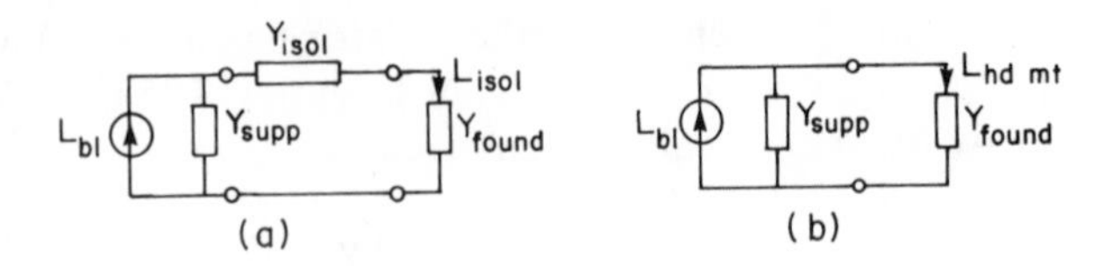

Figure 3.44 Mechanical mobility diagrams for isolated machine sitting on a nonrigid foundation.

obtain the force into the foundation:

$$L_{\text{found}} = L_{\text{bl}} \cdot \frac{Y_{\text{supp}}}{Y_{\text{supp}} + Y_{\text{isol}} + Y_{\text{found}}}. \tag{3.80}$$

We can find the force that will be put into the foundation if the machine were hard-mounted by simply setting $Y_{\text{isol}} = 0$. We thus obtain an expression for the ratio of the m.s. (mean square) force when the isolator is present compared to the m.s. force when the machine is hard-mounted:

$$\frac{\langle L^2 \rangle_{\text{isol}}}{\langle L^2 \rangle_{\text{hd-mt}}} = \frac{|Y_{\text{supp}} + Y_{\text{found}}|^2}{|Y_{\text{supp}} + Y_{\text{isol}} + Y_{\text{found}}|^2}. \tag{3.81}$$

Isolators are generally selected to have a bounce resonance frequency of about 10 Hz when combined with the weight of the machine; the foundation structure will begin to break up into resonant modes, and their mobilities will tend to become more like a dashpot at the higher frequencies. If the isolator mobility continues to rise, then attenuation will occur, as predicted by Equation 3.80.

CHAPTER 4

Vibration Transmission in Machine Structures

4.1 INTRODUCTION

This chapter explores ways that vibrational energy is transmitted from one part of a machine structure to another. It commonly happens that a part of a machine structure, such as an engine piston and connecting rod assembly, is directly excited by external forces, and other parts of the machine, such as the outer casing, receive their vibrational energy by vibration transmission. We want to understand what determines the spectrum of the transmitted energy and, in particular, how this depends on the junctions between the mechanical components. We shall find that there is a combination of experimental and analytical techniques useful for this purpose and can provide guidance in machine design to reduce vibration transmission. Because of our interest in energy, we emphasize the *magnitude* of response here. In Chapter 7, because of our interest in diagnostics, we shall be discussing the *phase* of the response.

4.2 ANALYZING VIBRATION TRANSMISSION BY TWO-PORT METHODS

This section introduces some of the basic ideas of vibration transmission of a small diesel engine. We examine each component as though it were a single input–output device, as shown in Figure 3.28. We noted in Chapter 3 that the input–output relations of a structural element can often be represented by a symmetric transfer matrix (e.g., the transfer mobility $Y_{12} = Y_{21}$). This is an example of the general principle of reciprocity, which applies to many structural systems.

We shall use reciprocity several times in this book. It has theoretical uses, but it also has important experimental applications. Sometimes it is convenient to measure force to acceleration in one direction in a structure and impractical in the other. For example, if we want to measure the ratio of acceleration of the engine block to a force applied to the top of a piston, it might be more convenient, because of the size of the sensors involved, to apply the shaker to the block and measure the response at the top of the piston with an accelerometer.

Figure 4.1 is a cross-sectional view of the four-cylinder engine that we will analyze. In Chapter 2, we studied combustion pressure as a source of engine structural vibration. We now examine the transmission of vibration through the engine structure using the transfer mobility matrices discussed in Chapter 3. Figure 4.2 is a sketch of the engine with the three major components of noise transmission: the piston and connecting rod, the crankshaft, and the engine casing. These components are shown in the figure as two-port networks labeled ①, ②, and ③.

Figure 4.3 shows the vibration magnitude at a location on the engine surface when various pistons are impacted. There is little difference between the vibration under these circumstances. We may infer that there is essentially the same transmission from any piston top to a typical surface location. The reason is that reverberation in the engine structure makes the vibration almost uniformly spread over the structure. Figure 4.4 is a similar set of spectra of surface vibrations when a piston is impacted just before, during, and just after TDC. Again, we see very little difference in the transmitted vibration, indicating that crank angle has little effect on vibration transmission, at least over this range.

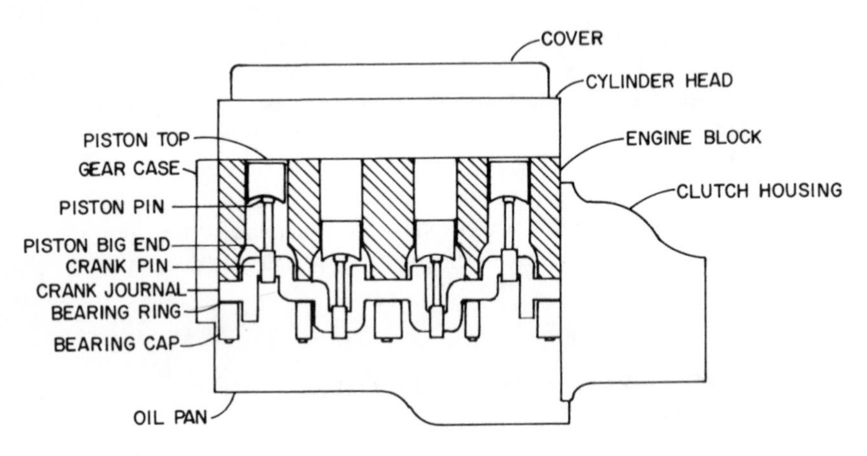

Figure 4.1 Cross-sectional view of diesel engine. The components of the transmission of vibration through the piston–connecting rod path to the engine casing are shown here.

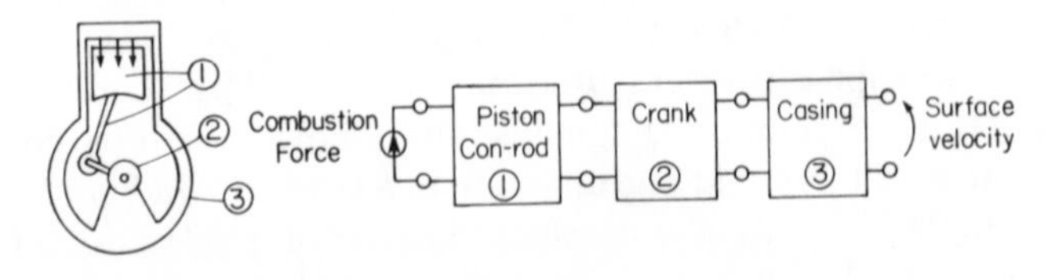

Figure 4.2 Division of the transmission path into three structural components. Overall transmission is determined from the joining of these components.

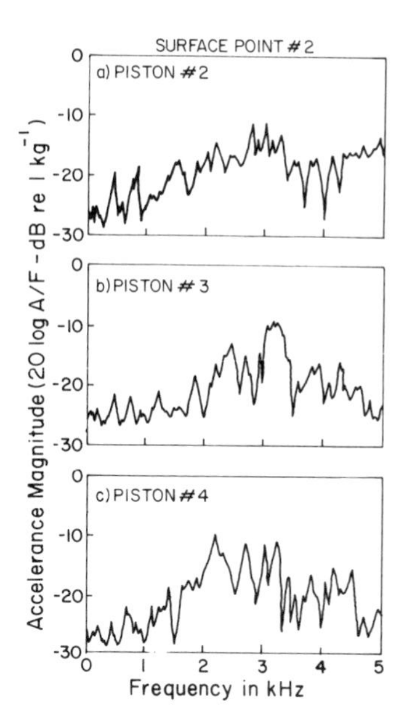

Figure 4.3 Vibration of casing due to impact of various pistons, all at TDC, shows small difference in transmitted level and therefore independent changing distance from impact.

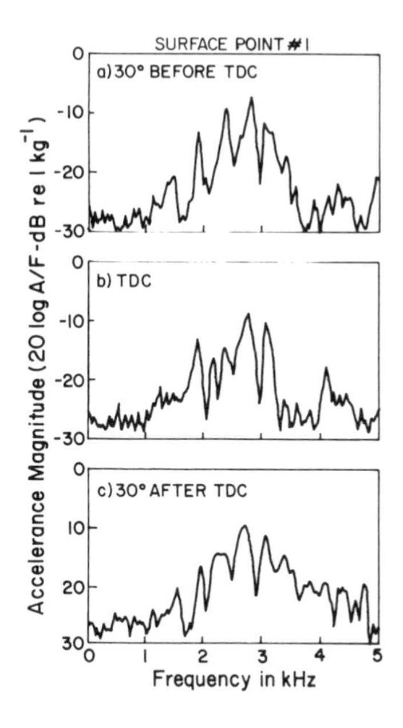

Figure 4.4 The transfer accelerance from piston top to engine casing shows relatively little change as a function of crank angle around TDC.

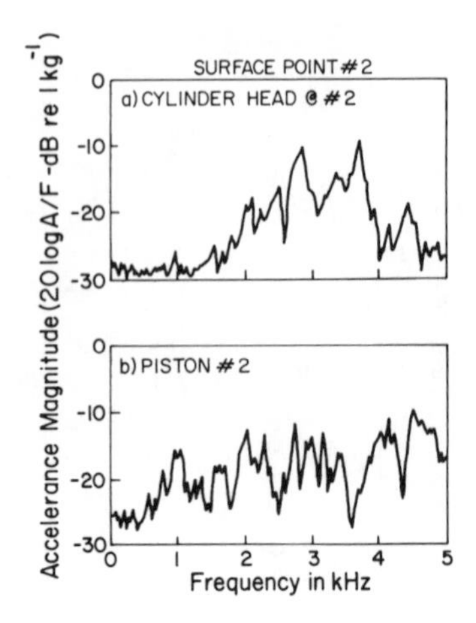

Figure 4.5 Comparison of transfer accelerance for transmission through head and through piston–connecting rod–crank to casing paths.

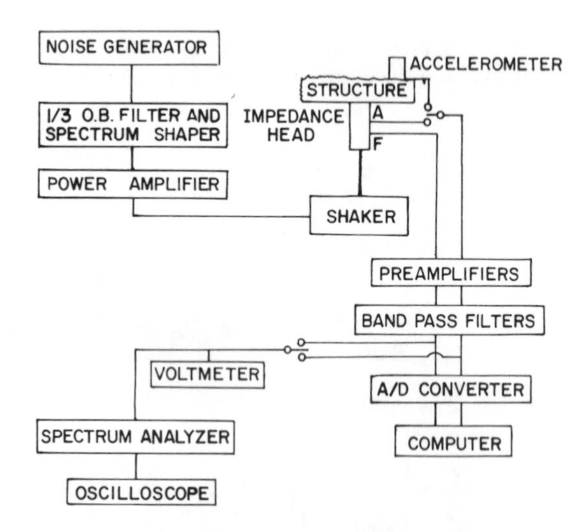

Figure 4.6 Mobility measurements are made using a minicomputer to calculate the Fourier transforms and cross spectra. Note the arrangement for monitoring the force and acceleration spectra.

The most interesting data are in Figure 4.5, which compares the vibration transmission when the piston–connecting rod combination is pushed down and the underside of the cylinder head is pushed up. We see that vibration below 3 kHz is much stronger when the force is applied to the piston, indicating that the piston, connecting rod, and crankshaft combination is the dominant path in this frequency

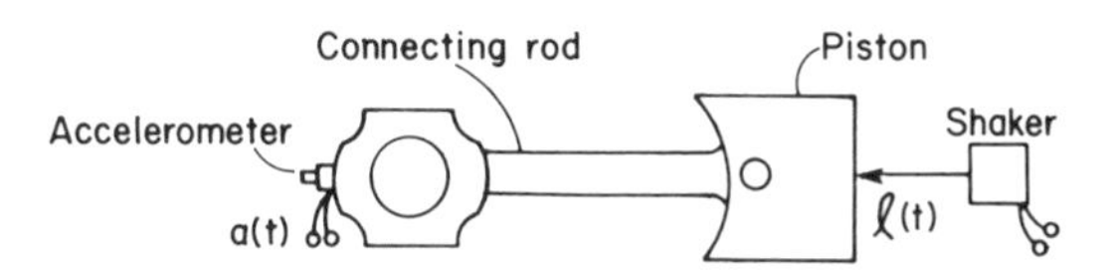

Figure 4.7 Arrangement for measuring the transfer accelerance from the piston top to the connecting rod big end.

range. This is also the frequency range that governs *A*-weighted sound levels for diesel engine noise.

Since the data in Figure 4.5 indicated that the transmission of energy from the combustion chamber to the casing of the engine occurred through the piston–connecting rod–crankshaft path for frequencies less than 3 kHz, the engine components in this path were analyzed separately, and the input and transfer mobilities for each component were determined. The transfer functions were measured using the system in Figure 4.6. The noise generator drives a dynamic shaker through an impedance head; a force gauge and accelerometer in the impedance head, or a roving accelerometer, are used to obtain signals for computing the input or transfer mobility. The signals from these sensors are sent to a two-channel A/D converter and then into a computer for calculating transfer functions. A real-time analyzer monitors the spectrum of the two signals to insure a reasonable signal-to-noise level for the measurements.

This piston and connecting rod assembly is shown in Figure 4.7. Figure 4.8 shows the drive point mobility looking into the top of the piston. The shaker is placed at the top of the piston, and the impedance head is used to measure the drive point mobility. We see a minimum in the mobility (antiresonance) at about 2.7 kHz, and a peak in the mobility (resonance) at about 4.3 kHz. At very low frequencies, the mobility is that of a mass, corresponding to translation of the entire assembly. The antiresonance corresponds to the system resonance when the drive point is held fixed—that is, if the piston top were sitting on an infinitely rigid surface, as shown, and the big end of the rod were vibrating against this rigid surface. The resonance at 4.3 kHz corresponds to a "dumbbell resonance" in which the piston and the big end move in opposite directions and the connecting rod acts as a spring.

To illustrate, we examine the results of two experiments on the piston and connecting rod assembly. In the first, the top of the piston is excited with a shaker and the acceleration is measured at the big end of the connecting rod, as shown in Figure 4.7. In the second experiment, the big end of the connecting rod is excited, and the vibration at the top of the piston is measured. If reciprocity applies, then these two experiments should give the same results.

Figure 4.9 shows the results of the experiment in which the piston top is driven and the big-end vibration is measured. Note that the resonance is still evident, but that the antiresonance is not. If we now carry out the reciprocal experiment by driving the big end and measuring the vibration at the top of the piston, we get the data shown in Figure 4.10. An overlay of these two curves in magnitude and phase shows that they are essentially identical and that reciprocity does apply.

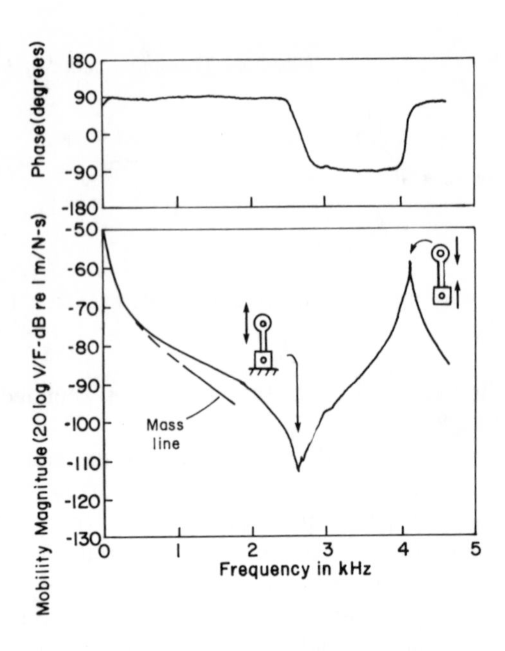

Figure 4.8 Drive point mobility at the top of the piston. Note the antiresonance at about 2.7 kHz and the resonance at about 4.1 kHz.

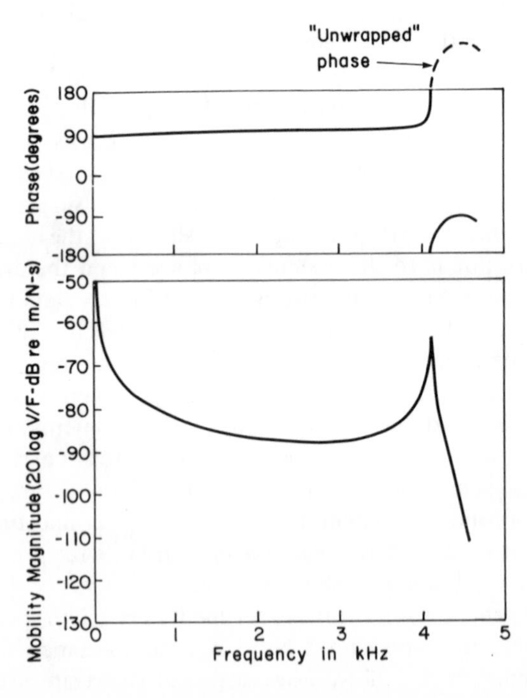

Figure 4.9 Transfer mobility from piston top to connecting rod big end. Note absence of antiresonance and that the interval between the poles at $f = 0$ and $f = 4.1$ kHz does not contain a zero for this transfer function.

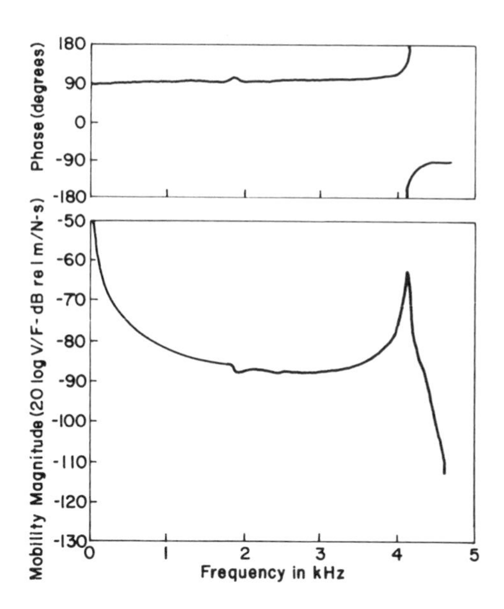

Figure 4.10 Transfer mobility from connecting rod big end to piston top. By reciprocity, this transfer function should be identical to that from the piston top to the con rod big end. The two functions are very nearly equal.

The mobility matrix for each element in Figure 4.2 was evaluated. These matrices can be combined by the methods of the next section to predict the overall transmission from combustion force to surface velocity. We therefore leave the present example to develop the needed theoretical relationship.

4.3 JOINING SYSTEM COMPONENTS

Suppose we have two systems, a and b, for which we know the mobility matrices separately. We wish to join them and determine the overall system transfer mobility. We label the terminal pairs of system a 1 and 2 and the terminals of system b 3 and 4, as shown in Figure 4.11.

The transfer mobility matrix is not the most convenient form for this joining process, and we use instead a matrix α_{ij} that relates the input and output drop and flow variables. For example, for system a, we have

$$V_1 = L_1 Y_{11} + L_2 Y_{12}, \qquad V_2 = L_1 Y_{12} + L_2 Y_{22}, \tag{4.1}$$

from which we can solve for V_1 in terms of V_2 and L_2 to get

$$V_1 = \frac{Y_{11}}{Y_{12}} V_2 + \left(Y_{12} - \frac{Y_{22} Y_{11}}{Y_{12}} \right) L_2. \tag{4.2}$$

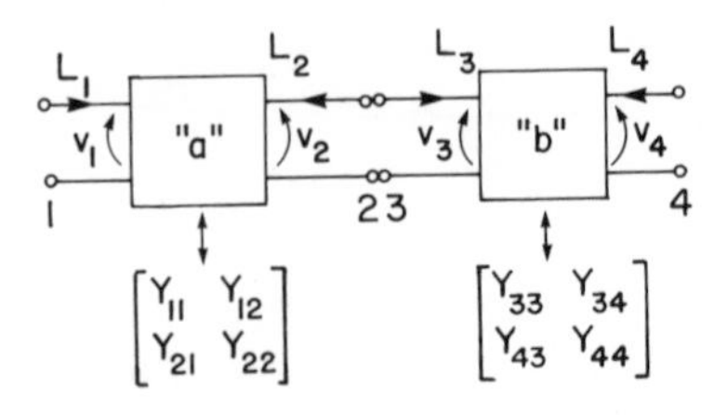

Figure 4.11 The joining of two systems requires compatibility of joining conditions on the port variables. In this case, they are $L_2 = -L_3$, $V_2 = V_3$.

Equation 4.2 and the second equation of 4.1 give us the desired matrix

$$\begin{pmatrix} V_1 \\ L_1 \end{pmatrix} = \begin{pmatrix} \alpha_{11} & \alpha_{12} \\ \alpha_{21} & \alpha_{22} \end{pmatrix} \begin{pmatrix} V_2 \\ L_2 \end{pmatrix}, \tag{4.3}$$

where

$$\begin{aligned} \alpha_{11} &= \frac{Y_{11}}{Y_{12}}, \qquad & \alpha_{12} &= \frac{Y_{12}^2 - Y_{11}Y_{22}}{Y_{12}}, \\ \alpha_{21} &= \frac{1}{Y_{12}}, \qquad & \alpha_{22} &= -\frac{Y_{22}}{Y_{12}}. \end{aligned} \tag{4.4}$$

One also has a similar matrix between the input (V_3, L_3) and the output (V_4, L_4) for system b. When a is joined to b, $V_3 = V_2$ and $L_3 = -L_2$. This allows us to write

$$\begin{aligned} \begin{pmatrix} V_1 \\ L_1 \end{pmatrix} &= \begin{pmatrix} \alpha_{11} & \alpha_{12} \\ \alpha_{21} & \alpha_{22} \end{pmatrix} \begin{pmatrix} V_2 \\ L_2 \end{pmatrix} = \begin{pmatrix} \alpha_{11} & \alpha_{12} \\ \alpha_{21} & \alpha_{22} \end{pmatrix} \begin{pmatrix} \alpha_{33} & \alpha_{34} \\ -\alpha_{43} & -\alpha_{44} \end{pmatrix} \begin{pmatrix} V_4 \\ L_4 \end{pmatrix} \\ &\equiv \begin{pmatrix} \alpha'_{11} & \alpha_{14} \\ \alpha_{41} & \alpha'_{44} \end{pmatrix} \begin{pmatrix} V_4 \\ L_4 \end{pmatrix}. \end{aligned} \tag{4.5}$$

The transfer mobility we seek is $1/\alpha_{41} = Y_{41} = V_4/L_1$ for $L_4 = 0$. We see from Equations 4.3 and 4.4 that the lower left component of the matrix is a reciprocal of the desired transfer mobility. We evaluate α_{41}, which, from Equation 4.5, is

$$\alpha_{41} = \alpha_{21}\alpha_{33} - \alpha_{22}\alpha_{43} = \frac{Y_{22} + Y_{33}}{Y_{12}Y_{34}}.$$

Taking the reciprocal gives

$$Y_{14} = \frac{Y_{12}Y_{34}}{Y_{22} + Y_{33}}. \tag{4.6}$$

Equation 4.6 tells us how to join two systems to obtain an overall transfer function from an input at a location on one component to the vibration at a location on the second.

4.4 APPLICATION OF THE JOINING RELATION TO A DIESEL ENGINE

The transfer mobilities measured for two of the elements in Figure 4.2 were combined according to Equation 4.6 to determine the overall transfer mobility from the piston top to the crankshaft journal. The result is shown in Figure 4.12. In addition, the vibration transmission measured with the piston and connecting rod attached to the crankshaft is shown in Figure 4.13. Comparing Figures 4.12 and 4.13 shows good agreement up to about 3 kHz, indicating that the simple force and translational velocity model of subsystem interaction represented by Figure 4.2 is adequate up to 3 kHz. Above that frequency, however, the moments and rotational motion at the junctions between system components become important.

One advantage of this model is that even though it is based on experimental data for the system components, it is a computational model, and therefore one can conceptually add other elements such as resilient pads to the system and predict their effect in reducing vibration transmission. Such a result is shown in Figure 4.14

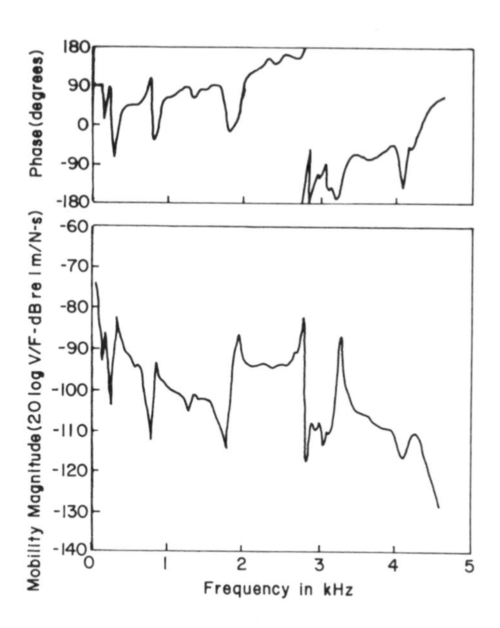

Figure 4.12 Transfer mobility from the top of piston 2 to crankshaft journal 3 calculated from experimentally determined structural component mobility matrices.

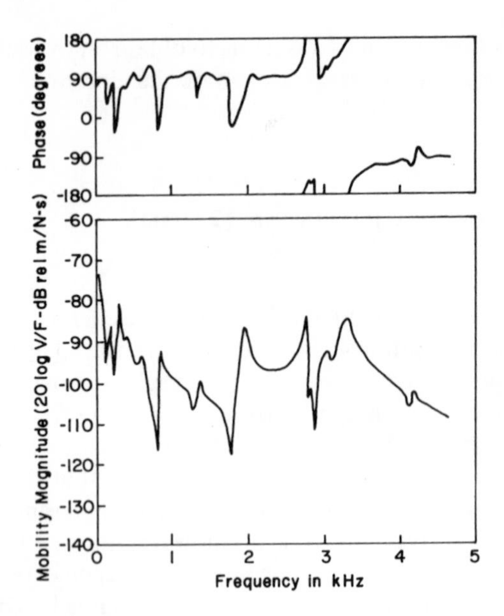

Figure 4.13 Measured transfer mobility from top of piston 2 to crankshaft journal 3 when piston and con rod are attached to crankshaft. Agreement with calculation is good up to 3 kHz.

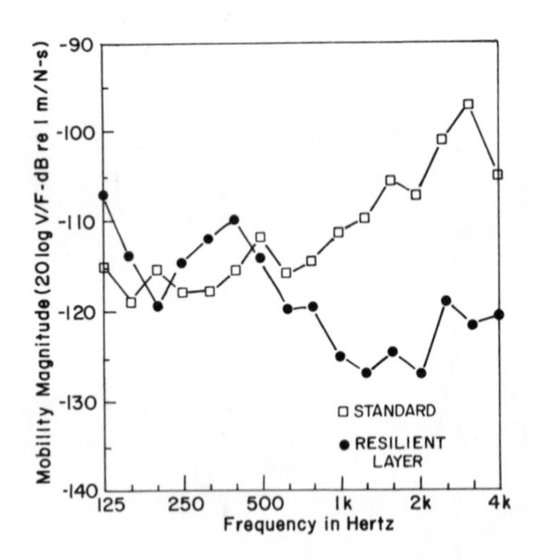

Figure 4.14 Reduction in transfer mobility from piston top to casing when the crankshaft is resiliently supported.

when a resilient layer is added to the crankshaft journal bearing support structure. Theoretical calculations predict a rather abrupt dropoff in transmission for frequencies greater than 500 Hz, with some amplification of transmitted vibration just below 500 Hz for this configuration. The engine was assembled with such resilient bearing support pads at the crankshaft journals, and the resulting vibration transmission is shown in the figure. A benefit from this modification is that the basic data on the components are useful for evaluating such a design modification. In general, if we have transfer mobilities, whether determined experimentally or analytically, we can use them as a design tool to see the effect of proposed changes.

4.5 NOISE TRANSMISSION THROUGH STRUCTURAL JUNCTIONS

We note from Equation 4.6 that the assembled system transfer mobility is not simply a product of two transfer mobilities, but that the mobilities at the joining parts are also important in the overall transfer function. For example, if we connect an isolator at input 3 of system b, then we would have the modified system shown in Figure 4.15(b). This would greatly increase the mobility Y_{33}, which, from Equation 4.6, would reduce the transfer mobility, as we would expect. This "isolator" might be an added component or a compliant connection in a structure.

The expression also shows that for an isolator to have any effect the mobility that it adds must be large compared to the mobility that was already present at the junction. We have already seen this effect in Section 3.17. Thus, an isolator which might work very well joining low-mobility components (high mass or stiffness) might not have as much effect in reducing the vibration transmission if that same isolator is used on a soft structure where one of the junction mobilities is large.

Let us now use Equation 4.6 to estimate the mean square response of the structure b when structure a is driven by a band of noise over some frequency interval Δf. Accordingly,

$$\frac{\langle v_4^2\rangle_{\Delta f}}{\langle l_1^2\rangle_{\Delta f}} = |Y_{14}|^2 = \frac{|Y_{12}|^2|Y_{34}|^2}{|Y_{22}+Y_{33}|^2}. \tag{4.7}$$

We represent $|Y_{12}|^2$ by the ratio $\langle v_a^2\rangle/\langle l_1^2\rangle$, where $\langle v_a^2\rangle$ is a typical free motion of structure a. From Equation 3.65, we can represent $|Y_{34}|^2$ as the ratio of the average drive point mobility G_b of structure b to $\omega\eta_b M_b$, or using $G_b = n_b/4M_b$ we obtain

$$\frac{\langle v_b^2\rangle}{\langle l_1^2\rangle} = \frac{\{\langle v_a^2\rangle/\langle l_1^2\rangle\}(1/\omega\eta_b M_b)}{|Y_{22}+Y_{33}|^2}\frac{n_b(f)}{4M_b}. \tag{4.8}$$

Note that in deriving Equation 4.8 we have ignored the loss of energy from system b to system a. We shall rederive this relation more carefully in Section 4.7.

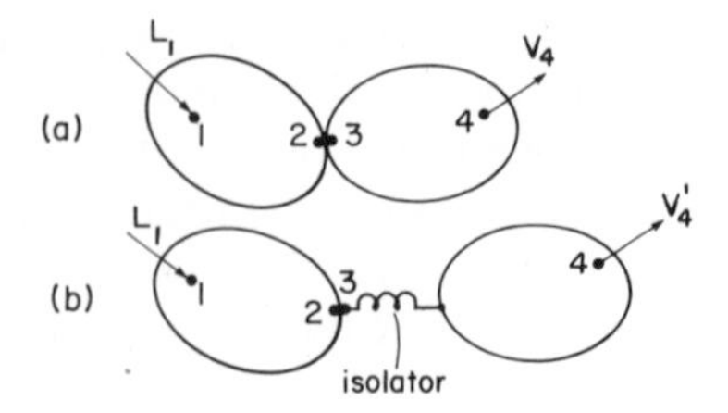

Figure 4.15 Two systems connected at a point are shown in (a). If an isolator is interposed between these systems, then one of the junction mobilities, Y_{33} in this case, is altered.

Equation 4.8 can be rewritten as

$$\frac{M_b\langle v_b^2\rangle}{n_b\Delta f} = \frac{M_a\langle v_a^2\rangle}{n_a\Delta f} \cdot \frac{\eta_{ba}}{\eta_b}, \tag{4.9}$$

where

$$\eta_{ba} = \frac{4}{\omega\eta_b(f)} \frac{G_a G_b}{|Y_{22} + Y_{33}|^2}, \tag{4.10}$$

which is a positive quantity. Equation 4.9 says that the average energy per resonant mode in the bandwidth Δf of the second system b is equal to that of the directly driven system a times the ratio of the quantity η_{ba}, called the *coupling loss factor*, to η_b, the damping of the second system.

The coupling loss factor in Equation 4.10 is determined by the modal density of the second system and the properties at the junction of the two structures. The loss factor η_{ba} represents the damping that structure a puts on structure b because of the connection between them. We notice from Equation 4.10 that $n_b\eta_{ba}$ is a symmetric quantity and is therefore equal to $n_a\eta_{ab}$. Thus, if we have calculated the coupling damping that structure a puts on structure b, then we can use this consistency relationship to determine the damping load that structure b puts on a by its connection.

For example, let system a be a wall panel in a room and system b the sound field itself. Then η_{ab} is the panel radiation damping produced by the sound field. If we are interested in calculating the damping that such panels provide to reverberant sound fields, we can use the relationship

$$\eta_{ba} = \frac{\eta_{ab} n_a}{n_b}, \tag{4.11}$$

which gives the room loss factor due to panel absorption when the radiation loss factor η_{ab} and n_a, n_b, the modal densities of the panel and the sound field, are all known.

4.6 TRANSMISSION LINE "FINITE ELEMENT" ANALYSIS

Some structures can be modeled as combinations of simple lumped or beam elements for which input and transfer mobility (or their inverse, impedance) matrices are known analytically. When this is the case, relatively simple calculations can predict system transfer functions in much the same way that experimentally determined mobilities were used in Section 4.4.

Each element, such as spring K, shown in Figure 4.16(a) has a force–velocity relationship that may be expressed as an impedance. A spring has two nodes, one at each end, and a force and velocity at each node. The impedance matrix of a spring, given the sign conventions of Figure 4.16, is

$$Z_{\text{spring}} = \begin{bmatrix} \dfrac{K}{j\omega} & \dfrac{-K}{j\omega} \\ \dfrac{-K}{j\omega} & \dfrac{K}{j\omega} \end{bmatrix} \tag{4.12}$$

where

$$\begin{Bmatrix} L_1 \\ L_2 \end{Bmatrix} = Z \begin{Bmatrix} V_1 \\ V_2 \end{Bmatrix} \tag{4.13}$$

V_i is the velocity of node i and L_i is the force applied to the node i. For the damper

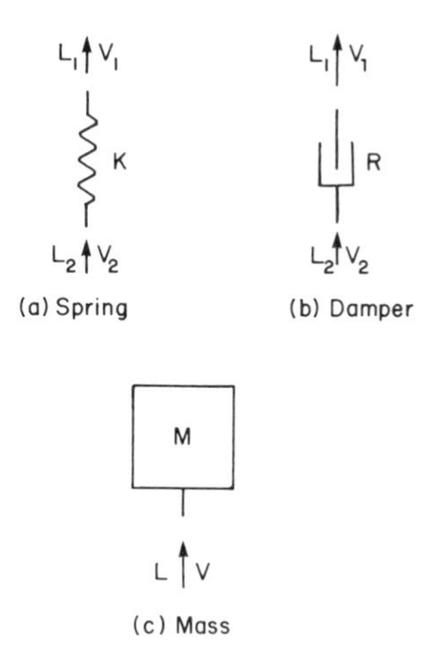

Figure 4.16 Lumped elements with force–velocity coordinates can be represented by impedance or mobility matrices.

shown in Figure 4.16(b), the impedance matrix is

$$Z_{\text{damper}} = \begin{bmatrix} R & -R \\ -R & R \end{bmatrix}. \tag{4.14}$$

Since the velocity of a mass shown in Figure 4.16(c) can be described by a single node,

$$Z_{\text{mass}} = j\omega M. \tag{4.15}$$

The flexural motion of a beam, shown in Figure 4.17, is described by translation and rotation at each end. The physical properties of the flexural beam element are length L, cross-sectional area A, density ρ, Young's modulus E, and cross-sectional area moment of inertia $I = A\kappa^2$. The impedance matrix Z is

$$\begin{Bmatrix} L_1 \\ M_1 \\ L_2 \\ M_2 \end{Bmatrix} = \begin{bmatrix} Z_{11} & Z_{12} & Z_{13} & Z_{14} \\ Z_{21} & Z_{22} & Z_{23} & Z_{24} \\ Z_{31} & Z_{32} & Z_{33} & Z_{34} \\ Z_{41} & Z_{42} & Z_{43} & Z_{44} \end{bmatrix} \begin{Bmatrix} V_1 \\ X_1 \\ V_2 \\ X_2 \end{Bmatrix}, \tag{4.16}$$

where
$$\begin{aligned}
Z_{11} &= iZ_0k(\cos kL \sinh kL + \sin kL \cosh kL)/\Delta, \\
Z_{12} &= iZ_0(\sin kL \sinh kL), \\
Z_{22} &= iZ_0k^{-1}(\sin kL \cosh kL - \cos kL \sinh kL)/\Delta, \\
Z_{31} &= jZ_0k(\sin kL + \sinh kL)/\Delta, \\
Z_{41} &= jZ_0(-\cosh kL + \cos kL)/\Delta, \\
Z_{42} &= jZ_0k^{-1}(-\sinh kL + \sin kL)/\Delta, \\
Z_{32} &= -Z_{41}, \\
Z_{33} &= Z_{11}, \\
Z_{43} &= -Z_{21}, \\
Z_{44} &= Z_{22}, \\
Z_0 &= \rho A c_1 \kappa,\ c_1 = \sqrt{E/\rho},\ \kappa^{-1} = \sqrt{A/I}, \\
k &= \sqrt{\omega/\kappa c_1}, \\
\Delta &= 1 - \cos kL \cosh kL, \\
Z_{ji} &= Z_{ij} \text{ (i.e., the matrix is symmetric).}
\end{aligned}$$

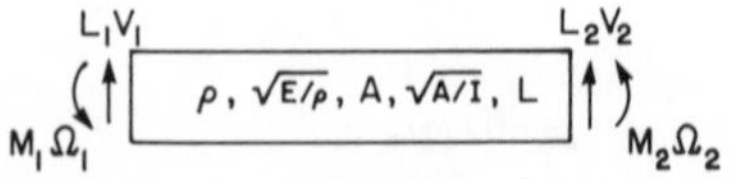

Figure 4.17 The bending beam element of length L used in sample analysis of a machine foundation.

Although the system impedance is generally complex, the matrices for these elements are typically purely real or purely imaginary. Suppose we wish to analyze the mounting system for a gearbox in a ship, as shown in Figure 4.18. The gear itself is modeled as a set of mass elements, and the case rail is the foundation structure shown in Figure 3.40. The isolator mounts are a parallel combination of spring and damper as shown, and the subbase in Figure 4.18 is a heavy beam structure that in turn rests on the hull elements, modeled as mounts with stiffness and damping. The case rail and submount are steel I-beams sketched in Figure 4.19. The parameter values for all these parts are given in Table 4.1.

The computer program uses the element transfer and input impedances in Equation 4.16 and assembles them into a complete system matrix by keeping the parts of interest and joining the subsystems at the numbered nodes while requiring the total force to vanish (or be equal to the external applied force, if any) at each junction and the velocities of all joining elements to be the same. This technique is the same as any finite element program, except that the transfer functions that define the element are functions of frequency.

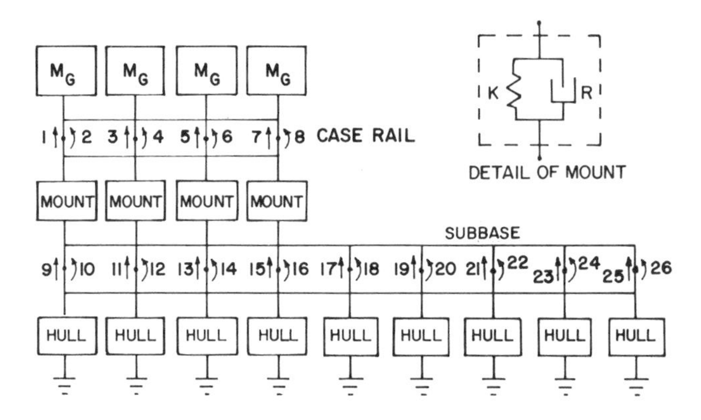

Figure 4.18 Model of reduction gear mounting system.

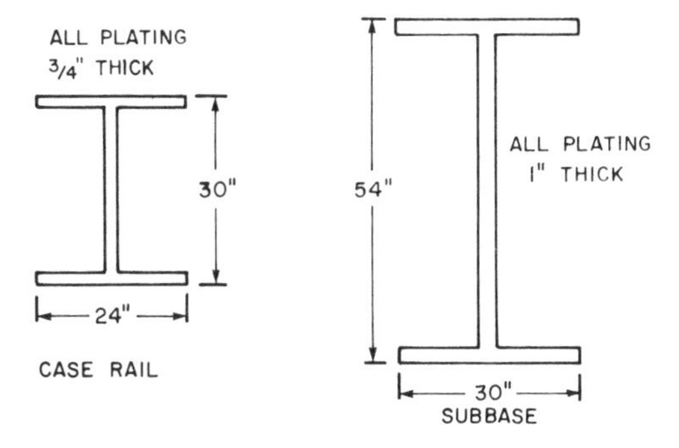

Figure 4.19 Cross sections of case rail and subbase.

Table 4.1 Baseline Parameter Values for Mounting System

Lumped mass of gear case (in addition to rail)	$M = 3700$ kg (for 4 per side)
Mount: stiffness	$K = 6.2 \times 10^7$ N/m
damping	$R = 4.9 \times 10^4$ N-sec/m
Hull: stiffness	$K = 10^9$ N/m
damping	$R = 2 \times 10^4$ N-sec/m
Material properties (for case rail and subbase): density	$\rho = 7850$ kg/m^3
bar wave speed	$c_L = \sqrt{E/\rho} = 5110$ m/sec
shear wave speed	$c_s = \sqrt{G/\rho} = 3180$ m/sec
Case rail: length	$L = 0.813$ m/element (for 3 elements)
cross-sectional area	$A = 0.037$ m^2
characteristic wave number	$\kappa^{-1} = A/I = 3.12$ m^{-1}
shear factor	$\gamma = 0.95$
Subbase: length	$L = 0.813$ m/element (for 8 elements)
cross-sectional area	$A = 0.0732$ m^2
characteristic wave number	$\kappa^{-1} = A/I = 1.80$ m^{-1}
shear factor	$\gamma = 0.95$

When the system matrix is assembled, various calculations can be made. Figure 4.20, for example, shows the power radiated by the hull—that is, the total power dissipated in the hull elements of Figure 4.18 due to a unit force at node 1. Not only does the model allow us to calculate quantities of interest in this way, but changes in material dimensions and properties can be made to assess their effect on the quantity of interest—in this case, radiated sound power.

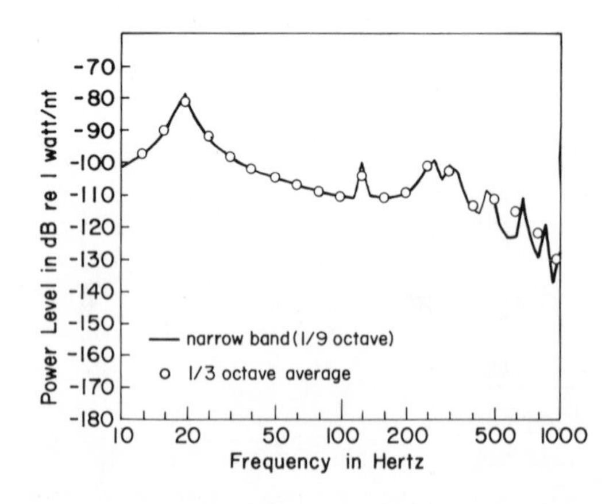

Figure 4.20 Radiated power for unit force at node 1, baseline parameter values.

4.7 STATISTICAL ENERGY ANALYSIS

In this section, we derive the statistical energy analysis (SEA) relationships by using the principle of reciprocity. We imagine two connected structures, a plenum and duct, labeled systems ① and ②, as shown in Figure 4.21. We first imagine that the plenum is excited by a point force that is band-limited white noise over the frequency interval Δf. The power spectral density of the excitation is $\langle l_1^2 \rangle / \Delta f$. The power injected into the plenum from the point source is

$$\Pi_1 = \langle l_1^2 \rangle G_1, \qquad G_1 = \frac{N_1}{4M_1 \Delta f}, \tag{4.17}$$

where G_1 is the average conductance of the plenum structure and N_1 is the mode count in the frequency interval Δf.

If the response of the plenum is linear, its mean square velocity is proportional to this injected power, or

$$\langle v_1^2 \rangle = \beta \Pi_1 = \beta G_1 \langle l_1^2 \rangle. \tag{4.18}$$

We will see that the parameter β does not have to be evaluated. We now continue to assert linearity and say that the mean square response of the duct will be proportional to the mean square response of the plenum through a parameter α. This is the quantity that we eventually want to know:

$$\langle v_2^2 \rangle = \alpha \langle v_1^2 \rangle = \alpha \beta G_1 \langle l_1^2 \rangle. \tag{4.19}$$

Thus,

$$\frac{\langle v_2^2 \rangle}{\langle l_1^2 \rangle} = \alpha \beta G_1. \tag{4.20}$$

We now imagine the reciprocal experiment in which we apply a band-limited white noise l_2' with the same spectrum as l_1 at the previous observation point on the

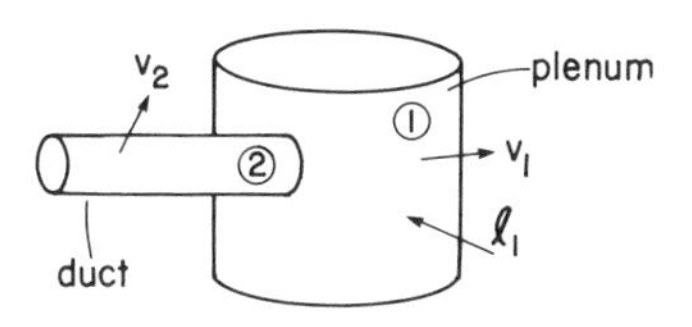

Figure 4.21 Transmission of vibration from a directly excited plenum to an attached duct. We are interested in the vibration response ratio $\langle v_2^2 \rangle / \langle v_1^2 \rangle \equiv \alpha$.

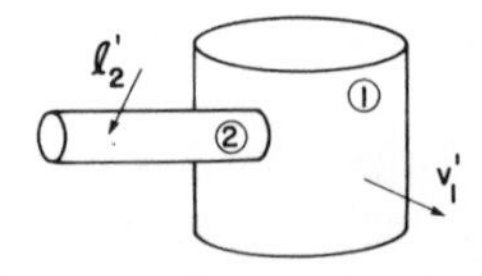

Figure 4.22 The reciprocal experiment in which the structure is driven with a force l'_2, where v_2 was measured previously and the response v'_1 is observed where the excitation force l_1 was previously applied.

duct, and we measure the response velocity v'_1 due to this excitation at the point where we have previously excited structure 1, as shown in Figure 4.22. The power injected into the duct by this force is

$$\Pi'_2 = \langle l'^2_2 \rangle G_2, \tag{4.21}$$

where $G_2 = N_2/4M_2\Delta f$.

We introduce the idea of a coupling loss factor at this stage by saying that a certain fraction of the power injected into the duct is transmitted into the plenum, and we treat the loss to the plenum as damping. If the coupling loss factor from the duct to the plenum is η_{21}, the fraction of the power that flows into the plenum is $\eta_{21}/(\eta_2 + \eta_{21})$. Because this ratio must be less than unity, we require $\eta_{21} > 0$.

The power into the plenum, structure 1, is therefore

$$\Pi'_1 = \Pi'_2 \frac{\eta_{21}}{\eta_2 + \eta_{21}}. \tag{4.22}$$

Again, if the mean square velocity of the plenum is proportional to the power injected into it, we have

$$\langle v'^2_1 \rangle = \beta \Pi'_1 = \beta G_2 \frac{\eta_{21}}{\eta_2 + \eta_{21}} \langle l'^2_2 \rangle. \tag{4.23}$$

This gives the ratio of mean square velocity of the plenum to mean square force acting on the duct as

$$\frac{\langle v'^2_1 \rangle}{\langle l'^2_2 \rangle} = \beta G_2 \frac{\eta_{21}}{\eta_2 + \eta_{21}}. \tag{4.24}$$

The two experiments just described can be represented by the diagrams in Figure 4.23. In the first experiment, a force l_1 was applied at one location, and the velocity v_2 was measured at a second; then a force l'_2 was applied at the previous location where velocity was measured, and the velocity v_1 was measured where the

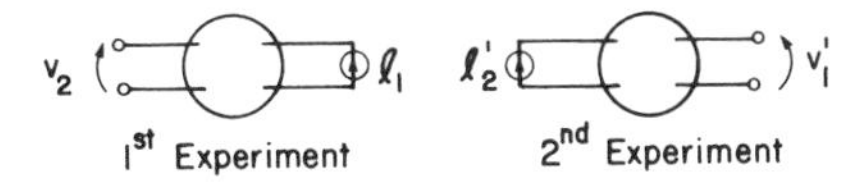

Figure 4.23 Input–output diagrams for the two experiments used in the reciprocity derivation of the response ratio α.

force was previously applied. Reciprocity says that $v_2/l_1 = v'_1/l'_2$ or, in bands of noise,

$$\frac{\langle v_2^2 \rangle}{\langle l_1^2 \rangle} = \frac{\langle v_1'^2 \rangle}{\langle l_2'^2 \rangle}, \tag{4.25}$$

where l_1 and l'_2 have the same spectral shape. Reciprocity requires that Equations 4.20 and 4.24 equal each other, or

$$\alpha = \frac{\langle v_2^2 \rangle}{\langle v_1^2 \rangle} = \frac{G_2}{G_1} \frac{\eta_{21}}{\eta_2 + \eta_{21}}. \tag{4.26}$$

We can convert Equation 4.26 into a ratio of modal energies by the definitions of the point mobilities:

$$\frac{M_2 \langle v_2^2 \rangle}{N_2} = \frac{M_1 \langle v_1^2 \rangle}{N_1} \frac{\eta_{21}}{\eta_2 + \eta_{21}}. \tag{4.27}$$

The left side of Equation 4.27 is the average modal energy of the duct in the band Δf, and the first factor on the right side is the average modal energy of the plenum in that band. If the coupling damping is large compared to the internal damping of the duct, then the ratio of loss factors is unity and we have modal energy equipartition between the modes of the duct and the modes of the plenum. If the coupling loss factor $\eta_{21} \ll \eta_2$, then we again get Equation 4.9.

The formula also shows that the internal damping of the duct must become comparable to its coupling damping into the plenum before any added damping will affect the vibration levels of the duct. This point was made in Chapter 1 with regard to the damping of machine covers when they are well coupled into the main structure of the machine. This is particularly true for in-plane vibrations, which are difficult to damp because of their high stiffness and because in-plane vibrations tend to be coupled well to connecting structures. This makes their response doubly difficult to reduce by adding damping.

Rearranging the terms in Equation 4.22 to collect the terms representing power dissipation in the duct on the left side gives

$$\omega\eta_2 M_2 \langle v_2^2 \rangle = \left\{ \frac{M_1 \langle v_1^2 \rangle}{N_1} - \frac{M_2 \langle v_2^2 \rangle}{N_2} \right\} \omega\eta_{21} N_2, \tag{4.28}$$

where the first factor on the right side is the difference in average modal energies between the plenum and duct, and the second factor is proportional to the coupling loss factor and represents an energy conductance between the two systems.

Since the coupling loss factor is positive, the power always flows from the structure of higher average modal energy to lower average modal energy. This fact is experimentally useful because we can interpret vibration levels in terms of average modal energy. Plotting the data in this way often allows us to infer the direction of power flow between two connected structures, a quantity that is often difficult to determine from an observation of the vibration levels. The structural intensity method described in Section 3.6 can be used to determine power flow.

The various parameters entering Equations 4.27 and 4.28 are called the statistical energy parameters:

$$\begin{array}{ll} \text{Mass} & M_i \\ \text{Loss factor} & \eta_i \quad (=Q_i^{-1} = 2\xi_i) \\ \text{Mode count} & N_i \quad (=n_i\Delta f) \\ \text{Coupling loss factor} & \eta_{ij} = \dfrac{\eta_{ji}N_j}{N_i} \end{array} \tag{4.29}$$

The first is the mass M of the structure or the subsystem. The second is the damping or internal loss factor η. Various damping measures are used in different technologies. The loss factor is the reciprocal of the Q (quality factor) used in electrical engineering and is twice the critical damping ratio widely used in mechanics. The third parameter is the mode count, N, which is a fundamental quantity in SEA and is the modal density times the bandwidth when modal density is a relevant quantity. These three parameters are relatively conventional and were used long before SEA was introduced. Loss factors are usually determined by experiment, since it is very difficult to calculate the damping of real structures. The mode count is normally determined by theoretical calculations, but can be measured under certain circumstances.

4.8 COUPLING LOSS FACTORS

The new parameter introduced by SEA is the coupling loss factor, which measures the power flow or damping effect that one system has when attached to another. It is the most difficult quantity to determine, but it can usually be found by measurement, calculation, or "guesstimate"—that is, the use of engineering judgment to extrapolate from known situations to estimate the value of the coupling loss factor for a new situation. A common and important situation is the interaction between a structure and the sound field. The loss factor from the structure to the sound field is the radiation loss factor, given by

$$\eta_{\text{structure-sound}} = \eta_{\text{rad}} = \frac{\Pi_{\text{rad}}}{\omega M \langle v^2 \rangle_{\text{struct}}}. \tag{4.30}$$

When two structures are joined at a point, we have shown that the coupling loss factor between them can be found from

$$\eta_{21} = \frac{2}{\pi N_2} \frac{\Delta f}{f} \frac{G_1 G_2}{\langle |Y_1 + Y_2|^2 \rangle_{\Delta f}}, \tag{4.31}$$

where Y_1 and Y_2 are the structural mobilities at the joining points. Equation 4.31 is an obvious variation of Equation 4.10.

For a line junction between two structures, Equation 4.31 can be modified to give the coupling loss factor per unit length of junction:

$$\eta_{21} = \frac{1}{k_2 A_2} \left[\frac{1}{1 + (k_2/k_1)^4} \right]^{1/4} \left\langle \frac{G_1 G_2}{|Y_1 + Y_2|^2} \right\rangle_{\Delta f}, \tag{4.32}$$

where k_1 and k_2 are the wave numbers for the appropriate wave types in the two joining structures, and Y_1 and Y_2 are the point mobilities. Often these are the moment mobilities, and we evaluate the terms in brackets by taking ratios of point mobilities for the two joining systems and adjusting by the other factors in Equation 4.32 to determine the line junction coupling loss factor. Indeed, we can carry the process a bit further. By associating the point mobilities with modal densities, we can express the coupling loss factor entirely in terms of mode counts or modal densities.

It is possible to deal with line junctions in a more fundamental manner than by simply applying Equation 4.32. Some situations of interest for vibration transmission are illustrated by the configurations in Figure 4.24. They include the "ell" [Figure 4.24(a)], the "tee" [Figure 4.24(b)], and the full "cross" [Figure 4.24(c)]. In every case, we consider waves of a particular type (flexure, in-plane longitudinal, or

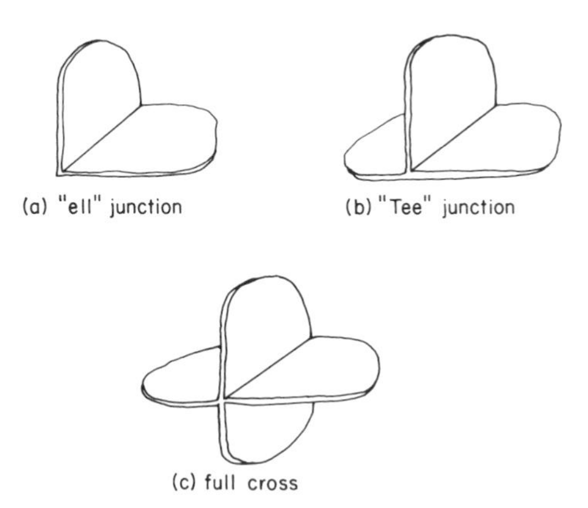

Figure 4.24 Plate connections of interest in built-up structures.

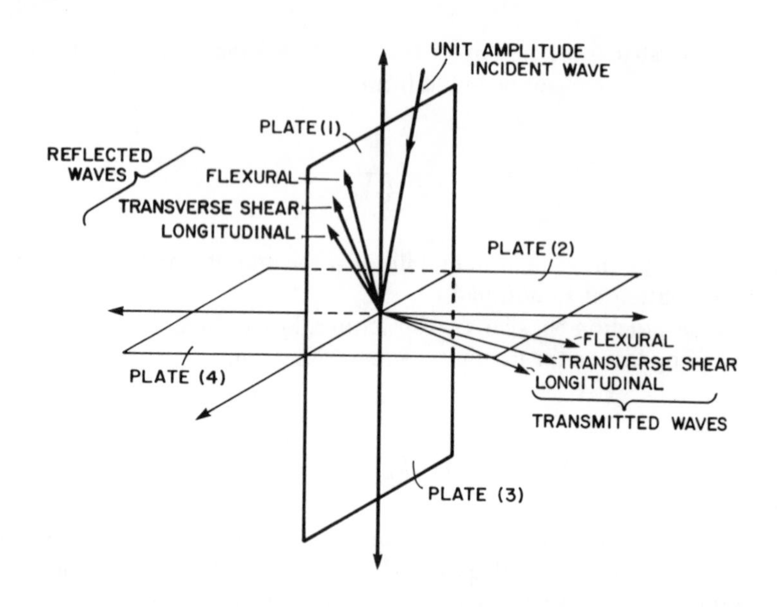

Figure 4.25 Incident, reflected, and transmitted waves at a structural junction.

in-plane shear) to be incident on the junction in one of the connecting plates. Continuity of displacements, rotations, forces, and moments at the plate junctions allows us to find the amplitudes of all the waves generated in the plate containing the incident wave and in all the plates connected to it, as shown in Figure 4.25.

A diffuse vibration field is incident on the junction, and for each angle of incidence the power in each wave emanating from the junction is calculated. This power is added for all angles of incidence and a transmissibility, averaged in angle of incidence, is computed for all the waves. Thus, for example, we obtain the transmissibility of energy from in-plane shear energy in plate 1 to flexural energy in plate 2, or the reflectivity of energy from flexure in plate 1 to in-plane longitudinal in plate 1. In an SEA calculation, the energy storage systems are mode groups, and it makes no difference whether these mode groups are in the same structural element or different but physically connected ones insofar as the procedure is concerned.

The transmissibilities are converted to coupling loss factors according to the energy decay rate formula

$$\omega\eta_{ij} = \nu\tau_{ij}, \tag{4.33}$$

where ν is the collision rate of the energy with the joining boundary of length L_{ij}, and τ_{ij} is the energy transmissibility of that boundary. If the energy (group) speed is c_i and the mean free path between boundary collisions is $d_i = \pi A_i/L_{ij}$, then the collision rate is $\nu = c_i/d_i = c_i L_{ij}/\pi A_i$, leading to an expression for the coupling loss factor in terms of the transmissibility. This expression is

$$\eta_{ij} = \frac{c_i L_{ij}}{\pi \omega A_i} \tau_{ij}, \tag{4.34}$$

where η_{ij} is the coupling loss factor and A_i is the area of the source plate.

4.9 EXAMPLE OF NOISE TRANSMISSION IN A SHIP STRUCTURE

We illustrate the use of SEA with some calculations and measurements carried out for a 1:2.5 scale model of the diesel propulsion engine foundation and hull structure of a frigate ship. A perspective sketch of the modeled components of this structure is shown in Figure 4.26. By symmetry the structure is further simplified to that shown in Figure 4.27, showing the 12 plate elements that were modeled—a series of models of varying complexity of 2 to 12 plates were studied. We shall only discuss results for the 7-plate and 12-plate models here.

The various dimensions of the plate elements are given in Table 4.2. As each model of increasing complexity was studied, elements were added to those already welded together to form the next model. The damping of a welded-up structure in the laboratory is somewhat lower than that of a constructed ship, due to various ancillary items such as piping, small motors, and so on, in the full-scale structure. To

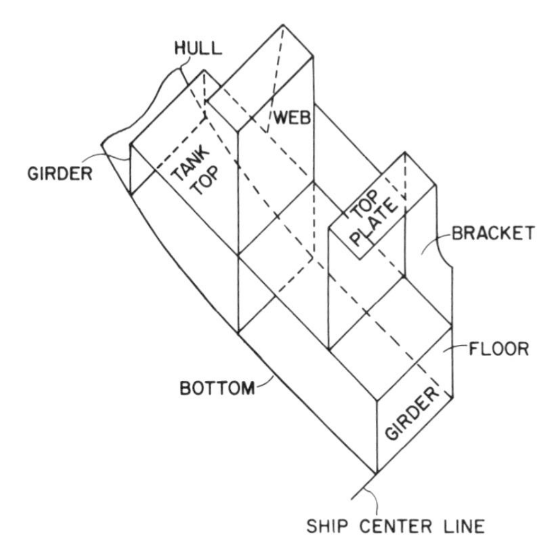

Figure 4.26 Perspective sketch of the engine foundation and fuel tank structure of a diesel-powered ship.

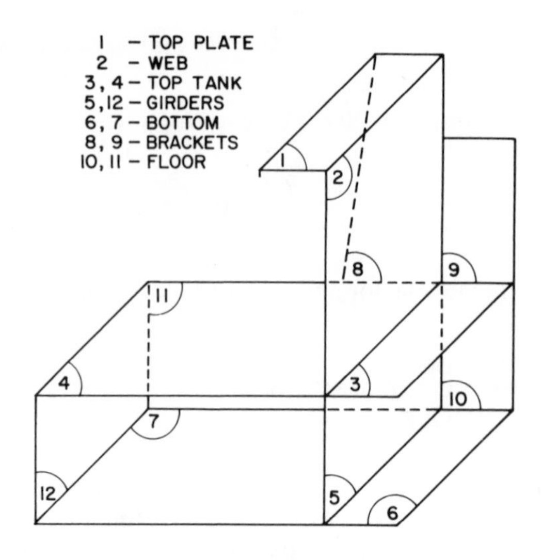

Figure 4.27 The 12-plate structural model used for analytical SEA and experimental studies.

Table 4.2 Basic Model-Plate Characteristics

Plate	*Thickness (mm)*	*Area* (M^2)	*Dimensions (mm)*	*Weight (kg)*	*Modal Density* $(\times 10^3)$ *Bend*	*In-Plane*
1 Top	8.2	0.007	178 × 432	5.335	3.1	—
2 Web	3.3	0.225	584 × 432	6.654	22.2	0.2×10^{-3} f
3 Tank top	2.6	0.088	203 × 432	1.698	11.2	0.8×10^{-4} f
4 Tank top	2.6	0.329	762 × 432	6.624	41.8	0.3×10^{-3} f
5 Long girder	3.3	0.143	330 × 432	3.717	14.1	0.12×10^{-3} f
6 Bottom	3.3	0.088	203 × 432	2.107	8.6	0.8×10^{-4} f
7 Bottom	3.3	0.329	762 × 432	8.154	32.3	0.3×10^{-3} f
8 Bracket	3.3	0.126	178 × 254 × 584	3.306	12.4	0.11×10^{-3} f
9 Bracket	3.3	0.077	381 × 203	2.027	7.6	0.07×10^{3} f
10 Floor	3.3	0.067	203 × 330	1.778	6.6	0.06×10^{-3} f
11 Floor	3.3	0.251	762 × 330	6.668	24.7	0.22×10^{-3} f
12 Long girder	3.3	0.143	330 × 432	3.717	14.1	0.12×10^{-3} f

f = frequency in Hz.

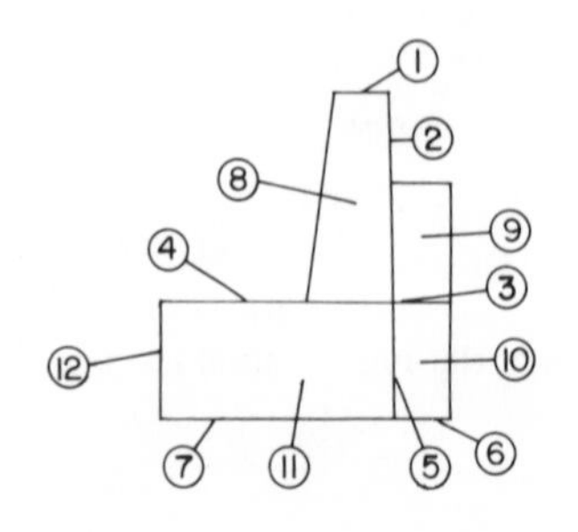

simulate this added damping, damping tape was applied to the elements to obtain the desired damping values.

The various modal densities for the plates were computed and are also presented in Table 4.2. The damping loss factors are also presented. Since each plate element was available for experimental study before its incorporation into the structural model, it was possible to measure its mode count and damping. In some cases, these experimentally determined values were also used in the analytical studies, particularly at lower frequencies where fewer modes are present. Generally, the agreement between experiment and calculation was improved when the experimentally determined parameters were used.

In an operational ship, the hull, tank floors, and side walls will be fluid loaded. Fluid loading effects have not been included in this study.

4.10 COMPARISON OF RESULTS FROM CALCULATIONS AND EXPERIMENTS

The results of a sample calculation of transmissibility as a function of incidence angle are shown in Figure 4.28. The variations are strong, but when averaged over incidence angle the transmissibility becomes a smooth function of frequency. Figure 4.29 shows such averaged transmissibilities for waves in elements that join in an ell configuration. Figure 4.30 shows elements that oppose each other at a cross junction of the type in Figure 4.24(c).

The transmissibility calculations were carried out under two conditions. In the first, we assume no in-plane degrees of freedom, and flexure is the only wave motion allowed. These calculations are denoted by F. In the second condition, flexure or bending (B), in-plane longitudinal (L), and in-plane shear (T) are allowed. By

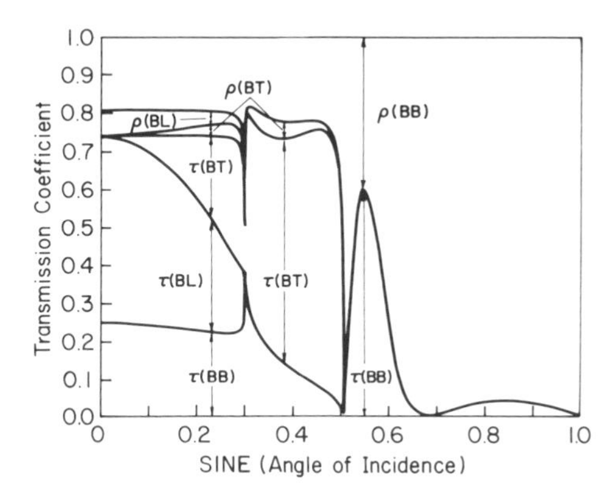

Figure 4.28 Variation of transmission coefficients τ and reflection coefficients ρ for a two-plate "ell" junction as a function of angle of incidence of the bending wave. The frequency ω and structural parameters are fixed in the calculation.

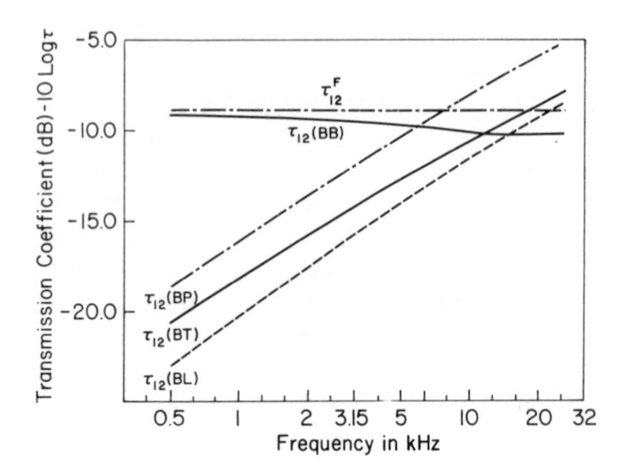

Figure 4.29 Transmissibility for the "ell" junction 1 → 2 for bending (B) wave incidence into longitudinal (L), in-plane shear (T), and bending (B). The transmissibility into combined in-plane waves is (P). When in-plane is ignored, the bending modes are labeled (F).

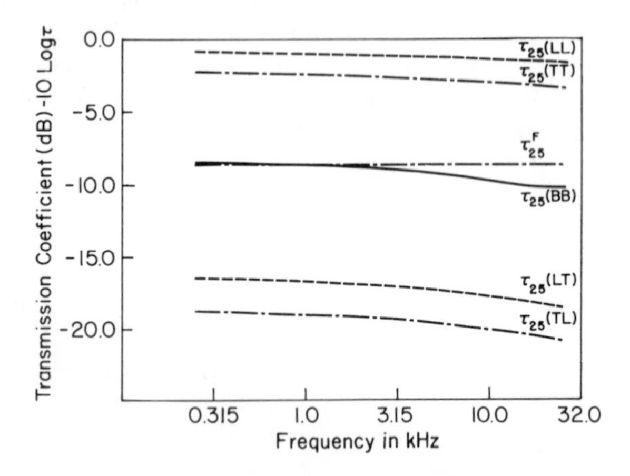

Figure 4.30 Transmissibility through the full cross junction 2-3-4-5 from selected wave types in plate 2 into various wave types in plate 5. Note the very high transmissibilities of in-plane waves for this stiff junction.

comparing these calculations, we can immediately see the (theoretical) effects of including in-plane degrees of freedom on the energy distribution in the system.

Calculations of transmissibilities, coupling loss factors, and energy distributions using SEA were all carried out on a computer. Power is injected into plate 1 of the model into flexural modes only. Figure 4.31 shows the computed results for the energy ratios of plates 1 and 2 of the seven-plate model for the two conditions. The plate energy combines all mode types, but it is always dominated by flexure, due to the much greater modal density for bending modes. For these plates, including in-plane modes makes only a slight difference in the prediction.

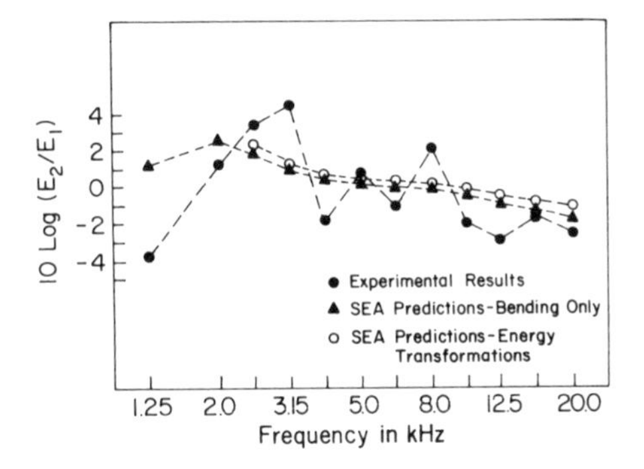

Figure 4.31 Comparison of calculated and measured vibrational energy ratios for plates 1 and 2 for a model consisting of plates 1 through 7 of the foundation model.

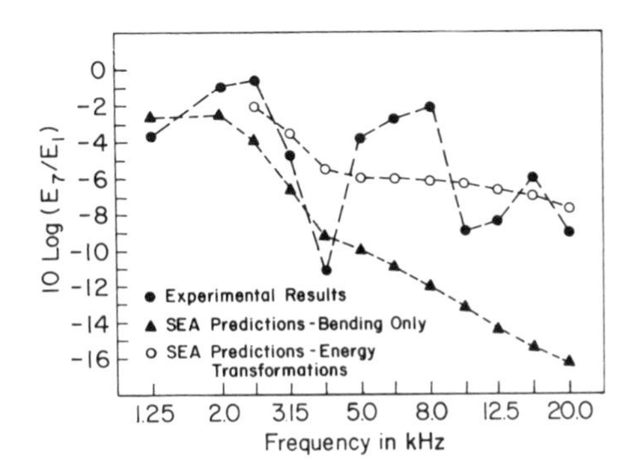

Figure 4.32 Comparison of calculated and measured vibrational energy ratios for plates 1 and 7 for the seven-plate model.

Examining plates that are farther apart, we get the results in Figure 4.32 for plates 1 and 7 of the seven-plate model. In this case, including in-plane motion makes a sizable difference in the predicted vibration of plate 7, increasing the level by several decibels.

The results are generally similar for the 12-plate model. Figure 4.33 contains results for plates 1 and 2; Figure 4.34 for plates 1 and 7. Again, there is little difference between the two calculations when the plates we study are near each other, but there is a large difference as we consider plates that are farther apart.

We also note that when there is a difference in the calculations, this difference begins at about 2 kHz (800 Hz full scale), which corresponds to a wavelength of 1.5 m (in-plane shear) or a half wavelength of 0.75 m, which is the condition for the lowest in-plane modal resonances to occur in the plate elements. Thus, we can say that as soon as the in-plane modes begin to appear, they dominate the vibration transmission.

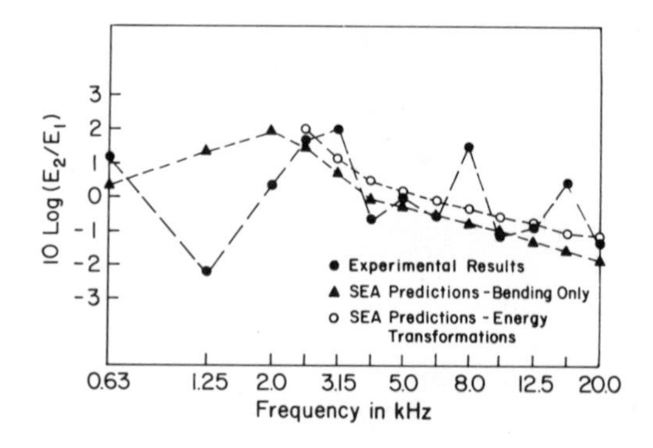

Figure 4.33 Calculated and measured vibrational energy ratios for plates 1 and 2 of the 12-plate model.

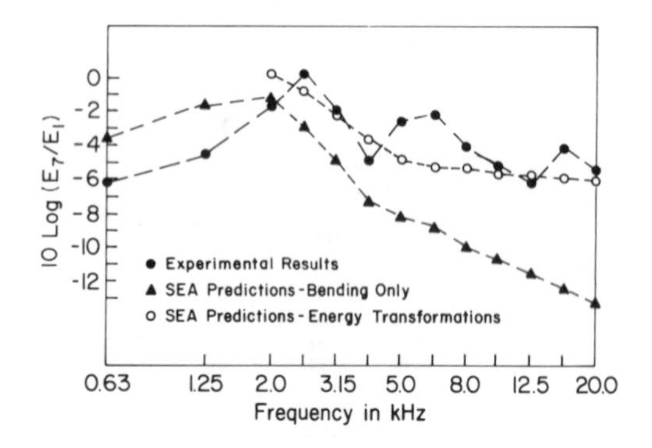

Figure 4.34 Calculated and measured vibrational energy ratios for plates 1 and 7 for the 12-plate model.

Experiments were carried out by applying a shaker perpendicular to plate 1 for each model at a series of locations, exciting the shaker with one-third octave bands of noise, and measuring the vibratory response at several locations on each of the plate elements. Again, assuming that flexure dominates the vibrational energy, we experimentally determine vibrational energy ratios and plot their values in Figures 4.31–4.34.

Comparing the predictions with these data, we see that the experimental results strongly support the inclusion of in-plane modes in the calculations. Since in-plane motions increase the energy transmission, they, in a sense, provide a "flanking path" to the bending motions, and if that flanking path is strong enough, we expect it to dominate the transmission.

We also note a pronounced frequency dependence, not predicted by the calculations, in the response that is present in all the data, particularly in Figure 4.32. This is probaby due to the equal depth of all the plate elements, causing a "pass-stopband" dependence that would probably be washed out by the variations in plate thicknesses, attachments, and so on, that are naturally present in a real structure.

When transverse elements 8–11 are added to the 7-plate model to produce the 12-plate model, this structure in the data is reduced, which can be seen by comparing Figures 4.32 and 4.34.

4.11 USE OF SEA IN ORGANIZING VIBRATION DATA

SEA is also a way of organizing vibration data so that it is more understandable. An example is data acquired on the small ship in Figure 4.35. Vibration levels were measured on this ship in the neighborhood of the various parts of the engine room and then at various positions around the deck house above. Vibration measurements were made on the girder that supports the engine, on a water pipe, and on various other components in the engine room. In the deck house region, measurements were made on the exhaust vent, the deck, and the starboard and port walls of the cabin. The data showed very little organization until they were expressed by modal energy,

$$E_m = \frac{M\langle a^2 \rangle}{\omega^2 N}, \tag{4.35}$$

which converts rms acceleration into modal energy.

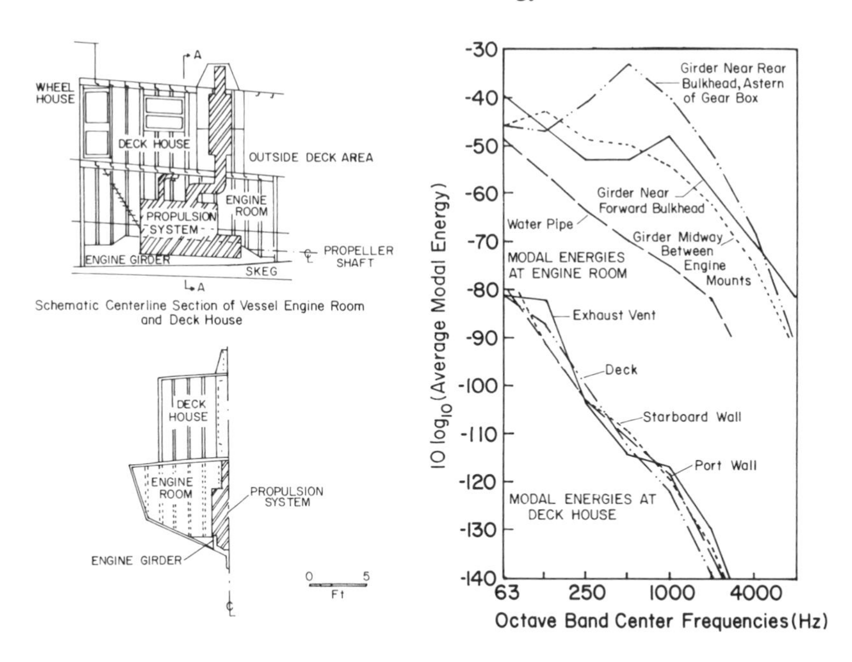

Figure 4.35 Sketch of small ship and graphs of modal energy of various structural components near the engine source and at remote locations in the deck house. Note tendency to equipartition away from the source.

Figure 4.35 shows that the vibration levels are quite inhomogeneous around the engine in terms of modal energy. When we go to a more remote location, such as the deck house, where the modes have a chance to interact with each other and come to equilibrium, we find a strong tendency toward equipartition of the various structural components. In addition, the modal energies in the deck house are significantly below those for the engine room, which we would expect because the excitation is in the engine room area. The inhomogeneity of levels near the engine is somewhat like the direct field—reverberant field behavior of a sound field in a room—whereas the uniformity in modal energy in the deck house is like the reverberant field in room acoustics.

4.12 UNCERTAINTY IN NOISE TRANSMISSION ESTIMATES

Calculations of vibration transmission by SEA are, by nature, estimates, and it is useful to ask how reliable they are. The magnitude and phase of transfer functions are highly variable functions of frequency that are smoothed out when we calculate the mean square response over a band of noise.

We evaluate the variability in response by calculating the variance, or standard deviation, of the response. The derivations of the formula for standard deviation are too complicated to be done here, but we can provide simple plausibility arguments for their form and the sources of the variability.

The variability in noise transmission is produced by the following effects:

1. The more mode pairs that interact, the less variability there will be in the response. If we imagine that each of the N_1 modes of structure 1 is "looking through" a rectangular filter of noise bandwidth $\pi\omega(\eta_1 + \eta_2)/2$, then each mode of system 1 will interact with $n_2(\omega)(\pi/2)\omega(\eta_1 + \eta_2)$ modes of system 2. Thus, the total number of interacting mode pairs is

$$N_{\text{pairs}} = N_1 N_2 \frac{\pi}{2} \frac{\omega}{\Delta\omega}(\eta_1 + \eta_2). \tag{4.36}$$

2. There are also geometric effects in variability. There will be more variation when we drive at a point and measure at a point than when we average the excitation and response over several locations.

3. The more "independent" junctions there are between structures, the less variability there will be. We have previously presented formulas for the coupling loss factors for point and line connections in Equations 4.31 and 4.32. A line junction of length L_{junct} may be thought of as a connection at N_{point} points, where

$$N_{\text{point}} = \frac{\pi}{8} \frac{L_{\text{junct}}}{\lambda_b/2}, \tag{4.37}$$

when λ_b is the bending wavelength. Thus, each half wavelength of junction acts as an independent point connection.

Combining these factors for response variability gives a normalized standard deviation

$$\frac{\sigma_{v_2}}{\langle v^2 \rangle} = \frac{1}{(N_{\text{pairs}} N_{\text{point}})^{1/2}} \frac{\langle \psi^4 \rangle^2}{\langle \psi^2 \rangle^4}, \tag{4.38}$$

where the second factor on the right side results from the geometric effects. Equation 4.38 has the form of a "central limit theorem," in which statistical variability is reduced by the square root of the number of independent contributors to a random process, in our case the transmission of vibrational energy to a receiving structure.

CHAPTER 5

Sound Radiated by Machines

5.1 INTRODUCTION

Now that we have studied how vibration is transmitted through a machine to its outer surfaces, we are ready to consider how that vibration is converted into radiated sound. We are concerned about how to calculate the radiation of sound, how to measure it, and how to use this information. By implication, we are also concerned about how people react to sound. That is discussed in Appendix B.

Sound radiation is inherently a complicated process. It turns out, however, that some fairly simple geometrical and dynamical parameters control sound radiation. These parameters allow us to make reasonably good estimates of sound radiation. We shall discuss estimation procedures in some detail. The measurement of sound radiation is also important in understanding where the sound comes from on a machine and in the QC procedures in machine manufacturing.

We shall discuss some procedures for measuring radiated sound. Current methods of measuring sound radiation favor those based on acoustical intensity and reverberant sound levels. We shall discuss these methods and give examples of their use, citing examples related to manufacturing quality control.

Last, we review some computer-based methods for calculating radiated sound and discuss various ways of reducing sound for noise control purposes.

5.2 PLANE-WAVE RADIATION AND RADIATION EFFICIENCYA

We begin our discussion of the radiation of sound—that is, compressional waves in air—by making an analogy with the one-dimensional longitudinal vibration of rods in Chapter 3. There we introduced the stiffness modulus $E = \rho_0 c^2$, where ρ_0 is the density of the medium and c is the speed of sound. In air, this modulus is γP_0, where $\gamma = C_p/C_v$ is the ratio of specific heats, and P_0 is the ambient pressure.

We consider a column of air captured in a tube of cross section A, as shown in Figure 5.1, which has a piston vibrating with velocity v at one end. For the elastic

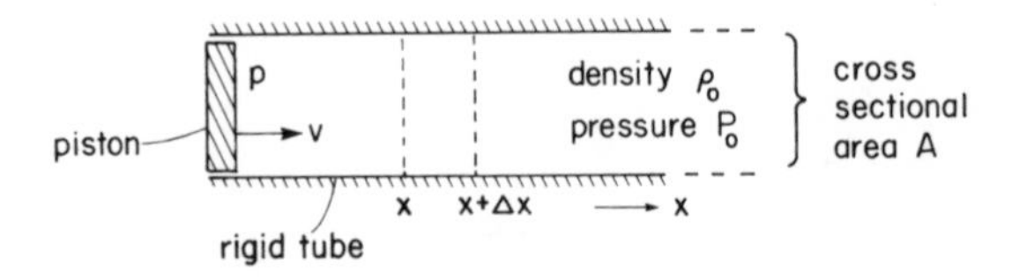

Figure 5.1 Generation of plane waves of sound by a vibrating piston at one end of an infinitely long tube.

column, we wrote for mass times acceleration the quantity

$$\Delta x \rho_0 A \frac{\partial^2 u}{\partial t^2} = \Delta x \rho_0 A \frac{\partial v}{\partial t}, \tag{5.1}$$

which was related to the gradient of stress. In a fluid, we use the pressure instead of stress, and since the pressure is the negative of mechanical normal stress, the relation between acceleration and stress gradient for a fluid is

$$\rho_0 \frac{\partial v}{\partial t} = -\frac{\partial p}{\partial x}. \tag{5.2}$$

The properties of waves produced in the column of Figure 5.1 are the same as those of compressional waves discussed in Section 3.2. In particular, Equation 3.5 shows that the ratio of stress (pressure) to particle velocity for a wave in the $+x$ direction is

$$\frac{p}{v} = \rho_0 c. \tag{5.3}$$

The speed of sound in air is about 345 m/sec, and ρc is about 407 mks units. We calculate the power radiated by the piston by taking the time average of the product of the force that the piston must exert on the fluid and the piston velocity:

$$\Pi_{\text{rad}} = \langle pAv \rangle_t = \langle v^2 \rangle_t \rho_0 c A = \langle v^2 \rangle R_{\text{rad}} \tag{5.4}$$

The coefficient of the m.s. velocity is denoted as the radiation resistance R_{rad}. We note that the radiation resistance is real and frequency independent, so the column of air acts like a viscous damper to the motion of the piston.

Equation 5.4 is used to define a radiation resistance for an arbitrary vibrating structure, shown in Figure 5.2. If we define a m.s. velocity averaged over time and over the surface of the machine, and if we measure the radiated power from the structure, we can relate the radiated power and the m.s. velocity by a formula similar to equation 5.4:

$$\Pi_{\text{rad}} = \langle v^2 \rangle_{x,t} A \rho_0 c \sigma_{\text{rad}}. \tag{5.5}$$

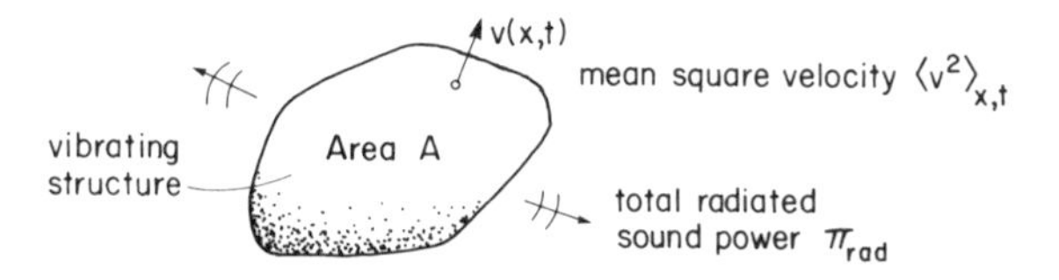

Figure 5.2 Sound radiation of a machine structure is expressed by its radiated power Π_{rad}, which is proportional to the space-time mean square velocity $\langle v^2 \rangle_{x,t}$.

We have added the radiation efficiency, σ_{rad}, to account for whatever differences there might be between the radiation of this machine and the radiation of a piston in a tube. We shall find that σ_{rad} is affected by the geometry of the structure and by the structural dynamics—that is, the kinds of modes of vibration that exist in the machine structure. When a structure vibrates in any particular frequency range, there are two characteristic dimensions of importance in sound radiation: the size of the structure and the wavelength of vibration. The critical factor, insofar as geometry is concerned, is how big the structure is compared to the wavelength of sound. The critical factor insofar as dynamics is concerned is how long the wave of vibration on the structure is compared to the wavelength of sound (or, equivalently, how fast the waves on the structure are compared to the speed of sound).

The plane-wave relationship of Equation 5.3 allows a useful first estimate of radiated pressure from any vibrating structure. Because the surface vibration of the structure equals the particle velocity in the air, we can multiply the structural velocity by ρc to get a first estimate of the pressure at the surface. If we reference the velocity level to 1 m/sec and use the usual reference pressure of $20\ \mu\text{P} = 2 \times 10^{-5}\ \text{N/m}^2$, then the relationship between velocity and pressure level is

$$L_p = L_v(\text{dB re 1 m/sec}) + 146. \tag{5.6}$$

This relation has been used to prepare Figure 5.3. Here we have plotted the velocity

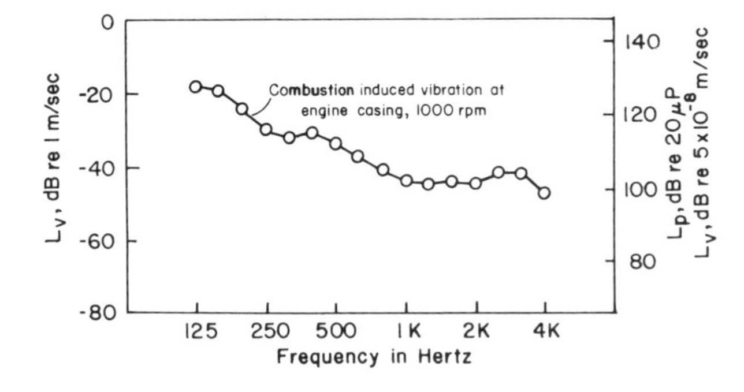

Figure 5.3 A vibrational velocity spectrum may be converted to a radiated sound pressure spectrum by the plane-wave relation $p = \rho c v$ or $L_p = L_v(\text{re 1 m/sec}) + 146$ in air.

level of the structure on the left-hand scale and the estimated near-field pressure, based on Equation 5.6, on the right-hand scale. If we select a velocity reference of 5×10^{-8} m/sec, then it turns out that the pressure and velocity levels are numerically equal in air. This convention has been used, particularly in Europe, to present structural vibration levels in situations where the noise radiated by the vibrating structure is to be estimated.

5.3 GEOMETRIC RADIATION EFFICIENCY: SOUND FROM A VIBRATING SPHERE

To see the effect of source size on radiation efficiency, we consider the radiation from the sphere in Figure 5.4. If the sphere radiates a certain amount of power, then that power must be present in the sound wave at all distances r from the sphere. Since power is the product of intensity and area, and the area increases as the square of the distance, then the intensity must decrease as the square of the distance. Since the intensity, in turn, is proportional to the square of the dynamical variables, we can assume that the pressure, for example, will vary inversely as the distance from the sphere. At a radius r, therefore, we write the pressure as

$$p = p_a \frac{a}{r} e^{j(\omega t - kr)}, \tag{5.7}$$

where p_a is the pressure in the wave at radius a, and we have included the factor e^{-jkr} to indicate that this is a propagating sound wave outward from the sphere.

We relate the velocity to this pressure by Equation 5.2, substituting the radial coordinate r for the coordinate x and using $\partial v/\partial t = j\omega v$:

$$j\omega \rho_0 v = -\frac{\partial p}{\partial r}. \tag{5.8}$$

Substituting Equation 5.7 into Equation 5.8, we can calculate a ratio of pressure to particle velocity at radius a. This is called the *specific acoustic impedance*, and at the surface of the vibrating sphere it is

$$\frac{p_a}{v_a} = \rho_0 c \frac{jka}{1 + jka}. \tag{5.9}$$

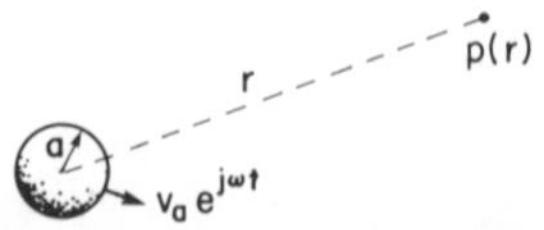

Figure 5.4 Radiation of sound by a pulsating sphere depends on the distance r, but not on the direction. The radiated pressure $p(r)$ is proportional to the velocity v_a.

Since $k = 2\pi/\lambda$, the product ka is the ratio of the circumference of the sphere to the wavelength of sound. If ka is large—that is, if we have a relatively large sphere or small wavelength—then the ratio of pressure to particle velocity in Equation 5.9 becomes ρc and the spherical vibrator has the same radiation resistance that a plane-wave piston radiator has. If ka is not large, then the complex fraction in Equation 5.9 can be rationalized to obtain

$$Z_{\text{rad}} = \frac{Ap_a}{v_a} = A\rho_0 c \frac{(ka)^2 + jka}{1 + (ka)^2} = R_a + jX_a, \tag{5.10}$$

where the pressure is multiplied by the area to convert to a mechanical impedance—the ratio of force applied to the fluid to the surface velocity of the sphere.

The real part of the radiation impedance in Equation 5.10 is the radiation resistance

$$R_a = A\rho_0 c \frac{(ka)^2}{1 + (ka)^2}, \tag{5.11}$$

and, referring to Equation 5.5, we can use Equation 5.11 to identify the radiation efficiency of a sphere to be

$$\sigma_{\text{rad}}(\text{sphere}) = \frac{(ka)^2}{1 + (ka)^2}. \tag{5.12}$$

This relation is graphed in Figure 5.5 and shows that the radiation efficiency of a sphere is almost unity when its circumference equals an acoustic wavelength. One finds for quite a variety of sources of differing shapes and vibration patterns that the radiation efficiency gets very close to unity when the distance around the object

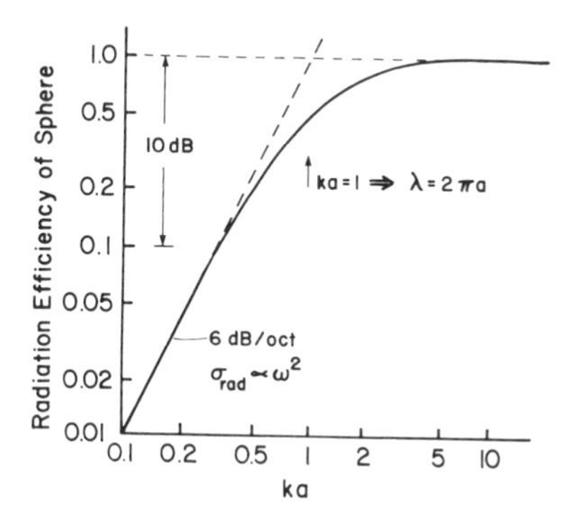

Figure 5.5 Radiation efficiency of sphere shows importance of radiator size to wavelength of sound.

becomes an acoustic wavelength. The ratio of a typical dimension of the radiator to the wavelength of sound is a useful parameter for estimating the effect of geometry on radiation efficiency.

5.4 OTHER ELEMENTARY SOURCES OF SOUND

The pulsating sphere in Figure 5.4 is a sound source that produces a changing mass flow into the medium. There are other elementary sources of sound that do not involve a net injection of flow into the medium. One example is the noise produced by an air jet when it impinges on a rigid obstacle such as a fan blade. When the turbulent flow produces forces on an obstacle, then, by Newton's law of reaction, the obstacle puts forces back on the fluid in the form of fluctuating lift and drag, resulting in sound radiation.

Both mass flow injection and fluctuating forces require external agents or objects placed in the fluid for sound to be produced. Sound can also be produced by paired forces transverse to each other or by longitudinally opposed paired forces that produce rotation or stretching of the fluid. Such paired forces occur in free turbulence and are primarily responsible for jet engine noise. In the turbulent mixing region of the jet, fairly near the engine nozzle, the transverse paired forces dominate the sound radiation, and in the downstream fully developed part of the turbulent flow the longitudinal stresses are more important in sound production.

For a slightly different interpretation of the simple source, let us look again at Equation 5.9, when the dimension of the sphere is very small compared to the wavelength of sound. The pressure at the surface of the sphere is

$$p_a = v_a \rho_0 cjka = \frac{j\omega v_a 4\pi a^2 \rho_0}{4\pi a} = \frac{\ddot{M}(t)}{4\pi a} \tag{5.13}$$

where, referring to the discussion of Fourier series in Appendix A, we have interpreted $j\omega$ as a time derivative. In this equation, $\ddot{M}(t)$ is the mass injected into the medium as a function of time, and equals the velocity of the sphere times its area times the density. The $j\omega$ in Equation 5.13 gives a second time derivative of the injected mass. Equation 5.7 indicates that the pressure at any distance is the ratio a/r times the version of the wave at $r = a$, but time delayed by the amount r/c. Consequently, we can write the pressure at any radius as

$$p(r, t) = \frac{\ddot{M}(t - r/c)}{4\pi r}. \tag{5.14}$$

As an example, consider the sound produced by the muzzle blast of a firearm, shown in Figure 5.6. The explosion injects a certain amount of mass into the medium within a time τ_s, as shown at the top of Figure 5.6. The mass flow rate, the product of volume velocity and cross-sectional area, is shown as the middle curve. The rate of change of mass flow, the second time derivative of the injected mass, is

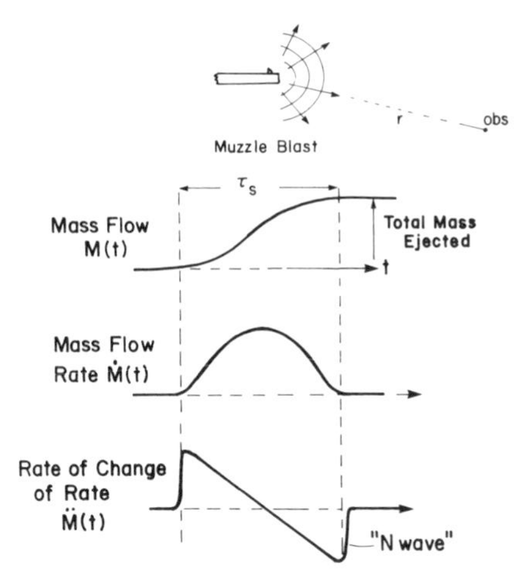

Figure 5.6 Sound radiation by mass injection into the medium, producing an "*N* wave" pressure pulse. This source is similar to the pulsating spherical source.

shown at the bottom of the figure. The figure shows how injection of a finite amount of mass into the medium leads to the generation of an "N" wave, which is a characteristic waveform for such sources.

Forces acting in a fluid can also produce sound. As an example, consider a force that acts on a small disk to produce the same velocity waveform as in the previous example. In this case, there is no net velocity injection, but local velocities are similar in the two cases. Such a velocity waveform requires the triangular force waveform shown at the top of Figure 5.7. This force results in an acceleration of the same shape. When the acceleration is integrated, the velocity is as shown. The "dipole" sound radiation from a fluctuating force in a medium is

$$p_d(r, t) = \frac{\dot{l}(t - r/c)}{4\pi rc} \cos\theta. \tag{5.15}$$

In this case, sound is produced by the rate of change of force, which is reasonable because we would not expect a steady force on the fluid to radiate sound.

As an example of a paired force, think of a torque applied to a small disk. Such a torque can be thought of as a transverse pair of forces operating on the fluid. If we want the disk to have the same velocity as in the dipole case, then the torque time dependence should also have the triangular waveform shown in the top of Figure 5.8. In this case, the sound radiation is proportional to the second time derivative of the applied force or torque, as shown in the figure:

$$p_q = \frac{\ddot{T}(t - r/c)}{4\pi rc^2} \cos\theta \sin\theta. \tag{5.16}$$

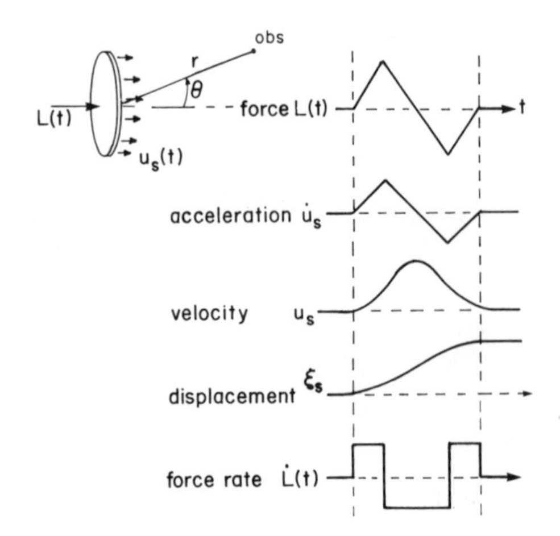

Figure 5.7 Sound radiated by a force $L(t)$ acting on the fluid medium. For a given velocity waveform, this source produces higher-frequency components in the radiated pressure than the pulsating sphere does.

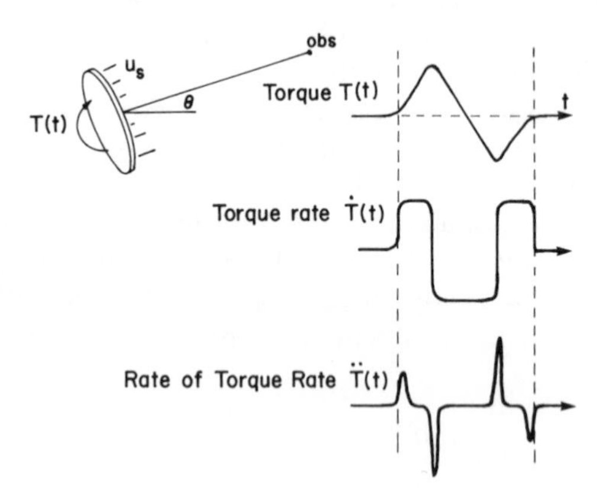

Figure 5.8 Sound radiated by a torque acting on the fluid medium. Even higher-frequency components are radiated by this process.

In this case, the second derivative of torque will produce the spiky pressure waveform from this "quadrupole" source.

The velocities for the three sources in Figures 5.6–5.8 all have the same waveform, but the sound radiation is increasingly dependent on higher-order derivatives of this waveform. One tends to get much more high-frequency sound radiation from dipole and quadrupole sources than from a monopole or mass-injection source.

5.5 SOUND RADIATION FROM COLLIDING PARTS

The elementary sources described in the preceding section do not usually dominate machinery noise. However, there are situations in which such radiation can be important, as when an air jet impinges on parts. Large-scale motions associated with structural vibration are usually much more efficient in radiating sound. There is an interesting example of sound radiation by elementary sources, however, in a popular child's toy called Clackers. This toy consists of two plastic spheres at either end of a string. One holds the middle of the string and causes the balls to collide and produce a very sharp sound. The question we want to ask is, "What is the source of this sound?"

Imagine a small sphere moving along a line at velocity v_0 (Figure 5.9) that strikes a second ball at rest. If the two spheres are elastic, then the conservation of momentum requires that the first ball stop and that the second ball continue on with the velocity v_0. Energy is also conserved since the kinetic energy prior to the collision was $(1/2)Mv_0^2$, and the second ball has the same energy after the collision. If the ball at rest is inelastic (made of putty, perhaps), then conservation of momentum requires that the new composite ball of mass $2M$ move with a velocity $v_0/2$. Now the kinetic energy is $(1/2)(2M)(v_0/2)^2 = (1/4)Mv_0^2$. Half the energy has been lost, and we say that the lost energy has gone into deformation of the inelastic ball. Note, however, that if the ball at rest had a spring with sticky tape on it and the two balls had stuck together, then the center of mass of the combined system after collision would still have a velocity $v_0/2$ and a kinetic energy $(1/4)Mv_0^2$. In this case, we would say that the missing $(1/4)Mv_0^2$ had gone into vibration of the two-mass system, but we note that in either case half of the energy goes into some form of internal motion or heating of the combined system.

Now we consider a sphere moving in a fluid at constant velocity v_0. The fluid motion around the mass has kinetic energy, which we normally represent through a so-called accretion mass M_e. Suppose that by some external agent we are able to stop the motion of the rigid ball in zero time. We apply force only to the ball; therefore, we have removed the kinetic energy $(1/2)Mv_0^2$, but we have not done

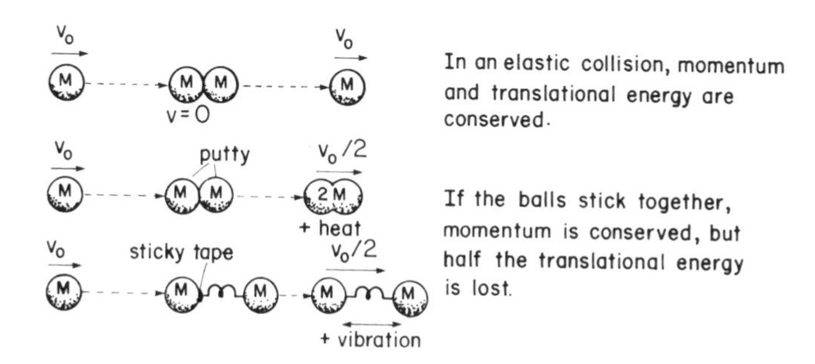

Figure 5.9 When an object collides with another object, some of the energy of translation may be converted to other forms if they are available.

anything about the kinetic energy in the fluid, $(1/2)M_e v_0^2$. What has happened to the energy of the fluid? Since the sphere has stopped, it is not possible for the fluid to do any work on the sphere, so the energy that was contained in the fluid must go into viscosity or sound generation. This is the mechanism by which the decelerating colliding spheres in the child's toy produce radiated sound. Such very abrupt deceleration can also cause radiated sound in certain machine components, such as the fly shuttle in a loom, which is very rapidly decelerated at the end of its travel.

Figure 5.10 shows the two spheres approaching each other. This motion can be thought of as producing a positive velocity source on the front face and a negative velocity source on the rear face. As the spheres rebound, the signs of these velocity sources change rapidly. A time waveform of the velocity in Figure 2.15 shows the change during rebound. Note that the velocity waveform is quadratic for a small time.

A quadratic velocity waveform will have a power spectrum that drops off at 18 dB/octave at high frequencies. When the acoustic wavelength is less than the circumference of the spheres, we expect the radiation efficiency to be unity, and, therefore, the radiated power should also drop off at 18 dB/octave. At low frequencies, however, the radiation efficiency of two spheres moving in opposite directions is that of a longitudinal quadrupole. For such a source, the radiation efficiency increases at 18 dB/octave. The force spectrum is flat at low frequencies (Figure 2.16), as is the acceleration, so the velocity drops at 6 dB/octave. Therefore, we expect in this frequency range the radiated sound power to increase at 12 dB/octave. Although such considerations will allow us to get some of the asymptotic behavior of the spectrum, the detailed spectrum is determined by the details in the velocity spectrum. The comparison between a more detailed theory and experimental data is shown in Figure 5.11.

We might assume that the sound in these collisions is due to the vibration of the objects. Figure 5.12 shows the pressure waveform trace from a single collision; the latter part of the trace has been amplified by a factor of 50 to show its details. The high-frequency oscillation is the natural frequency of vibration of the plastic spheres, in this case, resonating at about 50 kHz. This resonant vibration also shows up in the spectrum of the sound radiated by plastic spheres shown in Figure 5.13.

Note that the velocity waveform in this example is the time integral of the one in Figure 5.8; consequently, the pressure waveform is also the time integral of that in Figure 5.8. The pressure waveform is that of torque rate in this figure, which is seen

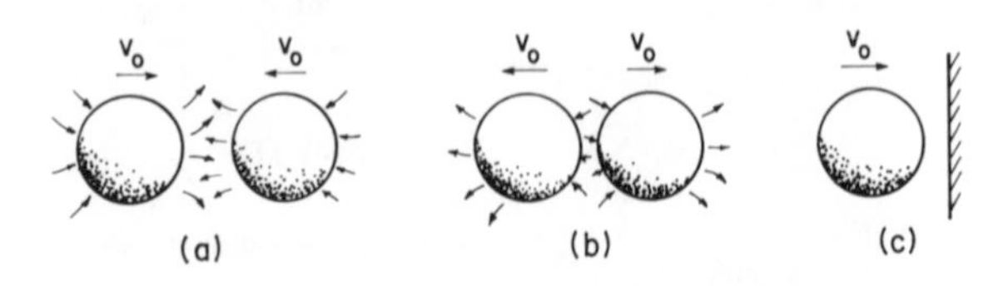

Figure 5.10 Two spheres that collide as in the Clarkers toy act as rapidly changing source distributions. Dynamically and acoustically, this is equivalent to the sphere rebounding from a rigid wall as shown.

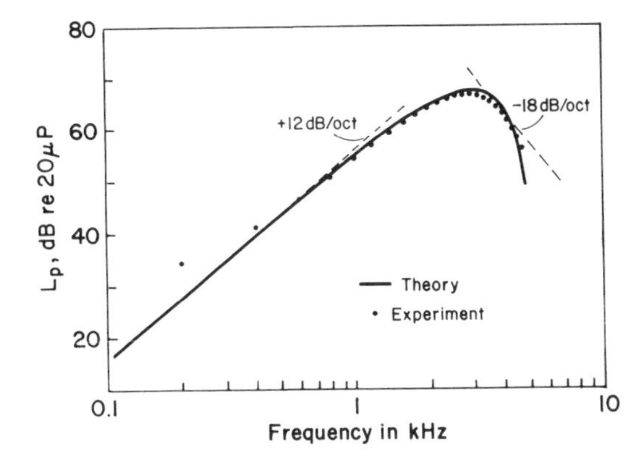

Figure 5.11 Calculated and measured pressure spectrum of radiated sound from impact of two steel spheres. Spectrum shape is determined by ratio of sound wavelength to size of spheres and contact stiffness.

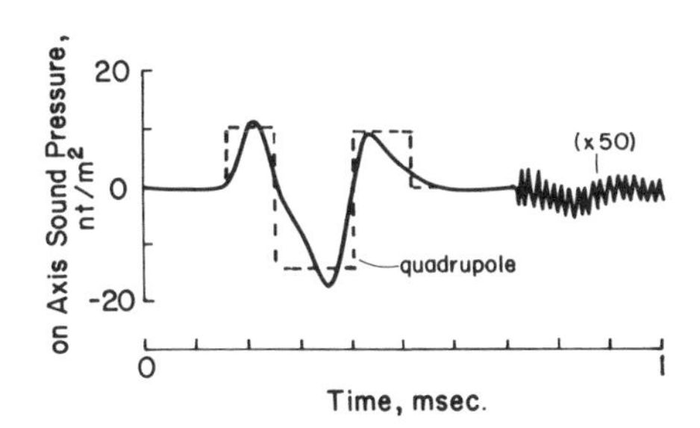

Figure 5.12 On-axis pressure waveform showing the low-amplitude radiation due to vibration.

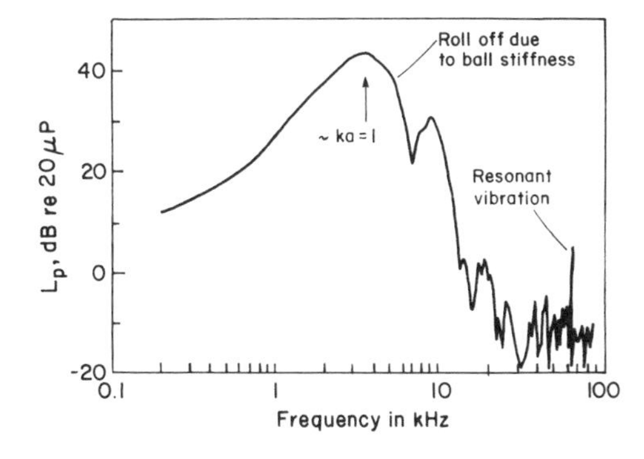

Figure 5.13 Radiated sound pressure spectrum for impacting plastic spheres. Note the notches in the spectrum, due to finite pulse duration and the peak due to resonant vibration at 50 kHz.

to be similar to the measured waveform in Figure 5.12. The Clackers toy is a longitudinal quadrupole, and the paired forces are the equal and opposite forces that the spheres exert on the fluid during collision.

This situation is very similar to a sphere impacting a rigid wall, because we could imagine a rigid wall at the plane of contact as shown in Figure 5.10. In the previous case, we found that the velocity made a transition from an incident velocity to the negative of that velocity for the rebound and the same thing occurs with the Clackers toy. The velocity spectrum of the sphere is determined in the same way the velocity was determined in Chapter 2 for the sphere during impact. The radiation is due to this change in velocity over what we might consider a single half-period of oscillation. This peak in the spectrum, however, is due not to the frequency of oscillation but to the size of the object with respect to the wavelength of sound. That is, the velocity spectrum is dropping off, and the radiation efficiency stops increasing when σ_{rad} becomes approximately 1. The spectrum of radiated sound begins to fall above this frequency.

The sound generated in machines can be very much affected by the impact process just described. The deceleration of the ball element of a typewriter will produce a very sharp peak in sound pressure. Impacting forces also produce a broad spectrum of vibration in the machine, and this represents another source of sound radiation. Generally, the sound energy produced by vibration will be greater, particularly for large machines that are resonant. For example, although there is direct sound radiation due to the deceleration of the impacting elements in a punch press, the major amount of sound usually comes from the impact-induced vibration and its subsequent radiation.

5.6 SPECIFIC ACOUSTIC MOBILITY AND SOUND INTENSITY

The particle velocity in a sound field is a vector, and therefore the specific acoustic mobility $\mathbf{y}$, which is the ratio of velocity $\mathbf{v}$ to pressure p, is also a vector:

$$\frac{\mathbf{v}}{p} = \mathbf{y}(\mathbf{x}, \omega) = \mathbf{G} - j\mathbf{B}. \tag{5.17}$$

Since the intensity of sound is

$$\mathscr{I} = \langle p\mathbf{v} \rangle_t = \tfrac{1}{2}\operatorname{Re}(p\mathbf{v}^*) = \tfrac{1}{2}|p|^2 \operatorname{Re}\mathbf{y}^* = \langle p^2 \rangle \mathbf{G}(\mathbf{x}; \omega), \tag{5.18}$$

any component of the intensity can be considered to be a measure of the specific acoustical conductance in that direction.

Let us now calculate the intensity component along a particular direction x. Assuming $e^{j\omega t}$ time dependence, we have, from Equation 5.2,

$$j\omega\rho_0 v = -\frac{\partial p}{\partial x}. \tag{5.19}$$

The velocity is found by integrating with respect to time:

$$v = -\frac{1}{\rho_0}\int \frac{\partial p}{\partial x}\,dt = \lim_{\Delta x \to 0}\left[-\frac{1}{\rho_0 \Delta x}\int (p_{x+\Delta x} - p_x)\,dt\right], \tag{5.20}$$

where we have expressed the pressure gradient in terms of the limit of its value between two adjacent locations, the usual definition of a derivative as a limiting process. For sinusoidal time dependence, Equation 5.20 becomes

$$v = \lim_{\Delta x \to 0}\left[-\frac{1}{j\omega\rho_0 \Delta x}\{p_{x+\Delta x} - p_x\}\right]. \tag{5.21}$$

To calculate the intensity, we multiply the velocity given by Equation 5.21 by the pressure, which we express by its average value between the locations x and $x + \Delta x$:

$$p = \tfrac{1}{2}(p_{x+\Delta x} + p_x). \tag{5.22}$$

Using Equation 5.18, we get

$$\begin{aligned} \mathscr{I} &= \langle pv \rangle_t = \tfrac{1}{2}\,\mathrm{Re}\, pv^* \\ &= -\frac{1}{2\omega\rho_0 \Delta x}\,\mathrm{Im}[|p_{x+\Delta x}|^2 - |p_x|^2 + p_x^* p_{x+\Delta x} - p_x p_{x+\Delta x}^*] \\ &= -\frac{1}{\omega\rho_0 \Delta x}\,\mathrm{Im}\{p_x p_{x+\Delta x}^*\}. \end{aligned} \tag{5.23}$$

The quantity $p_x p_{x+\Delta x}^*$ is the cross spectrum between the pressure at the two positions and is the basis on which two-channel spectrum analyzers are used to compute acoustical intensity.

As discussed in Section 3.12, the imaginary part of an impedance or mobility indicates that average values of kinetic and potential energies are not equal. For acoustic mobility, this means that there is a "sloshing" of kinetic or potential energy density past the test point in the x direction.

The imaginary part of the complex intensity $(1/2)(p\mathbf{v}^*)_x$ is, from Equation 5.23,

$$\begin{aligned} Q_x = \tfrac{1}{2}\,\mathrm{Im}(p\mathbf{v}^*)_x &= -\frac{1}{2\omega\rho_0 \Delta x}\{|p_{x+\Delta x}|^2 - |p_x|^2\} \\ &= -\frac{1}{2\omega\rho_0}|p|^2 \frac{\partial}{\partial x}\ln|p|^2, \end{aligned} \tag{5.24}$$

and the x component of the acoustic susceptance is therefore

$$B_x(\mathbf{x};\omega) = \frac{1}{\omega\rho_0}\frac{\partial}{\partial x}\ln|p|^2. \tag{5.25}$$

Therefore, locations where the sound pressure level is constant along all directions are places where the stored energy in the field is balanced between kinetic and potential.

5.7 MEASURING SOUND POWER USING ACOUSTICAL INTENSITY

The ability of dual-channel FFT analyzers and other analyzers using digital filters to quickly and accurately compute the cross spectrum between two microphone signals has been the basis for the very rapid growth in using acoustical intensity measurements to determine the sound power radiated by machines.

The usual measurement procedure is to surround the machine with a fixed array of dual-microphone probes or a traverse setup that sweeps over an area surrounding the machine. The example in Figure 5.14 is a dual-microphone probe moved about a piece of construction machinery. At a distance of 4 m, the probe is placed at 24 locations, and the total radiated power is found from

$$\Pi_{\text{rad}} = \sum_{i=1}^{24} A_i \mathscr{I}_i \tag{5.26}$$

where A_i is the area of the spherical segment assigned to each probe location, and $\mathscr{I}_i$ is the intensity determined for that location from Equation 5.24.

If the machine is located on a free-field, nonreflecting environment, then the sound intensity can also be determined by the relation

$$\Pi_{\text{rad}} = \frac{1}{\rho_0 c} \sum_i A_i \langle p_i^2 \rangle. \tag{5.27}$$

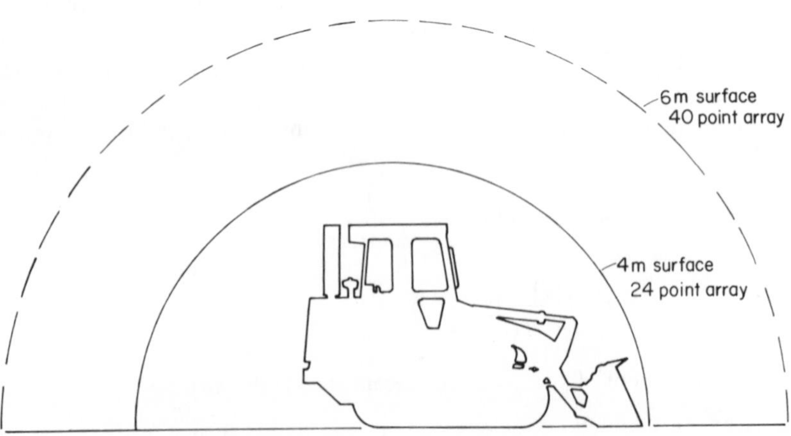

Figure 5.14 Array of positions for measuring radiated sound by a piece of construction equipment.

Figure 5.15 is a comparison of the sound power spectrums of a machine like that in Figure 5.14, determined by an acoustic intensity probe and Equation 5.26, and a single microphone, determined from Equation 5.27. Clearly there is a close relationship between the results of these two measurements, but the intensity measurement gives a slightly smaller value for the radiated power.

The acoustical intensity measurement has been useful only with the advent of digital processing, although it has been studied since the 1940s. Even so, there are still sources of error that affect the measurements, and these sources have been studied intensively:

1. Representation of a spatial derivative of pressure by the finite difference in Equation 5.21. Although there are more accurate ways to approximate the derivative, they require more microphone positions and greatly complicate the processing. The practical consequence is that we must put the microphones far enough apart to obtain a useful difference in the signal phases and, at the same time, keep their separation a small fraction of an acoustic wavelength.
2. Relative phase shifts between the two microphone channels, from the microphone diaphragm all the way through to the computation of the cross spectrum, must be kept very small. In early work, it was difficult to control the signal recording and processing, so that phase errors were too large. Digital recording and processing of the signals have now nearly eliminated this problem, but phase shifts in the microphones and the analog circuitry remain a problem, and are usually dealt with by a calibration technique that feeds as close a signal as possible to both microphone diaphragms.
3. Digital sampling introduces into the signal a random noise component and, if truncation of the digitization is used, a small bias.
4. Finite spatial sampling of the sound field can lead to errors in the overall sound power. Engineers do not consider this a serious deficiency, because they can correct it by choosing a finer grid size for the measuring surface.

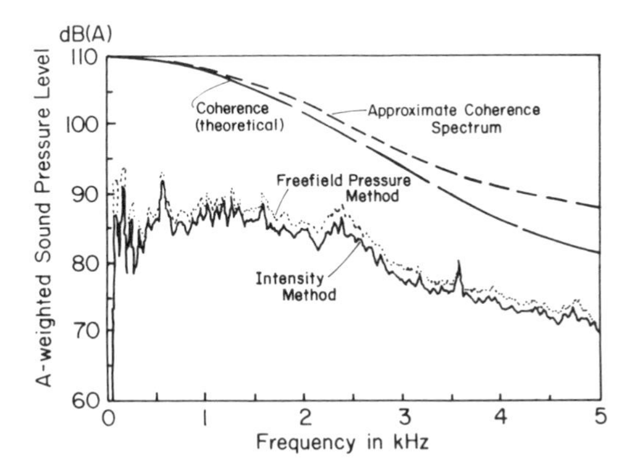

Figure 5.15 Comparison of intensity estimates by free-field pressure (108.3 dBA) and acoustic intensity (106.6 dBA).

5.8 REVERBERATION ROOM SOUND POWER

If the sound field in a room has linear dynamics, then the mean square reverberant pressure in the room is proportional to the sound power radiated into the space. In the reverberation room in Figure 5.16 the source radiates an amount of power Π_{rad} that results in a mean square pressure $\langle p_{rev}^2 \rangle$. If we follow a bundle of sound energy as shown, then its intensity $\mathscr{I}$ is reduced to an amount $\mathscr{I}(1-\alpha)$ after one reflection, where α is the average absorption coefficient of the interior surfaces of the room. After two reflections, the intensity has been reduced to $\mathscr{I}(1-\alpha)^2$; hence, after n reflections the intensity is

$$\mathscr{I}(n) = \mathscr{I}_0(1-\alpha)^n. \tag{5.28}$$

In time t the energy will have traveled a distance ct and the average distance traveled between collisions is the mean free path, given by

$$d = \frac{4V}{A}, \tag{5.29}$$

where V is the volume of the room and A is its interior surface area. The number of collisions in time t is then

$$n = \frac{ct}{d}, \tag{5.30}$$

so the intensity as a function of time becomes

$$\mathscr{I}(t) = \mathscr{I}_0(1-\alpha)\nu t, \tag{5.31}$$

where ν is the collision rate c/d. Expressing this in decibels, we get

$$10 \log \frac{\mathscr{I}(t)}{\mathscr{I}_0} = [4.34 \ln(1-\alpha)]\nu t \quad \text{(dB)}, \tag{5.32}$$

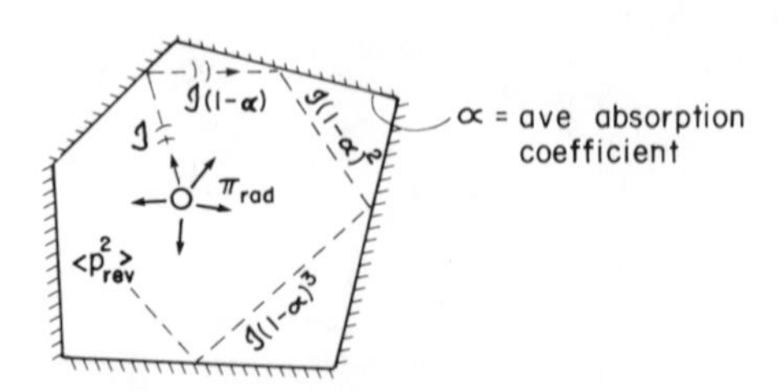

Figure 5.16 The sound produced by a source in a reverberant room rebounds from the walls many times, each time being reduced in intensity by a factor $(1-\alpha)$.

from which we can determine a decay rate

$$\mathrm{DR} = 27.3 f\eta = -4.34\nu \ln(1 - \alpha). \tag{5.33}$$

This allows us to relate a loss factor for the room to the surface absorption:

$$\eta = \frac{\alpha\nu}{\omega}. \tag{5.34}$$

Hence, $\omega\eta$ is the rate of energy decay and is equal to the fractional energy loss per collision (α) times the number of collisions per second with the wall (ν). Note the similarity between this discussion and that in Section 4.8.

The energy contained in the sound field can be related to the mean square reverberant pressure by noting that in 1 sec a wave travels a distance c, so the energy density times the volume of a cylinder of length c and 1 m^2 in cross section must be equal to the intensity. This means that the energy density is the intensity divided by the speed of sound:

$$\mathscr{E} = \frac{\langle p^2 \rangle}{\rho_0 c^2}. \tag{5.35}$$

The power into the reverberant field is the power that remains in the sound field after its first collision with the wall, and it must equal the energy lost from the reverberant field:

$$\Pi_{\mathrm{rad}}(1 - \alpha) = \omega\eta(\mathscr{E}V) = \frac{\omega\eta V}{\rho_0 c^2} \langle p^2 \rangle_{\mathrm{rev}}, \tag{5.36}$$

which therefore relates the reverberant pressure to the radiated power.

There is also a direct field—that is, the sound field that would exist in the absence of any reflections. From energy conservation, this is

$$\langle p^2 \rangle_{\mathrm{direct}} = \frac{\Pi_{\mathrm{rad}}}{4\pi r^2} \rho_0 c \mathscr{D}(\Omega), \tag{5.37}$$

where $\mathscr{D}(\Omega)$ is the directivity function. If we substitute Equation 5.34 into Equation 5.36, we obtain the m.s. reverberant pressure in terms of radiated power:

$$\langle p^2 \rangle_{\mathrm{rev}} = \Pi_{\mathrm{rad}} \rho_0 c \frac{4(1 - \alpha)}{\alpha A}. \tag{5.38}$$

The total mean square pressure is the sum of the components due to the direct and reverberant fields:

$$\langle p^2 \rangle = \langle p^2 \rangle_{\mathrm{dir}} + \langle p^2 \rangle_{\mathrm{rev}} = \Pi_{\mathrm{rad}} \rho_0 c \left[\frac{\mathscr{D}(\Omega)}{4\pi r^2} + \frac{4}{R} \right], \tag{5.39}$$

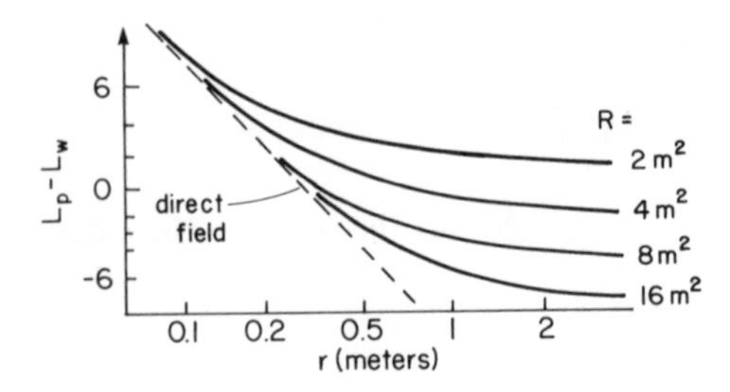

Figure 5.17 Transition from direct to reverberant field in a reverberant room depending on a room constant R.

where R is the room constant:

$$R \equiv \frac{\alpha A}{1 - \alpha}. \tag{5.40}$$

A graph of the sound pressure level resulting from Equation 5.40 is shown in Figure 5.17. For small distances, the direct field dominates, and the sound level drops off at 6 dB/double distance. A transition into an asymptotically constant amplitude region occurs as the reverberant field is entered. The reverberant level depends on the damping through R, which can be seen from Equation 5.37. The smaller the absorption, the higher the level in the reverberant field, and the closer the source is to the transition from the direct to the reverberant field.

A reverberant room will often use a rotating boom microphone to measure the m.s. pressure in the room, as shown in Figure 5.18. In addition, there may be a

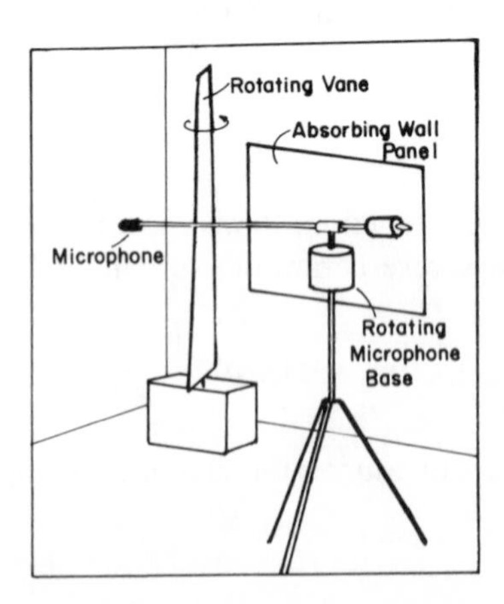

Figure 5.18 Interior of a reverberation room showing rotating microphone boom, low-frequency absorbing panels, and rotating vane. Room volume is about 50 m^3.

rotating vane that, in effect, changes the geometry of the room to produce a randomization of the modes of vibration. This is particularly useful at low frequencies. Panel absorbers on the walls are also used to increase low-frequency damping; they smooth out the response of the room at low frequencies.

In selecting positions for measurement, we attempt to avoid the direct field around the source so that the data are not contaminated by the possible areas of high directivity and, therefore, larger m.s. pressure than would truly represent the reverberant pressure. The absorption of the room can be found either by inserting a source of known sound power into the room and measuring the reverberant pressure associated with it or making decay rate measurements and determining the absorption or loss factor from the decay rate.

If a source of known power Π_k is used, then for a sound pressure $\langle p_u^2 \rangle$ due to the unknown source Π_u and one of $\langle p_k^2 \rangle$ due to the known source, we have

$$\langle p_u^2 \rangle = \langle p_k^2 \rangle \cdot \frac{\Pi_u}{\Pi_k} \tag{5.41}$$

or, in terms of pressure and power levels,

$$L_w\,(\text{unknown}) = L_w\,(\text{known}) + L_p\,(\text{unknown}) - L_p\,(\text{known}). \tag{5.42}$$

The methods indicated by Equations 5.39 and 5.42 are the usual ways of determining radiated sound power in a reverberation room.

5.9 USING RECIPROCITY TO DETERMINE SOUND RADIATION

A problem encountered some time ago was that of trying to estimate the noise produced by a small compressor B in a reverberant room when a larger compressor A was making even more noise in the same room, as shown in Figure 5.19. The larger compressor could not be turned off. The smaller compressor could be turned off, but there was little change in sound level because it was dominated by the larger compressor. Nevertheless, in this situation we can determine the sound radiation from the smaller compressor by using reciprocity.

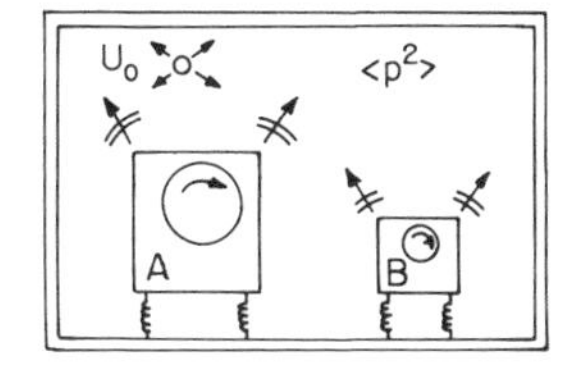

Figure 5.19 Large compressor A and small compressor B located in a reverberant room. The problem is to find the sound radiated by B if A cannot be turned off.

In preparation for this example, we need an expression for the amount of power produced by a simple source. Using Equation 5.11, if the source is small compared to an acoustic wavelength (i.e., $ka \ll 1$), then the radiated sound power from the source is

$$\Pi_{\text{rad}} = \langle v_0^2 \rangle \cdot 4\pi a^2 \cdot ka^2 \cdot \rho_0 c = \langle U_0^2 \rangle \frac{k^2 \rho_0 c}{4\pi}, \tag{5.43}$$

where the volume velocity $U_0 = 4\pi a^2 v_0$. The quantity $k^2 \rho_0 c/4\pi = R_0$ is called the point source acoustic radiation resistance.

Returning to our example, let us turn off compressor B and measure its vibratory response due to the sound produced by compressor A. We assume that the sound produced by compressor A is equivalent to that due to a simple source of volume velocity U_0, so the radiated sound is

$$\Pi_{\text{rad}} = \langle U_0^2 \rangle R_0 = \beta \langle p^2 \rangle \tag{5.44}$$

where we again assume that the m.s. pressure in the room is proportional to this radiated power. The parameter β is not evaluated because it will drop out of the calculation. Then the response velocity of compressor B due to this sound field is

$$\langle v_B^2 \rangle = \gamma \langle p^2 \rangle. \tag{5.45}$$

The parameter γ is determined from a measurement that we can carry out when compressor B is turned off. These two formulas allow us to determine a ratio between the velocity of B and the hypothetical point source volume velocity:

$$\frac{\langle v_B^2 \rangle}{\langle U_0^2 \rangle} = \frac{\gamma R_0}{\beta}. \tag{5.46}$$

This first experiment is shown in Figure 5.20(a).

The reciprocal experiment is shown in Figure 5.20(b). We apply a force to compressor B and measure the radiated pressure in the room, p'. When we excite the compressor with a drive force, we get a resulting mechanical vibration velocity of the compressor, which can be determined by experiment:

$$\langle v_B'^2 \rangle = \langle l'^2 \rangle \mu. \tag{5.47}$$

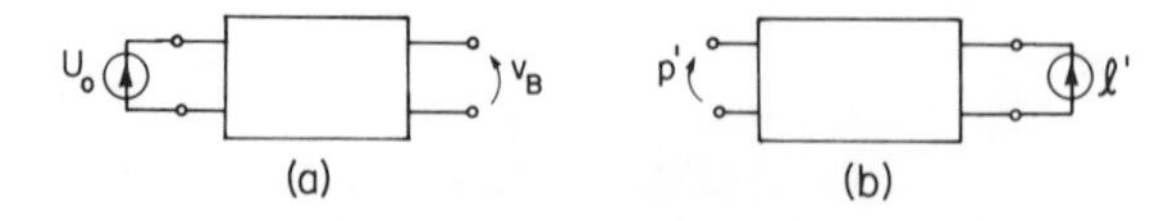

Figure 5.20 Diagrams for the reciprocal experiments used to determine the relation between the vibrational velocity of compressor B and its radiated sound power.

The radiated sound power is proportional to the m.s. velocity of the compressor and, in turn, determines the resulting reverberant pressure in the room due to the vibration of compressor B:

$$\Pi' = \alpha\langle v_B'^2\rangle = \beta\langle p'^2\rangle. \tag{5.48}$$

We want to know α because it relates the vibration of the small compressor to its radiated sound power. Combining Equations 5.47 and 5.48, we have a ratio between the radiated pressure and the driving force:

$$\frac{\langle p'^2\rangle}{\langle l'^2\rangle} = \frac{\alpha\mu}{\beta} = \frac{\langle v_B^2\rangle}{\langle U_0^2\rangle} = \frac{\gamma R_0}{\beta}, \tag{5.49}$$

where we have included the result from Equation 5.46 by reciprocity. Therefore, the desired quantity α is determined from the two measured parameters γ and μ and the known point source radiation resistance R_0:

$$\alpha = \frac{\gamma}{\mu}R_0 = \frac{\Pi_{rad}}{\langle v_B^2\rangle}. \tag{5.50}$$

Thus, if we know the vibration $\langle v_B^2\rangle$ of the compressor B when it is running, we can use α to determine its radiated sound power.

5.10 MEASURING PRODUCT SOUND WITH A SMALL REVERBERATION CHAMBER

A sewing machine company wanted to make more reliable measurements of the noise of sewing machines on the production line. A small semianechoic booth had been used, with an operator inside the booth operating the machine, testing its sewing properties and, at the same time, reading a small sound-level meter. If the meter read more than 72 dB during this sewing operation, the machine was to be rejected. The operator would write the sound level on a slip of paper and attach it to the machine.

Forty-eight machines were selected after this operation and the recorded sound levels on the slips of paper were noted. The distribution of these values is shown in Figure 5.21. One machine had a sound-level reading of 69, 5 read 70, 12 read 71, and 30 read 72. Since this did not appear to be a distribution that we would expect from a sample population, these same 48 machines were taken into a semianechoic chamber and the average sound levels over a hemisphere 5 ft from the machine were measured at five microphone locations and averaged. The distribution of sound levels measured in this experiment is shown in Figure 5.22. This distribution has much more of the bell-shaped form that we would expect from the measurement of any parameter over a population. Not only was there a difference in distribution between the two measurements, but there was almost no correlation

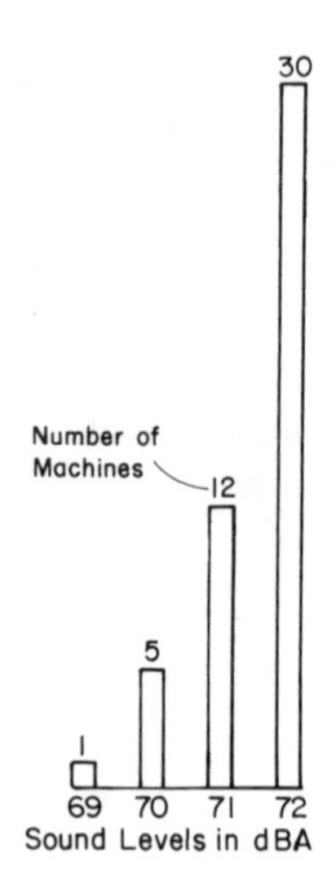

Figure 5.21 Distribution of production line sound levels as reported by operator. Strongly skewed distribution is atypical of expected distribution of sound from a population of machines.

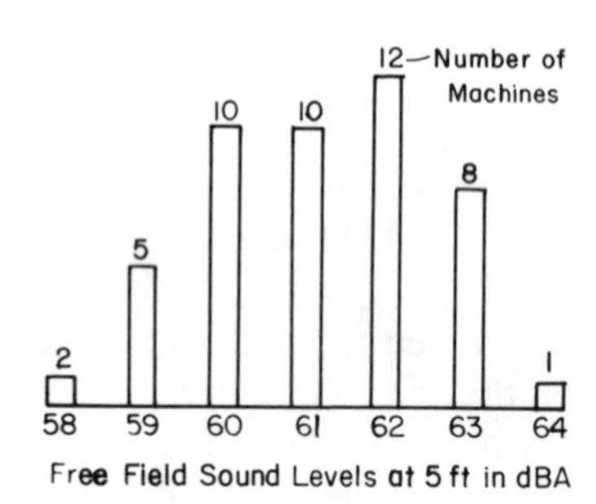

Figure 5.22 The distribution of sound levels for same population of machines, measured in a semianechoic room by laboratory personnel, shows the expected bell-shaped form.

between the values of sound pressure measured in the production test and those measured under controlled laboratory conditions.

The reason for the strangely skewed distribution in Figure 5.21 is likely to have been the pressure of the production line to pass machines. When we give the testing operator a lot of discretion in making the measurement and reporting the pass/fail condition, then unreliable quality control procedures are likely to occur.

Because of the amount of space needed and the long time for setting up the machine and doing the test, the laboratory measurement cannot be made on the production line. The design goal for an on-line production test was to be able to remove the machine from the line, place it in the test chamber, run it to speed, make

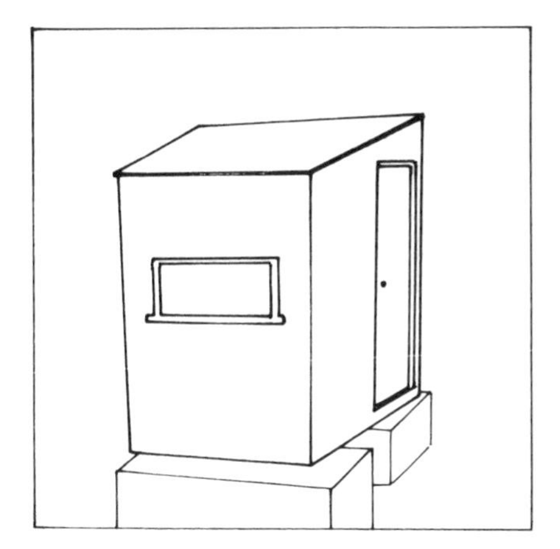

Figure 5.23 The small reverberant sound booth constructed to be used on the production line can be moved by a forklift.

the acoustic measurement, remove it from the chamber, and put it back on the production line within 1 min. A reverberation measurement was chosen for this process. Since the noises in a sewing machine that contribute to A-weighted sound are mostly above 300 Hz, it was decided to limit the measurement to frequencies 300 Hz and above. This allows us to make the reverberation room much smaller. A reverberation room that is limited to 300 Hz can be 1/27 the volume of a reverberation room intended to make measurements down to 100 Hz.

The reverberation room constructed for this application is shown in Figure 5.23. It is irregularly shaped and has two openings. One is a doorway for the QC supervisor to enter periodically and inspect the condition of the measuring equipment, carry out calibrations, and make other tests. The window, which is spring assisted, is used by the production line operator to insert and remove the test object. The booth is sturdily constructed so that it can be moved into position with a forklift. All construction is designed for a high transmission loss: double-glazed windows, double-leaf walls with resilient metal studs, and heavily gasketed door and window to eliminate air leaks.

A view of the interior of the booth is shown in Figure 5.24. We can see the humidity monitor and the low-frequency panel absorber, which are included to control sound absorption. Some of the five microphone booms and the preamplifiers can be seen, along with the acoustical calibrator used to set up the system. The photo pickup monitors the operating speed of the machine. The noise analyzer averages the squared outputs of the five microphones to develop a reverberant m.s. pressure estimate. The analyzer also measures the speed of the machine and generates a control signal to operate a powered variable transformer to bring the machine to the desired speed. The noise analyzer is normally located in a position so that the readings of sound level and speed cannot be read by the operator.

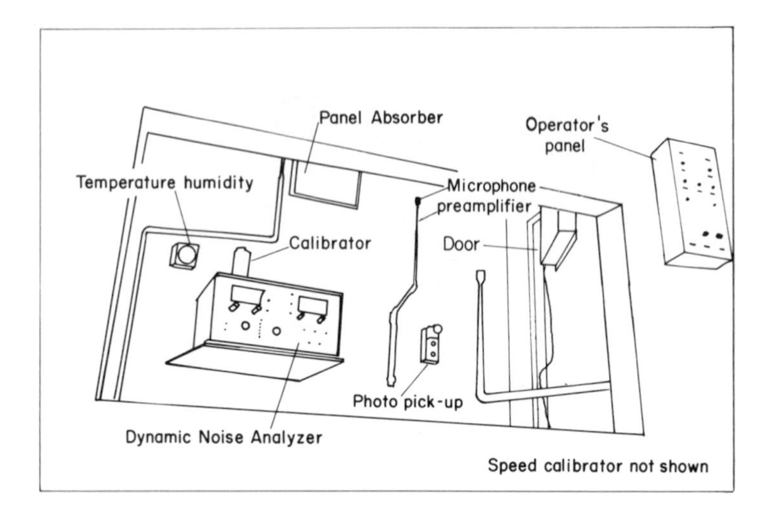

Figure 5.24 A view through the window of the noise booth shows several components of the measuring system. The noise analyzer is normally placed outside the operator's view.

The operator can only see the small control panel in Figure 5.24. This panel has a light that indicates when the test has started and the machine has come to speed. After it has come to and held speed for a certain time, the noise level is read; if it is within the acceptable limits, the "pass" light goes on for noise. If for any reason the machine cannot be brought to speed within a specified supply voltage region, then the machine also fails because of a low-speed condition. A picture of the window assembly and the machine in the test position is shown in Figure 5.25.

When used on the production line, the booth is placed next to the machine trolleys. Each machine is placed onto a lazy Susan arrangement and rotated into the test chamber as shown in Figure 5.25. The window is closed and the operator presses the start button on the control panel in Figure 5.24. When the test is over, a red or green light comes on to indicate whether the machine has passed or failed the test, either in terms of its ability to reach the prescribed speed or to meet the noise limits. The window is then opened; the machine is removed from the test chamber and put back onto the production line. This entire process takes about a minute.

A distribution of sound levels on the 48 machines described earlier, as measured in this booth, is shown in Figure 5.26. We see that the distribution is bell-shaped and is the sort of distribution we would expect. In addition, a cross-plot between the levels measured for the machines in this booth and the levels measured in the semianechoic room showed a high degree of correlation between the sound levels. In fact, measurements on a calibrated source indicated that the small reverberant booth makes more accurate noise power measurements above 300 Hz than the laboratory measurement does.

This example shows that the results from a quality control test can often be judged simply by looking at the data. An unrealistic distribution of results can signal

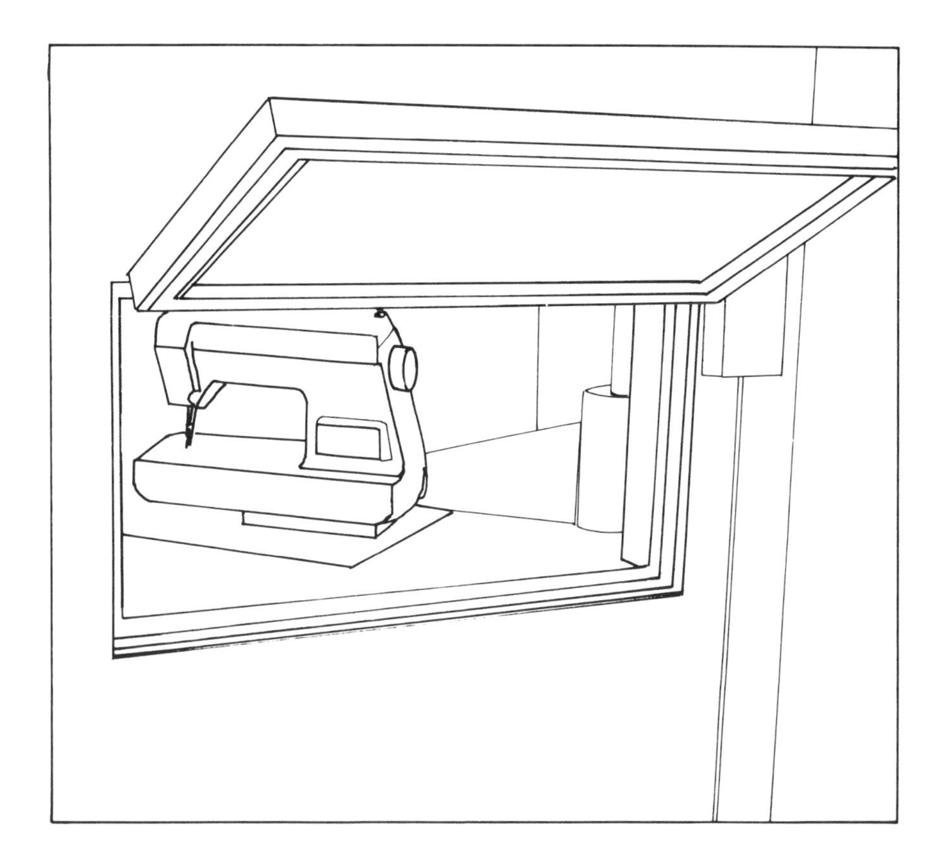

Figure 5.25 A machine under test has been placed on the lazy Susan platform and rotated into position for the measurement. The operator closes the window and pushes the "start" button to begin the test cycle.

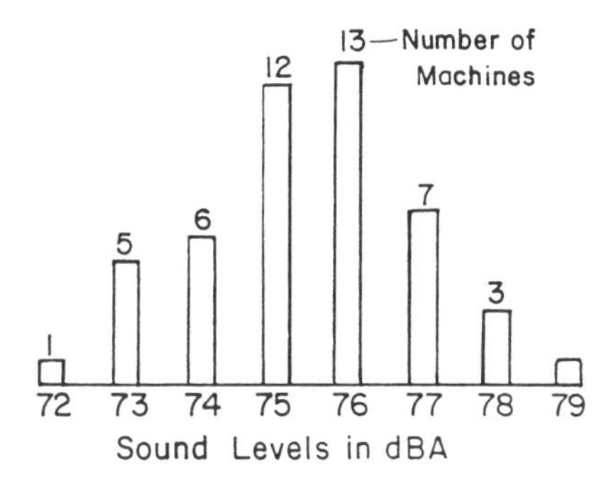

Figure 5.26 The distribution of sound levels measured by the production line booth for the control population shows the expected bell-shaped form also.

that something fairly fundamental is wrong with the measuring process. Often, production pressures will cause operators to pass machines that are above limits. The need to pass machines and send them through the QC process when production needs are great is understandable and, in an area like noise, perhaps acceptable. However, we would still like to catch the most noisy machines so that the need for relaxation of the noise standards in production does not eliminate the need to make reliable noise measurements. The measurement needs to be as accurate as possible, and the limits or criteria for passage can be influenced by production needs.

Setting QC limits is a management decision. The QC operator should only judge whether the machine has passed or failed the test. A good way to do this is to give the operator binary information through a red or green light or some similar output, which does not allow production pressures to influence the pass/fail decision.

5.11 SOUND RADIATION BY STRUCTURAL SURFACES

We now step away from the problems of measuring machinery noise to the question of how the vibrations that were calculated in Chapters 3 and 4 can cause sound radiation. We start with an infinite thin plate excited by a point force and then proceed to include the effects of boundaries on sound radiation.

We can use reciprocity to derive the radiation of sound by a thin sheet driven by a point force. This is our first example of radiation by structural vibration. In this case, we imagine that the large, thin sheet in Figure 5.27 is driven by a point force and produces a sound pressure p in the field. The reciprocal experiment will involve injecting a volume velocity U at the place where the pressure was measured previously and computing the response velocity v' of the sheet.

We will assume that the sheet is limp; that is, it has very little structural rigidity, and its response to the sound is primarily due to its mass reactance. We will also assume that at the frequency of interest, the mass reactance of the sheet is very large compared to the characteristic impedance of the air, $\rho_0 c$.

Figure 5.20 can be used to represent the input–output diagrams for the two experiments described in Section 5.9 (but note reversal in order!). In the first experiment, since the system is linear, the radiated pressure is proportional to the

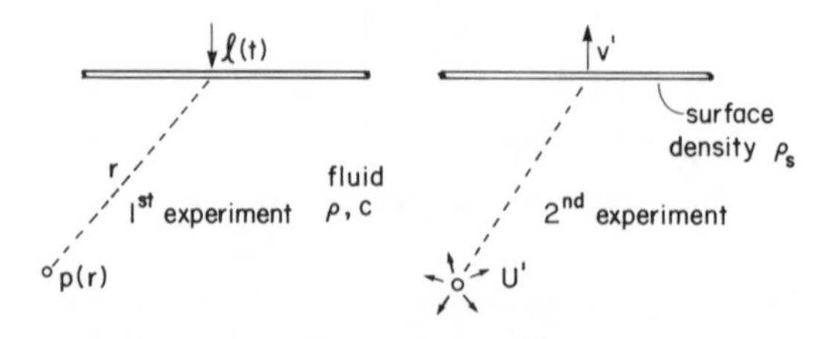

Figure 5.27 The reciprocal experiments used to calculate radiation from a limp plate.

applied force by a quantity θ:

$$\frac{p}{l} = \theta, \tag{5.51}$$

so θ is the quantity we want to determine.

Equation 5.14 allows us to relate the sound pressure produced in a free field due to a volume velocity U':

$$p' = \frac{j\omega\rho U' e^{-jkr}}{4\pi r} \equiv \alpha U', \tag{5.52}$$

where α is the coefficient of U'. When the sound impinges on the limp sheet that has a high mass reactance, there is a doubling of pressure due to the image of the sound field in the wall. This gives a blocked pressure at the surface

$$P'_{\text{bl}}\,(\text{pressure at wall}) = 2p'. \tag{5.53}$$

Now using the reactance of the wall due to its mass per unit area ρ_s, we can calculate the velocity due to this blocked pressure:

$$v' = \frac{P'_{\text{bl}}}{j\omega\rho_s} = \frac{2}{j\omega\rho_s}P' = \frac{2\alpha}{j\omega\rho_s}U'. \tag{5.54}$$

The reciprocity relationship is then

$$\frac{v'}{U'} = \frac{2\alpha}{j\omega\rho_s} = \frac{p}{l} = \theta = \frac{2j\omega\rho_0 e^{-jkr}}{4\pi r \cdot j\omega\rho_s} = \frac{\rho_0}{2\pi\rho_s r}e^{-jkr}, \tag{5.55}$$

which for an arbitrary waveform allows us to write

$$p(r,t) = l\left(\frac{t-r}{c}\right)\frac{\rho_0}{2\pi\rho_s r}, \tag{5.56}$$

since the factor e^{-jkr} in Equation 5.55 represents a time shift by the amount r/c.

A point force on a sheet therefore produces a pressure waveform that is a simple time replica of the applied force, weighted by the ratio of the density of air to the density of the sheet, and decaying as 1/distance away from the source. The force acts like a simple source and produces radiated sound of equal amplitude and waveform in all directions.

If an object strikes a large thin sheet, then there will be radiation from the point of impact, which is a time replica of the applied force. If the sheet is very thin, the vibrations generated in the sheet will have short wavelengths and will not be very efficient radiators of sound. However, when these waves arrive at the boundaries of the sheet and reflect, forces are applied to the sheet and additional radiation will be produced. There is a sound wave produced by this structural reverberation that generally has lower amplitude than the initial impulse but may actually radiate more sound energy because of its greater duration. We shall discuss the radiation from this structural reverberation a little later. Its actual importance relative to the initial radiated pulse will depend upon the circumstances.

Sometimes it is more convenient to measure the motion at the impact point with an accelerometer, for example, rather than the force, so we want to compute the radiated sound in terms of this local velocity. We saw in Chapter 3 that the drive point mobility of a large thin sheet is

$$Y_{\mathrm{dp}} = \frac{1}{8\rho_{\mathrm{s}}\kappa c_{\mathrm{l}}}, \tag{5.57}$$

so the local velocity at the drive point is

$$v_{\mathrm{dp}} = \frac{l}{8\rho_{\mathrm{s}}\kappa c_{\mathrm{l}}}. \tag{5.58}$$

This gives a pressure expressed in terms of the drive point velocity

$$p = \frac{l\rho_0}{2\pi\rho_{\mathrm{s}}r} = \frac{4\rho_0\kappa c_{\mathrm{l}}}{\pi r}v_{\mathrm{dp}}. \tag{5.59}$$

We can further relate this pressure to an effective volume velocity, which is the drive point velocity times the area of some small disk of effective radius r_{e} situated at the impact location:

$$U = \pi r_{\mathrm{e}}^2 v_{\mathrm{dp}}. \tag{5.60}$$

The pressure radiated by this volume velocity, accounting for the factor of 2 that results from the volume velocity of source being located at a high-impedance surface, is (using Equation 5.59)

$$|p| = \frac{\omega\rho v_{\mathrm{dp}}\pi r_{\mathrm{e}}^2}{2\pi r} = \frac{4\kappa c_{\mathrm{l}}\rho}{\pi r}v_{\mathrm{dp}}. \tag{5.61}$$

This allows us to calculate the effective piston radius:

$$r_{\mathrm{e}} = \lambda_{\mathrm{b}}\sqrt{\frac{2}{\pi^3}} = 0.254\lambda_{\mathrm{b}}. \tag{5.62}$$

The radius is about one quarter of a bending wavelength at the frequency being considered. The addition of damping can reduce the amount of sound produced by the reverberant flexural waves as they rebound in the structure, but damping will not reduce the radiation produced by the force or vibration at the drive point.

5.12 RADIATION FROM STRUCTURAL BENDING WAVES

Although radiation can occur from longitudinal and in-plane vibrations in general, the motion perpendicular to the surface resulting from these wave types is generally much smaller than that due to bending. Bending waves dominate the radiation of sound in most machine structures. Figure 3.3 shows a single bending wave of wavelength λ_b and at frequency f on the plate. The speed of bending waves is

$$c_b = \sqrt{\omega \kappa c_l} = 100\sqrt{h\,(\mathrm{mm}) \times f\,(\mathrm{kHz})} \quad \mathrm{m/sec}, \tag{5.63}$$

where we have used a longitudinal wave velocity of 5000 m/sec appropriate for steel, aluminum, or glass.

Equation 5.63 is a very simple way of estimating the bending wave speed if we know the thickness of the panel and the frequency. As an example, consider a panel that is 4 mm thick and vibrating at 1 kHz. This equation indicates that a bending wave on this panel will have a speed of 200 m/sec. This wave is slower than the speed of sound in air, and, as we know, such disturbances do not generate a sound wave but only a localized flow around the disturbing object.

We first consider supersonic waves that travel faster than the speed of sound. The frequency f_c at which bending waves become supersonic is

$$c_b = c = \sqrt{2\pi f_c \cdot \kappa c_l}, \tag{5.64}$$

where the critical frequency f_c is

$$f_c = \frac{c^2}{2\pi \kappa c_l} = \frac{13{,}000}{h\,(\mathrm{mm})}\ \mathrm{Hz}. \tag{5.65}$$

For example, a steel plate 1 cm thick has a critical frequency of 1.3 kHz.

When a projectile travels through air at supersonic speed, it generates a mach wave (see Figure 5.28) at an angle θ such that the speed of sound $c = v \sin \theta$. Similarly, if the wave on a bending plate is supersonic, then the angle of radiation θ

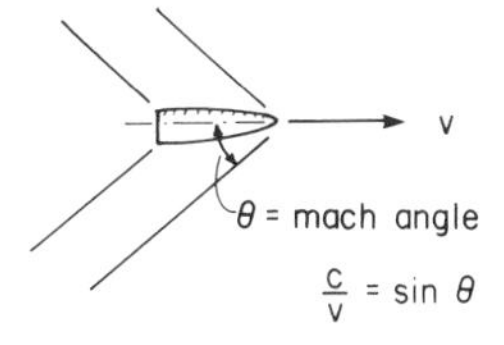

Figure 5.28 A body that moves faster than the speed of sound radiates sound waves at the mach angle $\sin^{-1}(c/v)$.

from this wave can be determined, as shown in Figure 5.29, to be

$$\frac{c}{c_b} = \sin\theta = \frac{c}{\sqrt{\omega\kappa c_l}} = \sqrt{\frac{f_c}{f}}. \tag{5.66}$$

Thus, it is the ratio of the frequency of the vibration to the critical frequency that determines the angle of sound radiation.

In Figure 5.29, we assume that the bending speed is greater than the speed of sound, which means that the bending wavelength is greater than the acoustic wavelength. In this circumstance, as the waves move away from the plate at the mach angle θ, the particle velocity is parallel to the wave front and is given by the pressure divided by $\rho_0 c$, the specific acoustic impedance. The plate velocity must equal the component of fluid velocity perpendicular to the plate surface and, therefore, must equal the acoustical particle velocity times the cosine of the mach angle or

$$v_{\text{plate}} = v_n = v\cos\theta = \frac{p}{\rho_0 c}\cos\theta. \tag{5.67}$$

This relation between the plate velocity and the pressure that it generates in a sound field is the radiation impedance (radiation resistance in this case), which is given by

$$R_{\text{rad}} = \frac{Ap}{v_{\text{plate}}} = \frac{A\rho c}{\cos\theta} = \frac{A\rho c}{\sqrt{1-\sin^2\theta}} = \frac{A\rho c}{\sqrt{1-f_c/f}}. \tag{5.68}$$

Thus, the radiation efficiency is

$$\frac{1}{\sqrt{1-f_c/f}} = \sigma_{\text{rad}} \qquad \text{(supersonic waves)}, \tag{5.69}$$

where the radiation efficiency is expressed in terms of the ratio of the critical

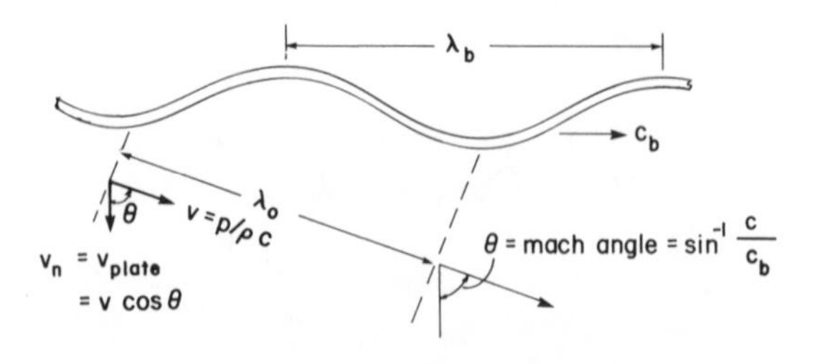

Figure 5.29 Radiation of sound by supersonic bending waves at frequencies greater than the critical frequency.

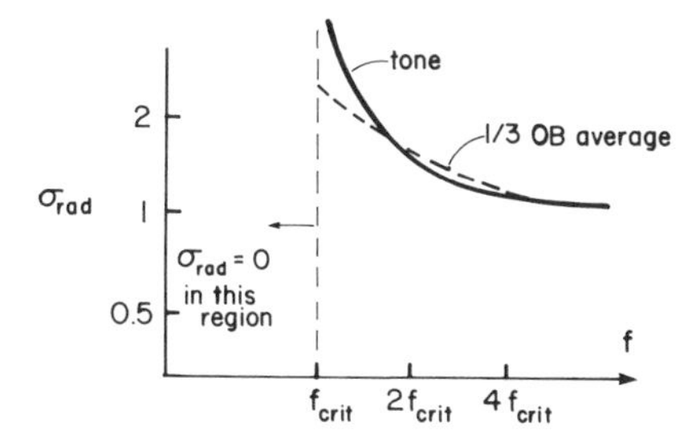

Figure 5.30 Radiation efficiency for bending waves on an infinite plate. Results for single frequency and third-octave averages are shown.

frequency to the frequency of vibration. Equation 5.68 implies that if the frequency is less than the critical frequency, then there is no radiation because the pressure and particle velocity are 90° out of phase and the fluid simply presents a mass reactance to the vibration.

The radiation efficiency is graphed in Figure 5.30. It approaches unity well above the critical frequency but has a singularity at the critical frequency. For this reason, it is not quite correct to refer to it as an efficiency, and in some textbooks it is called the radiation factor.

If the frequency is twice the critical frequency, then the radiation efficiency is $\sqrt{2}$, or 1.5 dB greater than its asymptotic value, and when the operating frequency is 2 octaves above the critical frequency, the radiation efficiency is about 0.25 dB. Thus the radiation efficiency approaches 0 dB fairly quickly above the critical frequency.

5.13 RADIATION BELOW THE CRITICAL FREQUENCY

In air, we normally deal with frequencies above and below the critical frequency, whereas in water the common situation is to be below the critical frequency. The speed of sound in water is 1500 m/sec, and if we compute the critical frequency, we obtain

$$f_c = \frac{250{,}000}{h\ (\text{mm})} \qquad \text{(water)} \tag{5.70}$$

which indicates that for most structures in water the major part of the frequency range will be below the critical frequency. We have already seen in the discussion of the point-driven thin plate that the radiation does not vanish for frequencies less than the critical frequency, but there is radiation directly from the drive point. Bending waves below the critical frequency are subsonic and, therefore, do not radiate as free waves. This situation is shown in Figure 5.31, where we have a vibration of wavelength λ_b that is smaller than the acoustic wavelength λ_0. The wave

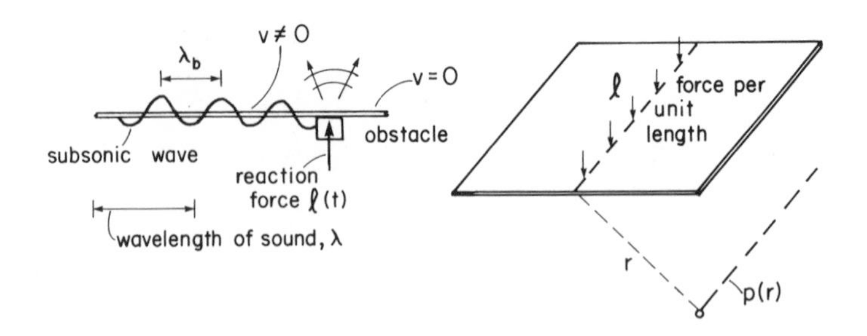

Figure 5.31 Generation of force by subsonic bending wave that impinges on an obstacle and the subsequent radiation due to this line force.

is subsonic and does not radiate unless it meets an obstacle that produces a reaction force on the plate. Then the reaction force may be thought of as causing radiation much as the point force produced radiation in Figure 5.27. The obstacle might be one of the boundaries of the plate, which produces a line force as shown in Figure 5.31.

As before, we assume that the pressure produced by this force at a distance r is proportional to this force per unit length by a quantity Γ. The reciprocal experiment is shown in Figure 5.32. A line of volume velocity at a distance r from the point where the force was placed now generates a sound field p'. The radiated sound due to the line of volume velocity is

$$p' = \pi \rho c U' \sqrt{\frac{k}{2\pi r}}\, e^{j\pi/4}. \tag{5.71}$$

Also as before, the blocked pressure at the plate surface is $2p'$. The velocity of the plate due to this blocked pressure is determined by the mass reactance:

$$v' = \frac{2p'}{j\omega \rho_s} = \frac{2\pi\rho c}{\omega \rho_s} \sqrt{\frac{k}{2\pi r}}\, U'. \tag{5.72}$$

We now use the reciprocity relation to obtain Γ:

$$\frac{v'}{U'} = \frac{p}{l} = \frac{2\pi\rho c}{\omega\rho_s}\sqrt{\frac{k}{2\pi r}} = \Gamma. \tag{5.73}$$

This allows us to calculate the intensity radiated by the force per unit length

$$\mathscr{I} = \frac{\langle p^2 \rangle}{\rho c} = \frac{\Gamma^2}{\rho_0 c}\langle l^2 \rangle = \frac{2\pi\rho_0}{\omega\rho_s^2 r}\langle l^2 \rangle. \tag{5.74}$$

Since this line force is radiating into a half-cylinder, the power radiated per unit

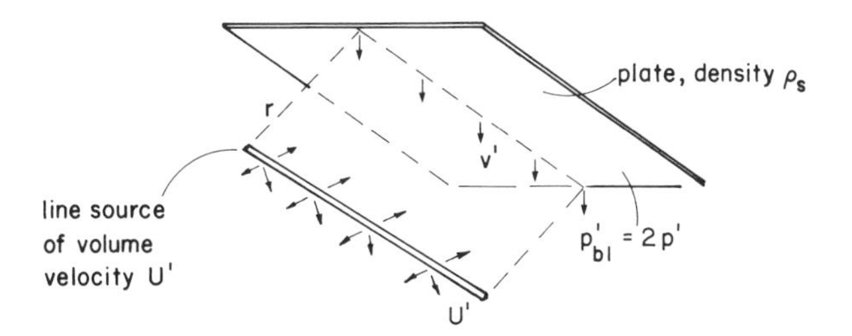

Figure 5.32 The reciprocal experiment used to compute the sound radiated by a line force on the plate.

length is

$$\Pi_1 = \pi r \mathscr{I} = \frac{2\pi^2 \rho_0}{\omega \rho_s^2} \langle l^2 \rangle. \tag{5.75}$$

This gives the power per unit force per unit length applied to the panel discontinuity, whether the force is due to an edge or some kind of loading member along a line on the panel, and corresponds to the result of Equation 5.56 for a point force.

To relate the edge force to the plate vibration, consider a wave traveling toward a discontinuity, as shown in Figure 5.31. We can think of the force produced by the discontinuity as the force required to generate a wave that will cancel out the incident wave in the region to the right of the location of the obstacle. The velocity is v, and the force is l, and they are related by the impedance of a bending wave:

$$Z_{\text{bending}} = \text{const. } \rho_s c_b. \tag{5.76}$$

where the constant is about 1.

The force per unit span of the wave is therefore

$$l = v \rho_s c_b. \tag{5.77}$$

We can place this result in Equation 5.75 to obtain the radiated power per unit length of edge (or loading member)

$$\Pi_1 = 2\pi^2 \rho_0 \kappa c_1 \langle v^2 \rangle. \tag{5.78}$$

From Equation 5.78, we see that the radiation resistance and, therefore, the radiation efficiency are independent of frequency. We have assumed, however, that the wave direction is perpendicular to the discontinuity. In a real plate, the waves are traveling in all directions, and only those waves that intersect the boundary so that the speed of the force disturbance along the boundary is supersonic will cause radiation. We must, therefore, calculate the fraction of waves that satisfy this condition to modify Equation 5.78.

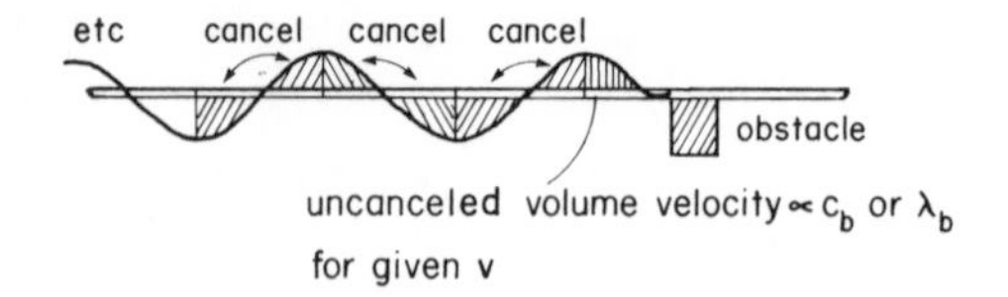

Figure 5.33 Illustration of the concept of uncanceled volume velocity, leading to sound radiation by an obstacle on a plate.

We note that in Equation 5.78 the power per unit length increases as the bending wave speed increases. As the bending wavelength increases, there is a greater amount of uncanceled volume velocity that radiates near the discontinuity because the width of this volume velocity strip will increase as κc_1, as shown in Figure 5.33.

The dependence of the radiation on boundary conditions ranges over a factor of 2 from a supported edge to a clamped edge. Differences of 3 dB are generally quite difficult to verify experimentally in practical structures. We shall not worry about such details in this discussion.

Figure 5.34 shows a top view of the panel and the discontinuity. When a wave is incident on the discontinuity, it produces a force disturbance along the discontinuity. Over some angular region less than a critical angle θ_c, the travel speed along the discontinuity will be supersonic and the wavelengths of force along the discontinuity will be greater than the acoustic wavelength. Therefore, only waves that are incident on the discontinuity within the angular region less than θ_c will cause sound radiation.

The condition for θ_c is that the trace speed be equal to the speed of sound, which is

$$\sin\theta_c = \frac{c_b}{c} = \sqrt{\frac{f}{f_c}}. \tag{5.79}$$

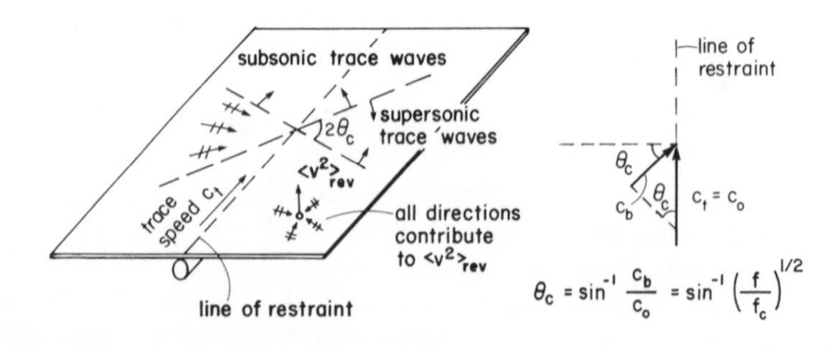

Figure 5.34 Identification of the fraction of the reverberant vibration that contributes to supersonic trace waves along a line of restraint.

The total angle occupied by waves that meet the supersonic trace conditions is $2\theta_c$, whereas the angle occupied by all waves that generate the mean square reverberant velocity is 2π. Therefore, the fraction of all waves that produce supersonic trace waves is θ_c/π. The mean square effective velocity that produces radiation is therefore

$$\langle v^2 \rangle_{\text{eff}} = \left[\frac{1}{\pi}\sin^{-1}\left(\frac{f}{f_c}\right)^{1/2}\right]\langle v^2 \rangle_{\text{rev}}. \tag{5.80}$$

Putting Equation 5.80 in Equation 5.78 gives the power per unit length:

$$\Pi_l = 2\pi^2 \rho_0 \kappa c_l \langle v^2 \rangle_{\text{eff}} = \left[2\pi\frac{\kappa c_l}{c}\sin^{-1}\left(\frac{f}{f_c}\right)^{1/2}\right]\langle v^2 \rangle_{\text{rev}}\rho_0 c. \tag{5.81}$$

Multiplying by the total edge length or perimeter to get the total radiated power, we have

$$\Pi_{\text{rad}} = \langle v^2 \rangle_{\text{rev}} \rho c A_p \sigma_{\text{rad}}, \tag{5.82}$$

where the radiation efficiency below the critical frequency for the panel of perimeter P and area A_p is

$$\sigma_{\text{rad}} = 2\pi\frac{P\kappa}{A_p}\frac{c_l}{c}\sin^{-1}\left(\frac{f}{f_c}\right)^{1/2}. \tag{5.83}$$

We note that the radiation efficiency now depends on frequency. At very low frequencies, it varies as $\sqrt{f}$ or 1.5 dB/octave and tends to rise somewhat more rapidly as the critical frequency is approached.

This radiation efficiency is graphed in Figure 5.35, which shows its behavior below the critical frequency for various values of Ph/A. This graph assumes steel or

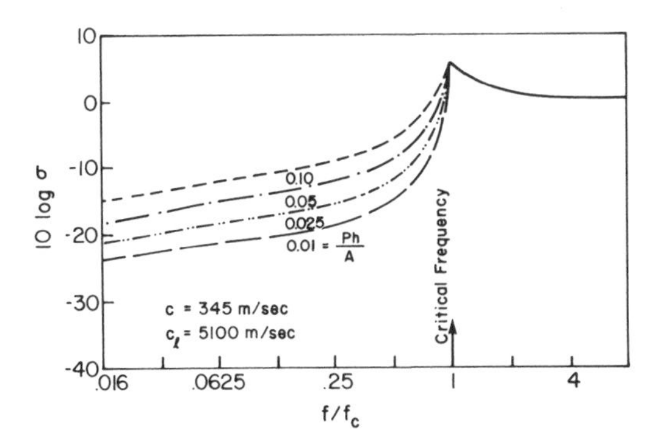

Figure 5.35 Calculated radiation efficiency for a simply supported plate, averaged in third-octave bands.

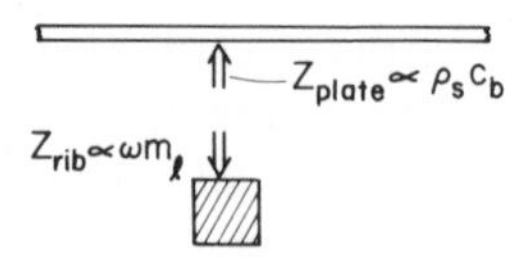

Figure 5.36 Comparison of impedances of plate and rib to determine if rib acts as an obstacle to flexural waves.

aluminum structures, so the longitudinal wave speed is about 5000 m/sec and the speed of sound in air is 345 m/sec. The calculations are for an average over one-third octave bands, so the singularity at the critical frequency becomes +6 dB at f_c.

A summary of theoretical evaluations of radiated damping is as follows:

- Supported, clamped, other b.c. (boundary condition) plates: Value depends on f_{crit} and perimeter
- Plates with reinforcing ribs: Depends on f_{crit} and impedance ratio of rib to plate
- Cylinder: Depends on f_{crit} and f_{ring}
- Cone: Depends on f_{crit}, cone angle, max and min diameters

The radiation efficiency for flat plates has been fairly well worked out. Various kinds of boundary conditions will result in different forces applied to the edges for a given vibration amplitude. The radiation efficiency depends on the critical frequency and the perimeter of the plates. In plates with reinforcing ribs, we must decide how much of a restraint the reinforcing rib is. This depends on the impedance ratio between the plate and the rib of mass m_l per unit length, as shown in Figure 5.36. If the

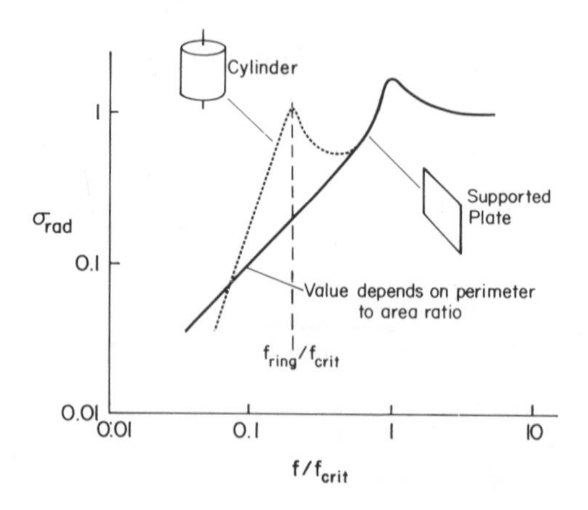

Figure 5.37 Comparison of radiation efficiency of a supported flat plate and for the same plate rolled into a cylindrical shape.

impedance of rib ωm_l is very small compared to that of the plate $\rho_s c_b$, then the rib does not represent a discontinuity and radiation efficiency from such discontinuity will be fairly small.

As discussed in Chapter 3, a curved structure is dynamically a flat plate above the ring frequency, and the flat-plate radiation efficiency can be used. When the ring frequency is below the critical frequency, the group of modes that produces the peak in the modal density tends to be supersonic, and therefore the peak in modal density is accompanied by a peak in radiation efficiency, as shown in Figure 5.37. When the ring frequency is above the critical frequency, the radiation frequency is already unity and there is very little effect of the ring frequency on radiation efficiency. In cones, the cone angle and the maximum and minimum diameters play a role, but the formulas become very complicated.

5.14 NUMERICAL METHODS FOR SOUND RADIATION

Computational methods are an alternative to experimental or analytical methods of determining sound radiation. The computer has not been extensively used for sound radiation calculation from machines because of their complicated shapes and the broad range of frequencies that they radiate. Nevertheless, the increasing power and lower computational costs have increased interest in computer-based methods for determining sound radiation.

Commonly used are the finite element and boundary element methods. Finite element methods are used primarily for enclosed spaces, particularly at low frequencies because of the need to space element modes at distances less than a quarter wavelength apart. When the wavelength becomes too short, the required number of nodes becomes very large, computation times increase, and inaccuracies in the calculations increase. Finite element methods are quite limited in their application to sound radiation. Their application to acoustical problems is essentially a straightforward extension of their use in structural mechanics.

The second method is the boundary element method. It is more successful because it uses finite elements over the surface of the structure only, and the sound field itself is represented by analytical expressions. The boundary element method is most appropriate for computing radiation into free space. This procedure is indicated by the sketch in Figure 5.38. Each surface element Q is assigned a velocity v_Q, area A_Q, and volume velocity $2v_QA_Q$.

A volume velocity U_0 produces a radial velocity v_r in the sound field given by (see Equation 5.9)

$$v_r = U_0 \frac{1 + jkr}{4\pi r^2} e^{-jkr}, \tag{5.84}$$

which produces an outward normal velocity component at element S (see

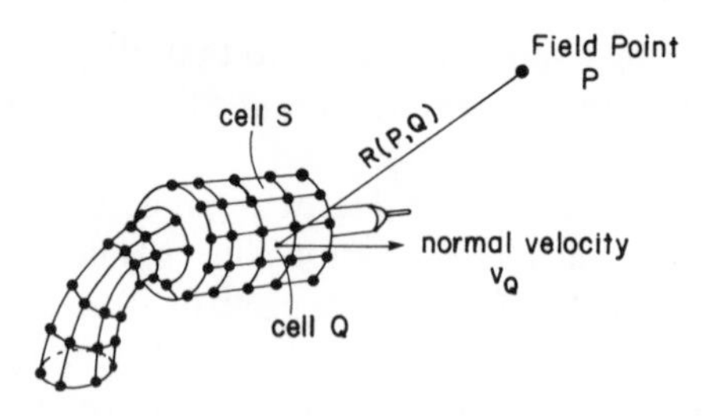

Figure 5.38 Geometry of surface elements defining structure for boundary element sound computation.

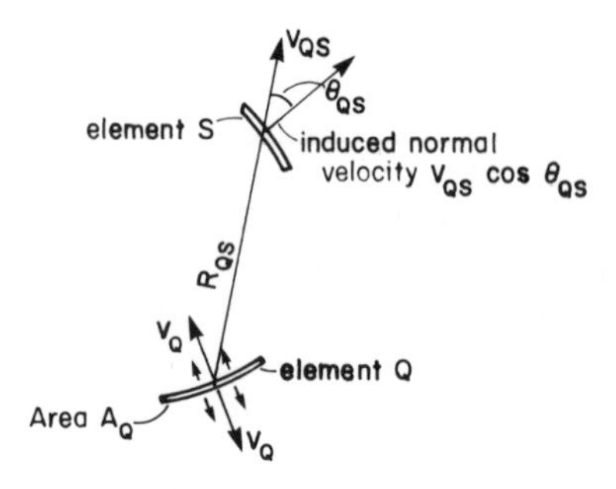

Figure 5.39 Interaction of sources in free space that must combine to produce the actual velocity distribution of a vibrating structure.

Figure 5.39), given by

$$V_{QS} = \frac{V_Q A_Q}{2\pi R_{QS}^2}(1 + jkR_{QS})e^{-jkr_{QS}} \cos\theta_{QS} \equiv V_Q T_{QS}, \tag{5.85}$$

where the matrix T is defined by the geometry of the model and the wavelength of sound. The total outward normal velocity at S is then

$$v_S + \sum_Q V_Q T_{QS} = v_{S,\,\mathrm{meas}}, \tag{5.86}$$

which must equal the actual measured (or assumed) normal velocity of the machine structure.

The required assigned velocity distribution is found by solving Equation 5.86 for $\mathbf{v}$ in terms of the known velocity $\mathbf{v}_{\mathrm{meas}}$:

$$\mathbf{v} = (\mathbf{T} + \mathbf{I})^{-1}\mathbf{v}_{\mathrm{meas}}, \tag{5.87}$$

where **I** is the identity matrix. When **v** is found, we can compute the sound pressure radiated to any point P by adding the combinations from all volume velocity elements corresponding to the surface elements of the machine structure:

$$p(P) = \frac{j\omega p}{2\pi} \sum_{Q} \frac{V_Q A_Q}{R_{PQ}} e^{-jkR_{PQ}}, \tag{5.88}$$

where R_{PQ} is the distance from element Q to point P.

The operations involved in finding **v** from Equation 5.87 and then solving for p in Equation 5.88 can be readily handled by most computers. Nevertheless, the large number of frequencies of interest in noise and diagnostic problems means that these calculations can be extensive. Also, the input data $\mathbf{v}_{\text{meas}}$ must be accurate in both magnitude and phase, which can impose quite a burden on experimental procedures.

CHAPTER 6

Diagnostics Using Signal Energy

6.1 INTRODUCTION

The vibratory or acoustical signals picked up by a sensor and used for diagnostics can be processed in many ways. The direct vibratory signal is very rarely used. Instead, the overall mean square or a power spectrum is likely to be used. More elaborate procedures are being studied, but they remain mostly in the research phase. We shall discuss these more advanced procedures in subsequent chapters.

This chapter considers methods that use signal energy, that is, the mean or integral square of the signal, resolved into either narrow- or broadband frequency intervals. These methods employ commonly available instrumentation and can therefore be implemented in factories or laboratories.

6.2 ENERGY VERSUS TIME ANALYSIS OF ROTATING MACHINE VIBRATION

In Chapter 5, we described a reverberant chamber used to measure the noise output of sewing machines on the production line. The purpose of that system was to provide a reliable and accurate acoustical measurement of the sound output of the machine as part of the QC procedure. The disadvantage with this form of test is that for the sound measurement to be reliable the machine must be completely assembled, with all covers in place and all operational features functional. The machine must also be very well isolated from the factory noise environment, a difficult condition to achieve unless very special measures are taken to provide adequate noise insulation. This noise measurement is also not able to identify the various noise sources in the machine; hence, if the machine does not pass the noise test, additional diagnostic procedures must be used to determine the source of the excessive noise and how the repair should be effected.

Vibration tests that can be correlated with the noise have significant advantages over an acoustical test on the production line. A list of these advantages is presented in Table 6.1. Since an accelerometer can be placed at various locations on the machine, it is possible to place it close to sources of vibration that may be of

Table 6.1 Why a Vibration Test?

Sensor can be placed close to source of vibration
Allows identification of noisemaker
Discriminates against other sources
Insensitive to environmental interference
Ambient vibration is mostly at low frequencies
Ambient sound inefficient in exciting machine
Vibration measurement can be made earlier in production cycle
Far enough along line to be representative
Early enough for easier repair
Interaction effects must be preserved

special interest to determine whether or not a particular noisemaker is a problem in that machine. Such identification also gives a much more accurate indication of the repair procedure needed to quiet the machine.

It is much easier to isolate a machine from factory vibration than from acoustical noise. The ambient vibration is generally at low frequencies, and a foam or rubber pad may be able to isolate the machine from vibrations, particularly at higher frequencies that are correlated with radiated sound. In addition, the ambient acoustical noise is very inefficient at exciting the machine and generally will produce background vibrations that are very small compared to the vibrations produced by the mechanisms when the machine is operating.

The greatest advantage is that the vibration measurement can be made earlier in the production cycle. When a particular set of mechanisms is installed in the machine and can either be operated under their own power or externally driven, a vibration measurement can indicate potential noise sources or other problems. This means that the repair will be more effective. However, the vibration measurement will not be reliable unless the various interaction mechanisms among the machine components are preserved, which means generally that the machine must be properly loaded so that backlash or other sources of spurious vibration excitation will not affect the test.

Figure 6.1 is a diagram of the instrumentation used in a particular vibration diagnostic system. A trigger pulse is generated by a photoelectric cell beam reflected from the shaft (handwheel) of the machine. Each analysis is carried out on a one-machine-cycle basis. The accelerometer is attached at particular locations on the machine, and the signal is broadband filtered over a frequency range selected to correlate well with the radiated sound. The signal is passed through a Waveburst Processor (Waveburst Processor is a registered trademark of Grozier Technical Systems, Inc.) that gives a short-time envelope and an integrated energy during the cycle. The results are presented on an oscilloscope.

Figure 6.2 shows the correlation of broadband bed shaft bushing vibration to the noise as measured in the reverberant booth of Figure 5.23 for a group of machines. The correlation is fairly good, with a standard deviation of about 1 dB. There is reason to believe that there should be a high degree of correlation between

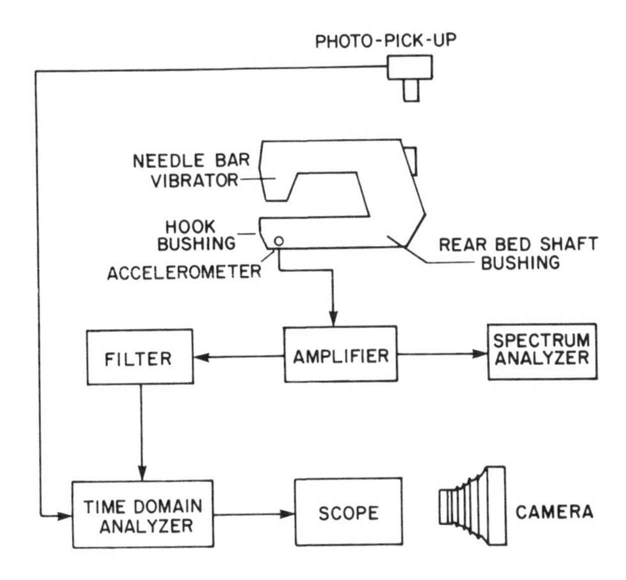

Figure 6.1 Setup for generating energy–time curves for repeated cycles of machine operation. The processor has parallel integrated and averaging outputs.

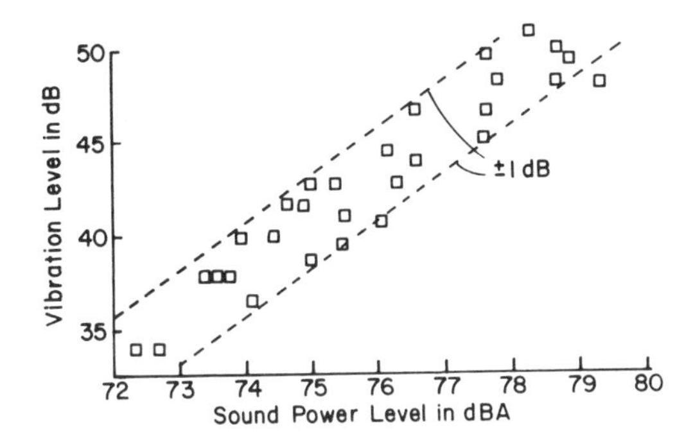

Figure 6.2 Correlation of broadband integrated acceleration signal energy with noise output of machines in the reverberant booth.

the vibration of the bed shaft, which is driven by a timing belt from the arm shaft, and the noise produced by this machine: an earlier experiment showed that when a large flywheel was attached to the arm shaft, the machines generally were significantly quieter. The flywheel causes the arm shaft to turn at a much more constant angular velocity, and this produces much less vibratory excitation of the bed shaft via the timing belt.

Vibration traces from this system of 10 machine cycles of a machine judged quiet are shown in Figure 6.3. We see some clean and repeatable impacts in the needle bar, the hook area is very quiet, and the bed shaft, although it shows more

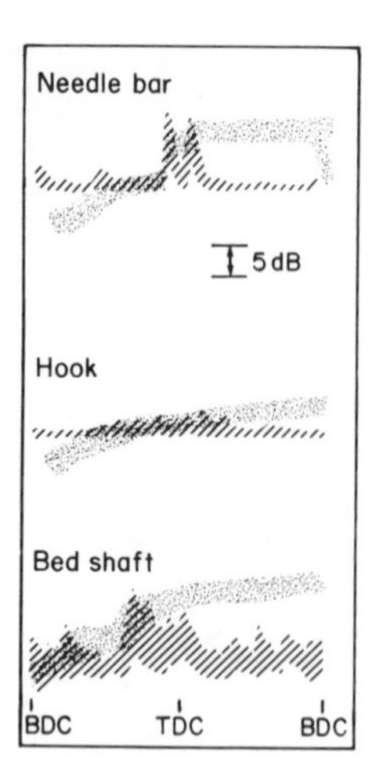

Figure 6.3 Vibration levels for a machine judged "quiet." The jagged line is short-time average (envelope). Ten cycles of machine operation are overlaid.

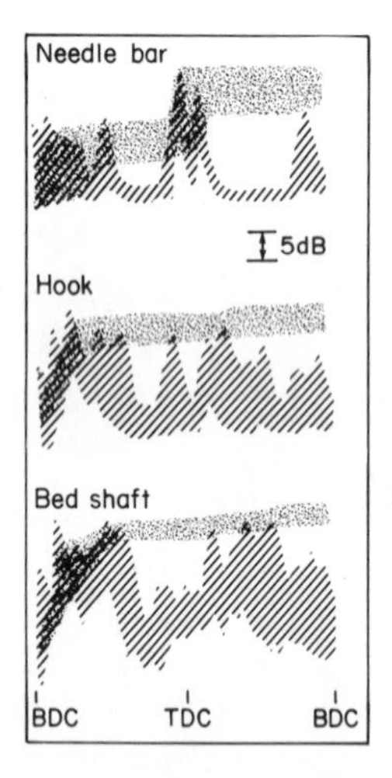

Figure 6.4 Vibration levels for a machine judged "noisy." The envelope levels are higher and less repeatable from one cycle to another.

fluctuation and higher vibration levels than at other locations, is relatively repeatable on a cycle-by-cycle basis, both for the energy and short-time average (envelope) presentations.

These data are to be contrasted with the vibration patterns for a "noisy" machine as shown in Figure 6.4. Not only are the levels significantly higher, but there is a lot more variation in the short-time average and the integrated energy traces. The contrast between the vibration patterns in Figures 6.3 and 6.4 indicate that the vibration test is clearly able to select machines that are noisy and in need of adjustment.

The 10 machine cycles overlaid in these figures also allow us to see the variability from one cycle to another. In some cases, we can associate this variability directly with impacts that occur in some cycles and not in others. These random impacts are particularly troublesome because even if they are relatively low in sound level, they give a subjective impression of a machine that has rattles and other irregularities in its operation and is therefore of lower perceived quality (see Appendix B).

The system described here has been put on a production line and used to determine noise levels and to identify machines that require additional repair before their assembly is completed. The system, as installed, has some features similar to the noise booth in that the test sequence is automatically controlled and a light indicates to the operator whether a particular transducer output is beyond acceptable limits.

6.3 ENVELOPE METHODS USING A HIGH-FREQUENCY RESONANCE

Faults in rotating machines may, as in Figures 6.3 and 6.4, involve an impact each cycle of rotation. The spectrum of such an impact is a set of lines, separated by the rotation (or cycle) rate, extending to very high frequencies. The lower-frequency lines may be masked by other sources, such as imbalance or misalignment, but the higher-frequency lines, which may include ultrasonic frequencies, are not so masked. A high-frequency resonance of a transducer or of the structure itself will have enough bandwidth to encompass several spectral lines.

To illustrate, suppose we have a machine turning at 3000 rpm. If there is one impact per revolution, then the line spacing is 50 Hz. Suppose we have a resonance at 20 kHz that has a loss factor $\eta = 0.02$. This gives an effective rectangular bandwidth for the mode of 628 Hz, which will include about 12 lines, spaced at 50 Hz. This signal will be like a 20-kHz carrier, amplitude modulated with a pulse waveform.

The periodicity and waveform of this modulation can give important information regarding the fault. For example, consider the ball bearing shown in Figure 6.5. If the inner shaft rotates with angular velocity Ω and the outer race is fixed, the shaft surface speed is $\Omega d_i/2$. This linear speed on one side of the ball gives it a center speed $\Omega d_i/4$ and a rotational speed

$$\Omega_{\text{ball}} = \Omega \frac{d_i}{d_0 - d_i}. \tag{6.1}$$

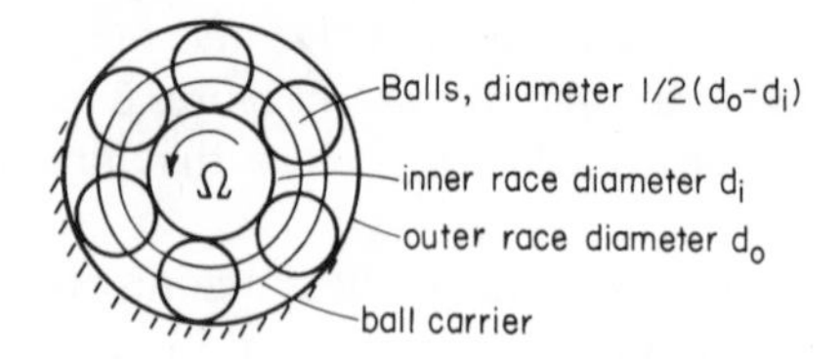

Figure 6.5 Ball bearing with fixed outer race and rotating inner race.

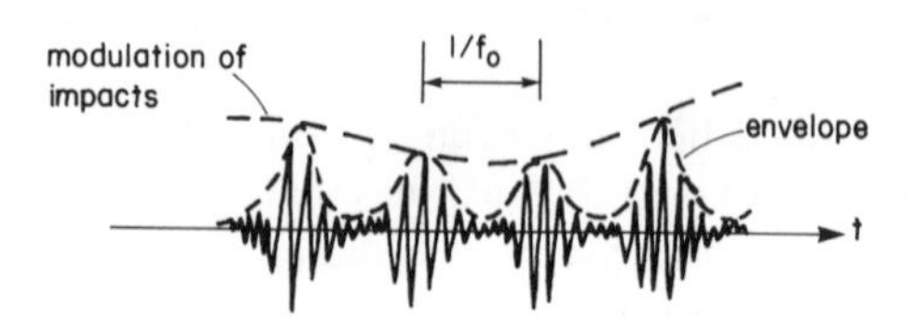

Figure 6.6 Response of high-frequency resonance to bearing impacts due to outer race defects.

Thus, the angular velocity of the ball carrier is

$$\Omega_{\text{cage}} = \frac{1}{4}\frac{\Omega d_i}{d_i + 1/4(d_0 - d_i)} = \Omega\frac{d_i}{d_0 + 3d_i}. \tag{6.2}$$

The balls travel along the outer race at a speed of

$$\Omega_{\text{ball}} \cdot \tfrac{1}{4}(d_0 - d_i) = \tfrac{1}{4}\Omega d_i, \tag{6.3}$$

and since each ball will encounter a defect on the outer race at distances πd_0 apart, the frequency of such impacts will be

$$f_0 = \frac{1}{4\pi}\frac{d_i}{d_0}\Omega n_B, \tag{6.4}$$

where n_B is the number of balls in the carrier.

There are similar formulas for impacts for inner race impact rates and for situations in which the outer race rotates. In any case, when these impacts excite the 20-kHz resonance, the response signal will appear as in Figure 6.6. The envelope of this signal may be obtained with an analog processor, like that shown in Figure 3.15, or the Hilbert transform, as discussed in Section A.13.

Some investigators report that a frequency spectrum of this envelope function itself may reveal important information in case there is modulation of the impacts. Modulation can occur, for example, if the defect is on the ball and successive impacts occur at locations where the shaft loading changes.

6.4 DETERMINING GEAR TRANSMISSION ERRORS FROM NOISE DATA

We discussed how gear errors of various kinds can produce fluctuating mesh forces and mechanical vibration in Section 2.8. The line spectra of such gear noise can provide detailed information about the kinds of errors inherent in a gear, but the analyses involved are very complicated. In this section, we shall discuss how noise spectra can be used to obtain a transmission error spectrum that will give a general indication of gear quality and a model for a gear system as a noise source in a structure.

In Figure 2.36, we showed the configuration of a double reduction marine gear with two high-speed, turbine-driven input shafts and a single, low-speed output shaft to drive the ship's propellor. This gear set is housed in a structure, as sketched in Figure 6.7. The vibration spectra as measured on the support structure is a very dense set of lines caused by the various multiples of the three shaft speeds in this gear set, with particular peaks occurring at the frequencies associated with meshing at the low-speed output gear and meshing at the high-speed input pinions. A typical gear noise spectrum is shown in Figure 6.8. Clearly, a gear analysis that will predict the amplitudes of all these lines must be very complicated.

The vibration spectral lines have been grouped into third-octave bands, as shown in Figure 6.9. For this gear, the low-speed mesh is in the 1.6-kHz band, and

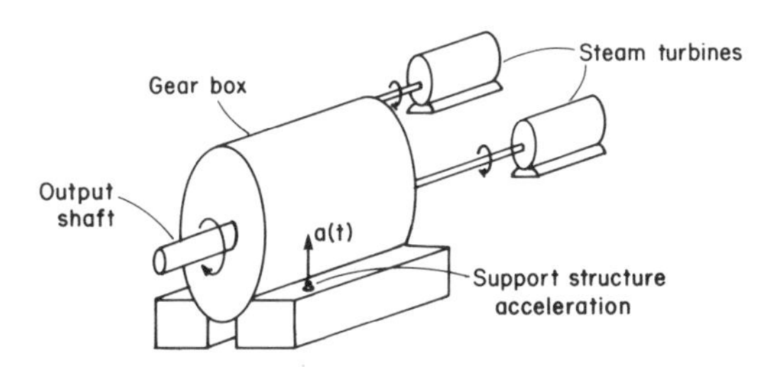

Figure 6.7 Layout of turbine-driven gear set showing location of acceleration measurement.

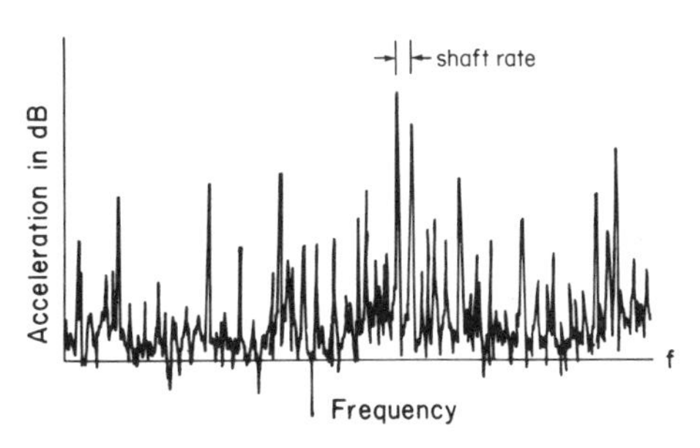

Figure 6.8 Typical gear noise spectrum of a complex gear set like that discussed in Chapter 3.

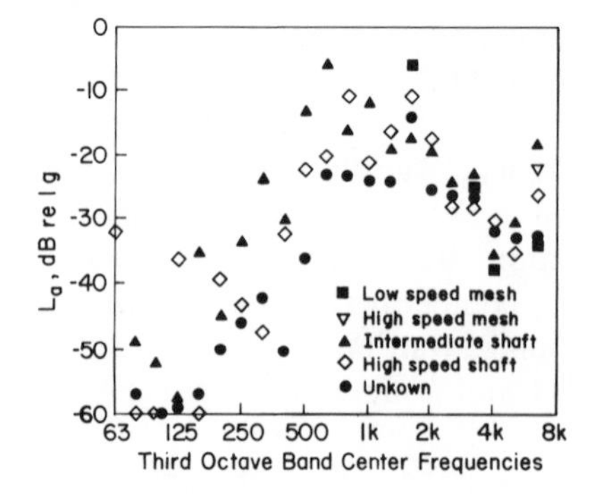

Figure 6.9 Third-octave band acceleration levels corresponding to various rotation and mesh rates, calculated from narrowband spectra.

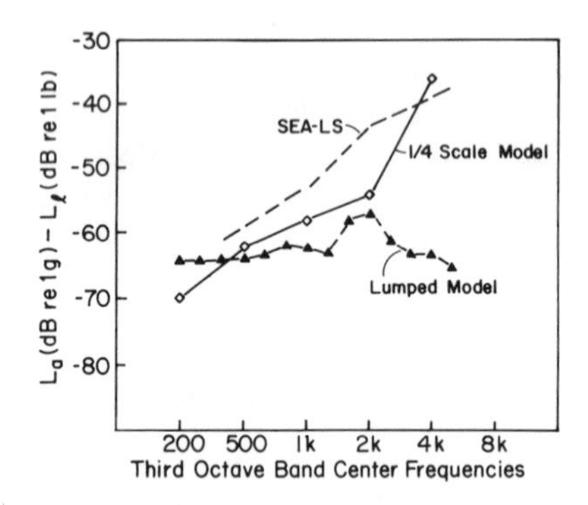

Figure 6.10 Analytically and experimentally determined transfer functions from output (low-speed) mesh to gear housing.

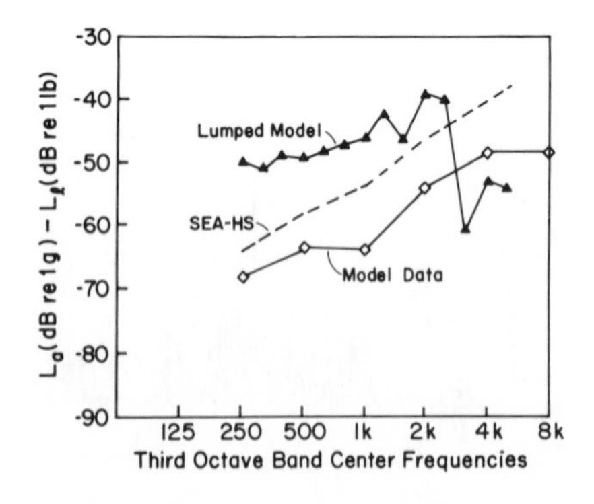

Figure 6.11 Analytically and experimentally determined transfer functions from input (high-speed) mesh to gear housing.

the high-speed mesh is in the 6.3-kHz band. Other contributions to the vibration at various orders of the intermediate shaft and high-speed shaft are also shown in the diagram. These vibrations can be associated with transmission error by using a transfer function that relates transmission error to structural vibration. Such transfer functions may be determined experimentally or analytically, as described in Chapter 4. Some transfer functions based on experimental and analytical techniques from the low-speed mesh to the housing of the gear set are shown in Figure 6.10, and similar transfer functions from the high-speed mesh to the housing are shown in Figure 6.11.

Using the observed vibration and these various estimates of the transfer functions, we can infer an error spectrum as in Figure 6.12. Concentrating on the solid triangles, for which the SEA transfer function has been used, and the vibration data associated with low-speed mesh, we see that the error spectrum peaks in the 500–1000-cycle bands and the overall rms error is about 1 μ rms. When we consider that this is a mesh-attenuated error, as discussed in Chapter 2, this means that individual tooth error is about 5–10 μ rms. An unattenuated error of this magnitude is larger than we would expect from machining processes. The conclusion is that tooth bending is the principal source of transmission error in this gearing system. Consequently, the system has to be designed at a certain load point. The error will increase away from that load point, however, because the bending will have a value other than that accounted for by the design.

If we examine the vibration associated with the high-speed mesh and use the high-speed mesh-to-casing transfer functions, then the error spectra in Figure 6.12 for the inverted triangles and small circular dots result. In this case, the overall rms error is about 1 μ using both transfer functions and the unattenuated tooth-to-tooth error would be 2–10 μ.

The vibration data in Figures 6.8 and 6.9 show strong sidebands. The high-speed and low-speed mesh frequencies, although high in amplitude, do not dominate the spectra. This means that there is considerable modulation of these mesh forces as

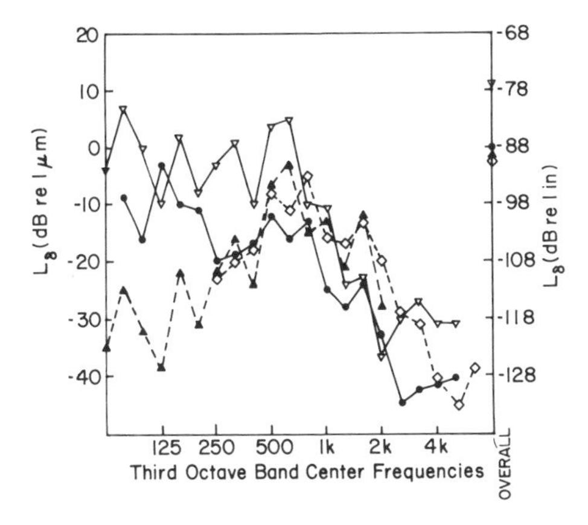

Figure 6.12 Inferred mesh-attenuated transmission error for gear set based on vibration data and transfer functions used as inverse filters.

we go around the gear during a meshing cycle. This variation causes sideband production and indicates that there is a fair amount of variability from tooth to tooth in the transmission error. Some gear sets show much stronger vibration at the mesh frequency and smaller sidebands, which means there is more uniformity in bending rigidity and in tooth form than is the case for this gear set.

Thus, the vibration spectrum can provide a good indication of the magnitude of the mesh transmission error and its cause. If we track this over time, we can use gear-induced vibration as a monitoring tool because changes in lubrication and tooth finish, tooth bending, wear, and chipping all produce detectable changes in vibration. More advanced signal processing can actually pick out the modifications in the transmission error produced by the chipping of a single tooth. We shall discuss procedures to do this in Chapter 8.

6.5 USING THE POWER CEPSTRUM IN DIAGNOSTICS

As discussed in Appendix A, the power cepstrum is the even or cosine transform of the log power spectrum. It is not possible to recover the initial waveform from the power or real cepstrum because the inverse transform of the real cepstrum only allows us to compute the magnitude of the Fourier transform or the power spectrum. Inverse time transformation of the power spectrum reproduces the correlation function, not the initial waveform, because the phase of the signal has been lost.

One application of the cepstrum is noted in Appendix A. The contributions of the path and source are additive, and time windowing can be used for separating them. The separation of source and path from each other can be illustrated by imagining that the structural path is a simple longitudinal rod, as shown in Figure 6.13, in which we have the response due to an impact at one end of the rod and an accelerometer picking up vibration at the far end of the rod. The impact produces a stress wave in the rod, which produces a pulse at the accelerometer every time the stress wave comes to the far end of the rod.

The transform of the impulse response $h(t)$ has a magnitude $|H(\omega)|^2$. The transform of log $|H|^2$, the power cepstrum $C_h(\tau)$, has the same time periodicity as the impulse response, but the peaks die away more quickly. The input $x(t)$ has the spectrum $|X|^2$. The power cepstrum of the accelerometer output is sketched at the bottom of this figure. It shows separately the combinations of the source and the sequence of peaks due to the transmission path. Because of the equally spaced and repeated nature of the transmission path peaks in frequency in this case, the path cepstrum can be extracted from the composite cepstrum, leaving only the power cepstrum of the source $x(t)$, which allows us to reconstruct the power spectrum of the source function. We cannot determine the source waveform, but we can recover the spectrum of the source, which may be sufficient for diagnostic purposes.

A second application of the real cepstrum is to provide a smoothing procedure by extracting the low-time components of the cepstrum. This might be done to

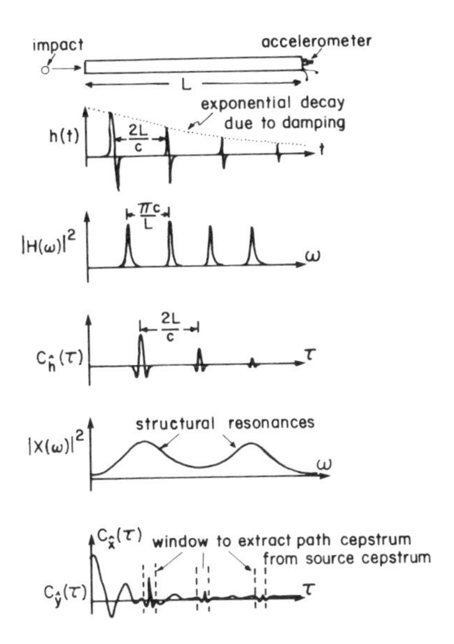

Figure 6.13 Illustration of separation of path and source cepstra by a simple time windowing.

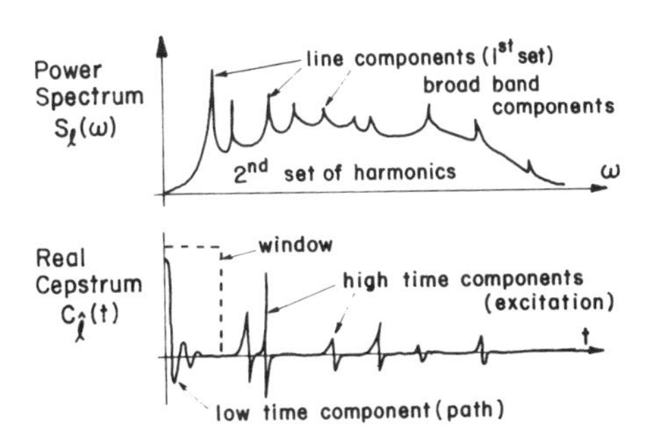

Figure 6.14 Power spectrum of force and real or power cepstrum showing relation between frequency and time domains.

extract a smoothed version of the source spectrum or of the transfer function or of both. To see this, note that the cepstrum in Figure 6.14 has "low-time" components that correspond to the "background" or continuous part of the power spectrum. By windowing out or eliminating the high-time components in the cepstrum and taking the Fourier transform, we can extract this smoothed spectrum.

The third application is the separation of different frequency multiples in a complicated spectrum. This technique has particular application to rotating machinery where there are various shaft frequencies, each of which might be contributing a set of independent harmonics to the overall response spectrum of the machine. Such separation can then allow us to identify varying amounts of noise excitation of the mechanisms driven by these shafts. The power spectrum in Figure 6.14 shows broadband noise on which is superimposed a sequence of harmonic peaks. This noise might be due to random and narrowband components of a small gear. The broadband components form a smooth part of the spectrum that, when transformed, produces the contribution to the cepstrum at low values of the time variable.

In some cases, there might be a second set of harmonics that appear with a different frequency spacing in the spectrum, perhaps produced by a second gear train in the system. This additional set of harmonics in the power spectrum will produce a separate set of time peaks in the real cepstrum, separated by a time that is inversely proportional to the frequency spacing. Since they are equally spaced, however, we can window out or extract each set of cepstral peaks to form a separate cepstrum for each equally spaced set of frequency peaks. This cepstrum has an inverse Fourier transform that will generate a power spectrum for the spectral components associated with each basic rotation speed. Using the power cepstrum to extract different orders of harmonically related frequency components has often been done in gear noise analysis.

6.6 DIAGNOSTICS OF A HORIZONTAL CENTRIFUGE

Vibration spectra of the horizontal centrifuge in Figure 6.15, as measured on the grounded gearbox pinion and on the feed-end bearing cap, are shown in Figure 6.16. The power cepstra for these spectra are shown in Figure 6.17. A low-time window is applied to these cepstra to separate the low-time part of the cepstra associated with the path or structural resonances of the centrifuge from the periodic excitation by the rotating elements.

The "path" spectra for these two locations, derived from low-time windowed cepstra, are shown in Figure 6.18. We note that the spectrum for the pinion data is

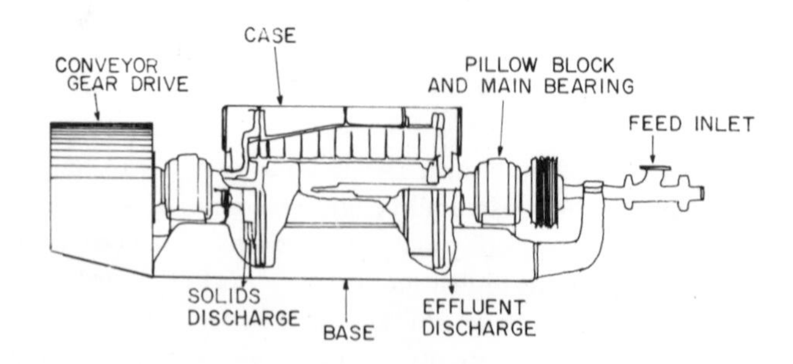

Figure 6.15 A horizontal centrifuge has drum rotation at 20 Hz and conveyor rotation at 19 Hz. Vibration is measured at the grounded gearbox pinion and on the feed-end bearing support.

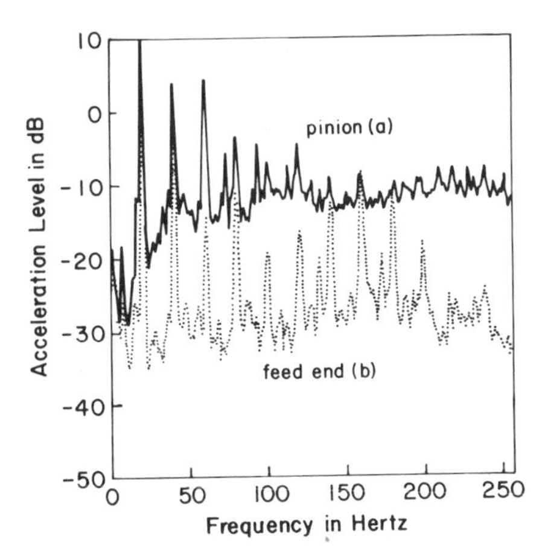

Figure 6.16 The acceleration spectra at the pinion (a) and the feed end (b) show a strong 20-Hz periodicity corresponding to drum rotation.

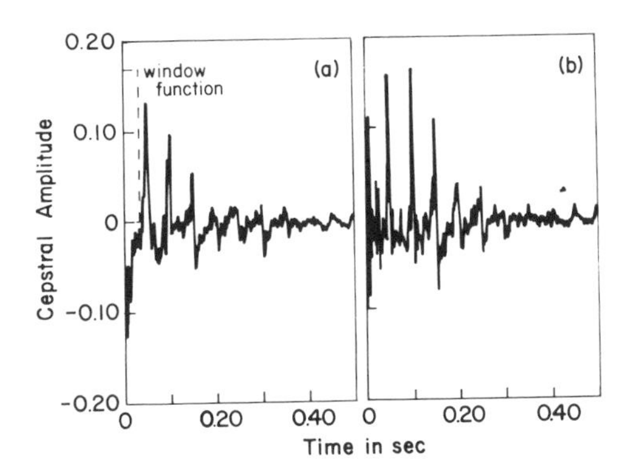

Figure 6.17 The cepstra for the 256-Hz spectra show peaks at 1/20 sec periodicity corresponding to the 20-Hz lines. A low-time window of these cepstra can reveal path information.

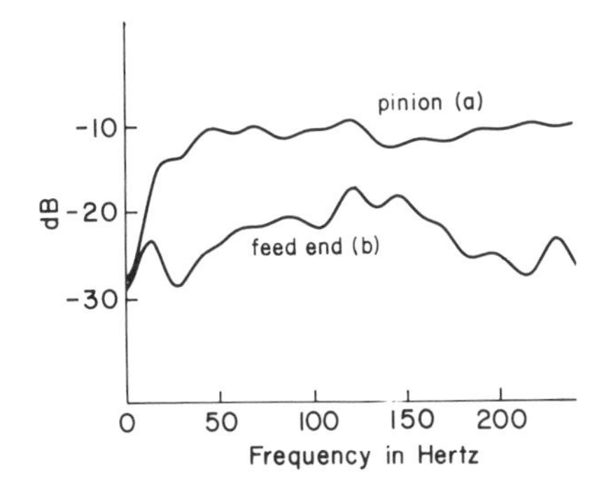

Figure 6.18 Spectra of the low-time windowed cepstra indicate a smoother path spectrum for the pinion data than for feed-end data, indicating that the unknown source is closer to the pinion.

fairly flat, with little frequency structure, whereas that for the feed end shows a resonant structure. The number of peaks in this path spectrum is consistent with the number of resonant modes of the centrifuge that might be expected for such a structure. We interpret the path spectra of Figure 6.18 to indicate that the excitation is close to the pinion and remote from the feed end. This occurs because local response to excitation can be expected to show a less resonant behavior.

A finer frequency resolution of the data in Figure 6.17(a) is shown in Figure 6.19(a). We can see periodicities at 20 Hz (the drum rotation rate) and at 1-Hz spacings (the difference between the drum and conveyor rotation rates). We also see that there is a 3-Hz periodicity in that every third line of the 1-Hz periodicity is enhanced in amplitude. The cepstrum in Figure 6.19(b) also shows this periodicity in that it has peaks at multiples of 0.05 sec, 1 sec, and 1/3 sec.

If we window and retain only the cepstral pulses at multiples of 1/3 sec, we generate the cepstrum and the corresponding frequency spectrum in Figure 6.20. This represents the excitation spectrum of the centrifuge and shows that something important is occurring at a rate of three times the conveyor/bowl relative rotation rate. Since there is no structural component in the system that could generate this periodicity, it is likely to be a nonlinear oscillation of the system for which factors of 3 in the periodicity are possible.

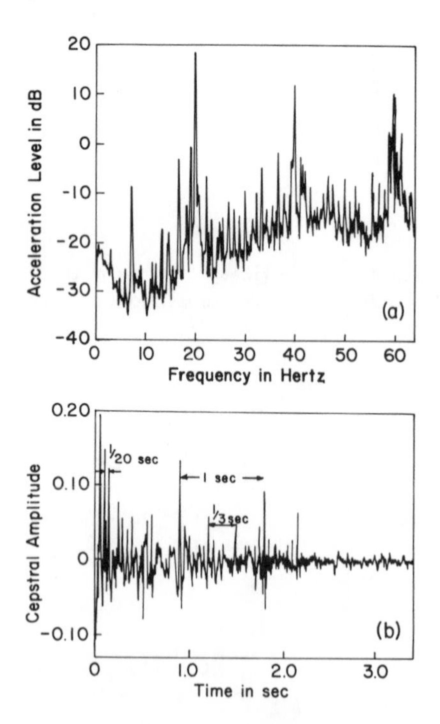

Figure 6.19 A finer scale spectrum of the pinion vibration data in (a) shows important frequency periodicities at 1, 3, and 20 Hz with corresponding peaks in the cepstrum in (b) at reciprocals of these values.

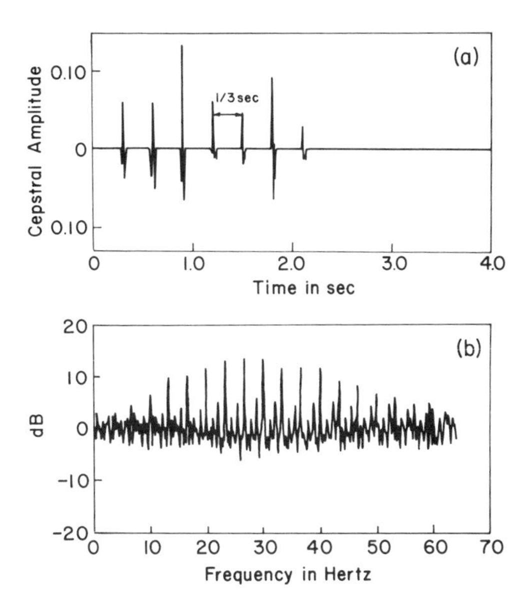

Figure 6.20 The pinion cepstrum is windowed at multiples of 1/3 sec in (a) to produce the spectrum in (b), which extracts the 1- and 3-Hz components from the 64-Hz spectrum.

6.7 DIAGNOSTICS OF VALVE AND VALVE SEAT IMPACTS

When a valve comes to its seat during the operation of an engine, there is an impact around the periphery of the valve head. If the valve and its seat are well mated, then this impact will be relatively uniformly distributed around the head. In this circumstance, we might expect the longitudinal resonances of the valve to be fairly strongly excited. On the other hand, if the valve strikes one side of its seat before the other, either because of the misalignment or an irregularity in the shape of the valve seat due to machining or deposit buildup, we might expect relatively more bending modes to be excited. Thus, the relative strength of excitation of longitudinal and bending modes in the valve structure is a potential diagnostic signal about the condition of the valve seat.

To determine valve and seat force signatures, we constructed the rig in Figure 6.21, which is very much like the system in Figure 6.13. A valve seat is machined into one end of a long steel rod, and a series of strain gauges are applied to the rod at some distance from this seat. The force between the valve and the valve seat is converted to strain according to

$$\varepsilon = \frac{\partial u}{\partial x} = \frac{\sigma}{E} = \frac{l}{EA} \tag{6.5}$$

where A is the cross-sectional area of the rod, and E is Young's modulus. The gauges

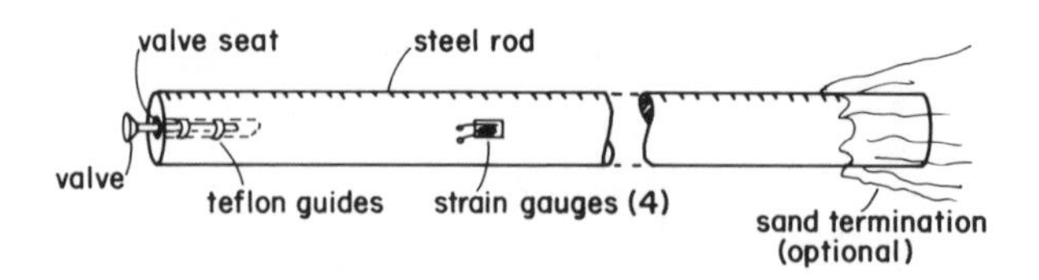

Figure 6.21 Steel rod with strain gauges and machined valve seat used to measure the force between a valve and its seat.

are placed around the rod and summed so that differences in strain across the rod do not generate an output signal, and bending motions are discriminated against. Originally, a damping termination using sand was applied to the end of the rod, but this was abandoned because it was not effective in absorbing the stress waves. Without a termination, a time history of the impact strain is shown at the top of Figure 6.22. The periodic sequence of pulses results from the reflection of stress waves in the rod and, as discussed before, can be regarded as a convolution of the impacting force and the periodic impulse response of the rod, sketched in Figure 6.13. The real or power cepstrum of this signal is shown in the bottom of Figure 6.22. We can see the periodic components in the power cepstrum due to the

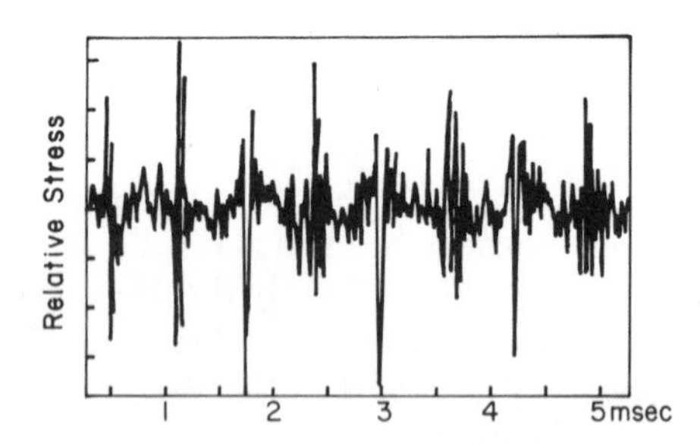

(a) Time history of impact, showing reflections from end of bar.

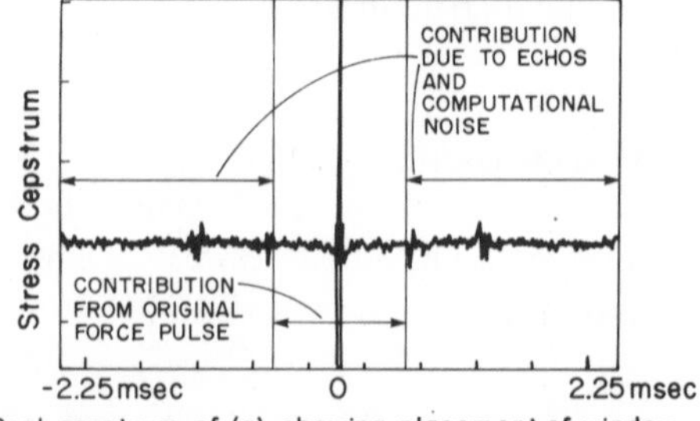

(b) Real cepstrum of (a), showing placement of window.

Figure 6.22 Removal of reflections of path effects from response of beam by windowing power cepstrum.

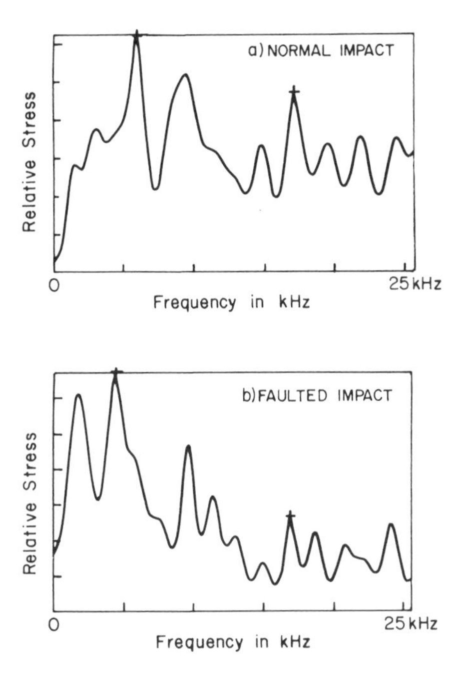

Figure 6.23 Source power spectra recovered from low-time cepstral windowing for normal valve–valve seat impact and an impact faulted by a small piece of tape placed on seat.

impulse response of the finite rod. If we window the cepstrum below the earliest of these pulses, a low-time cepstrum, due to the excitation force, is obtained, from which a power cepstrum of the dereverberated pulse can be determined.

Figure 6.23(a) is constructed from the cepstrum in Figure 6.22 when the impact is due to a well-aligned valve. We see that two resonances at 16 kHz and 5 kHz are very prominent in the vibration spectrum in part (a). If we put a piece of tape under one location on the valve seat to simulate a fault, then the recovered power spectrum of the vibration is shown in Figure 6.23(b). We see that the 16-kHz peak in the spectrum has dropped significantly.

6.8 ANALYSIS OF VALVE RESONANCES

In Section 3.14, we discussed the vibration of a structure in terms of modal response. The response of the entire structure could be represented as the sum of responses of individual modes, each mode acting as a simple resonator with its own resonance frequency and damping. We can better estimate transfer functions and interpret resonance frequencies if we know the mode shapes for the structure being studied. These mode shapes can be experimentally determined by procedures based on the analysis in Section 3.14.

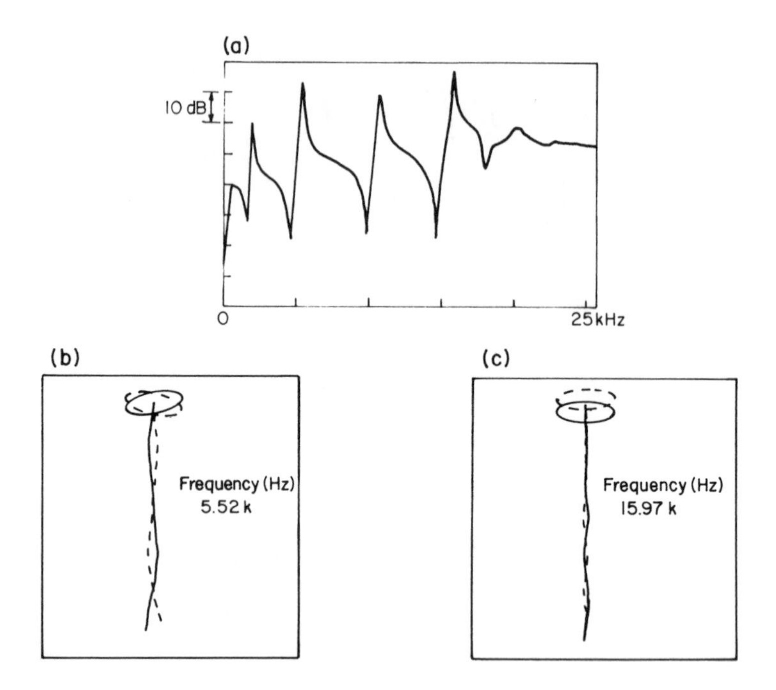

Figure 6.24 Output of modal analysis system showing a typical transfer function and the derived mode shapes at 5.5- and 16-kHz resonances.

The output of a modal analysis program is shown in Figure 6.24 for the inlet valve of an automotive engine. The upper part of the figure shows a typical transfer function between a point on the valve where the force is applied and a second location where the acceleration is measured. The peaks in this transfer function are the resonances of the system, and the minima (or zeros) tend to be defined by the locations at which the force and the acceleration are measured, as discussed in Section 3.3. It turns out, in this case, that the resonances at about 5.5 kHz and at 16 kHz are of special interest.

The mode shapes as measured for these two modes are shown in the lower part of the figure. The 5-kHz resonance is a second bending resonance of the stem of the valve. This mode has a critical damping ratio of 0.12% or a loss factor of 0.24%. The other mode at 16 kHz is a longitudinal vibration of the stem combined with a dishing motion of the valve, and it has a critical damping ratio of only 0.03% or a loss factor of 0.06%. We are especially interested in these resonances since they appear in the data in Figure 6.23.

6.9 THE USE OF INVERSE FILTERS

When the vibratory signal due to valve–valve seat impact travels through the engine structure, its spectrum will not remain as in Figure 6.23 but will be strongly modified

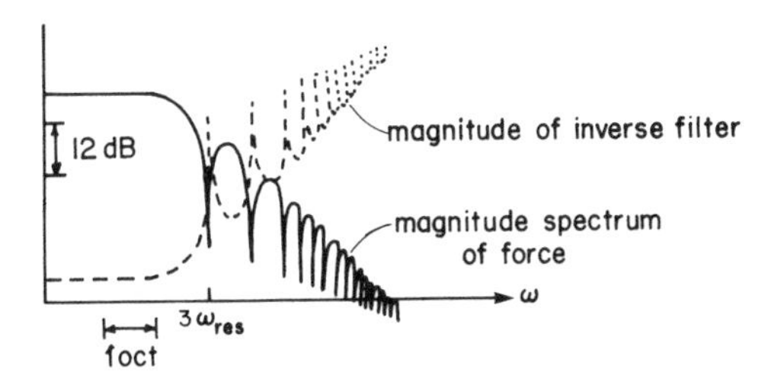

Figure 6.25 Magnitude of force spectrum produced by impact of a resilient sphere on a rigid surface, and the magnitude spectrum of a filter that will convert the force to an impulse.

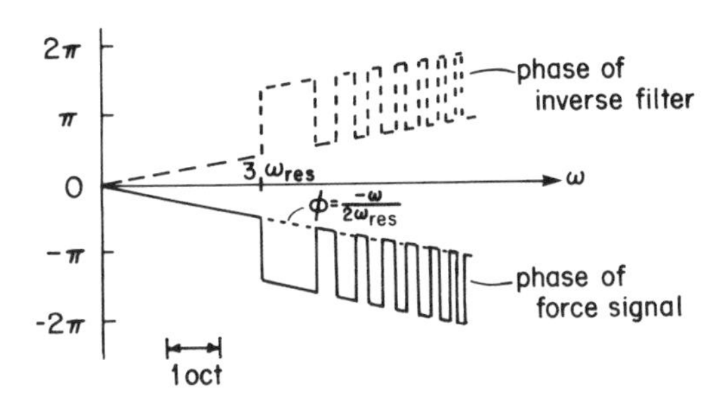

Figure 6.26 Phase of force spectrum due to impact on a rigid surface and the phase of its inverse filter. The "drift" in phase is due to the choice of zero time in the force record.

by structural resonances. An inverse filter (see Chapter 1) can be applied to the received signal to undo these spectral distortions due to propagation.

As an example, recall the impact of the elastic body against a rigid surface, as illustrated in Figure 2.14, and note that it generated the force pulse against the surface in Figure 2.15. This time waveform of force generated the energy spectrum in Figure 2.16, which is reproduced in Figure 6.25. In addition, the phase spectrum of this force pulse is given by Equation 2.6 and is sketched in Figure 6.26.

An inverse filter has a gain that is the negative of the log magnitude signal spectrum and a phase that is the negative of the signal phase. If the inverse filter is applied to the measured signal as in Figure 1.5, the resulting output will be a delta function. We can see from Figures 6.25 and 6.26 that the inverse filter is, in effect, a mirror image of the signal Fourier transform about the horizontal frequency axis. It may be extremely difficult to construct this inverse filter in terms of a real network, but its implementation as a computer program is not difficult.

The experiment in Figure 6.21 determined that the force produced by a valve when it impacts against its seat is due to internal resonances of the valve in much the same way as the internal resonances were responsible for the impact spectrum of piston slap discussed in Section 2.6. Having determined the spectral signature of well-aligned and faulted valve–valve seat impacts, we now proceed to look at the recovery of these impact signatures in an engine. The engine studied is the same one used for the study of piston slap. The transfer function we need relates a ring of force applied to the valve seat to vibration at a single location on the engine casing. Since the transfer functions we can measure are due to point forces, we must coherently add the transfer functions due to a series of point forces around the valve seat periphery to obtain the effective transfer function for a ring of force. This transfer function is shown in Figure 6.27(a) and is to be used as the basis of an inverse filter to recover the excitation or force spectrum from the observed acceleration.

Figure 6.25 shows that the extremely sharp minima that occur in Figure 6.27(a) will become extremely sharp peaks in the inverse filter. Any errors in the measured vibration spectrum (which tend to be greatest at response minima) will be greatly magnified at these peaks. To eliminate this problem, we have slightly smoothed this transfer function by applying a running three-point average to the data [see Figure 6.27(b)], and this latter spectrum is used to generate the inverse filter used for recovery of source spectra.

If we now use this transfer function to infer the spectrum of applied force at the valve seat from the measured vibration on the engine casing, we obtain Figure 6.28. For a properly aligned valve, the longitudinal force is slightly greater than that due to bending excitation. For the faulted case, for which one side of the valve seat has been raised by a thin strip of tape, the force excitation of the bending resonance has been increased and the excitation for longitudinal response for the longitudinal

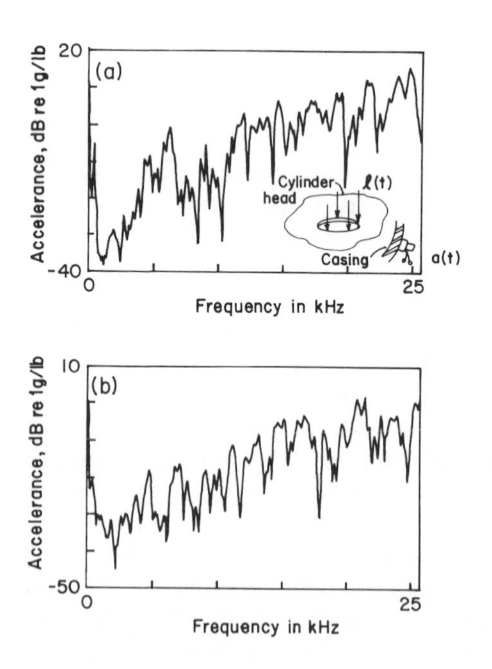

Figure 6.27 Measured transfer accelerance from valve seat force to engine casing. Ring force is approximated by addition of four locations on valve seat. Lower curve is smoothed by three-point running average to eliminate sharp minima.

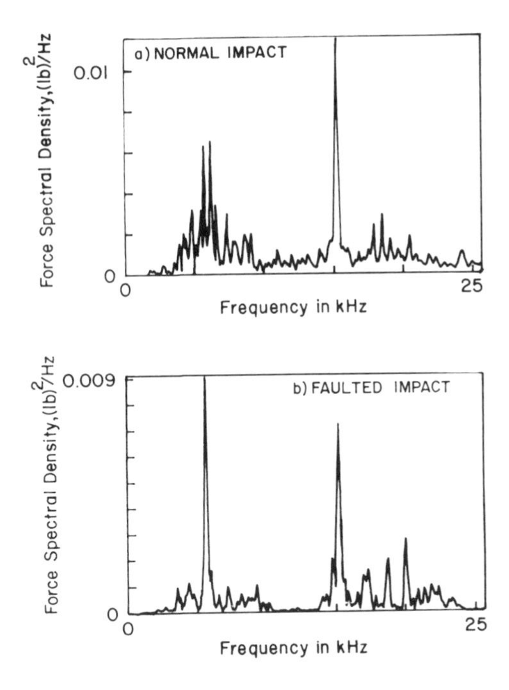

Figure 6.28 Reconstructed valve seat interaction force spectra for normal and faulted impacts in the engine. The relative excitation of the longitudinal and bending valve modes are as expected.

mode at 16 kHz has been reduced. This is the condition we might expect, based on our intuition and on the data in Figure 6.23.

Although it is possible to determine a faulted impact from the vibration spectrum in this case, we must remember that there is only one valve operating and, therefore, we do not have the interference from other noise sources in the experiment just described. Real engines have many valves and many other sources of vibration, and valve–valve seat impacts are not terribly strong sources of vibration compared to some of these other events. To recover weaker events such as valve and seat interactions, we must use an array of sensors and more advanced signal processing techniques to recover valve–valve seat impact signatures. We shall discuss some of these more advanced techniques in Chapter 8.

6.10 UNCERTAINTY IN THE VIBRATION TRANSMISSION

As we have just seen, when a signal is measured at some location at a distance from the source, then the desired inverse filter will depend on the properties of the propagation path as well as the source waveform. If one is trying to recover an energy or power spectrum of the source, it is not necessary to be concerned about the phase of the inverse filter. When we want to recover a waveform to describe the source, the phase is not only important, but may be much more important than is the amplitude spectrum. We shall discuss this topic in some detail in Chapter 7.

In general, we regard the inverse filter as a product of two inverse filters—one inverse to the source, and the second inverse to the transmission path. We saw earlier that the primary propagation effects that change signal spectrum and the waveform are dispersion and reverberation. Dispersion changes the waveform because different frequencies travel at different wave speeds, and, therefore, they get out of step as the wave progresses. The energy in any frequency range is not modified significantly by dispersion, so the energy spectrum is preserved. Thus, there is no particular interest in eliminating the effects of dispersion for recovery of power or energy spectra.

Reverberation is the effect of multiple reflections and propagation along different paths through the structure. These interfering paths produce very sharp variations in the magnitude and phase parts of the spectrum. Examples of the magnitude of structural transfer functions were shown in Figures 3.39, 4.12, and 6.27. Extremely sharp minima occur in all these transfer functions. Since the inverse filter is the mirror image of the magnitude function, these sharp minima become very sharp peaks in the transmission and are potential sources of error in spectral reconstruction. One must be very careful in designing inverse filters to provide a bit of smoothing at these sharp peaks so that spurious effects in the waveform or the spectrum are avoided, as was done in Section 6.9.

We have seen that there is a great deal of variability in the magnitude of the transfer function of a particular structure along the frequency axis. An equally important issue is how much variability exists between different structures made according to the same blueprints in the same factory. We have had occasion to measure transfer functions from a point on the head of a diesel engine to a point on the casing for a large number of engines on the production line. The mean values and standard deviations of the magnitude of the transfer function are shown in

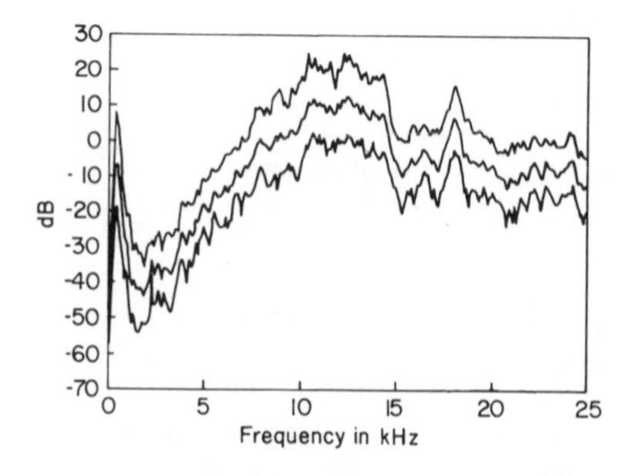

Figure 6.29 Mean and mean $\pm\ \sigma$ for magnitude of transfer function from head to casing for a population of diesel engine structures.

Figure 6.29. We note that the standard deviations in magnitude are about 8 dB at any frequency, which is a fairly substantial uncertainty in the vibration transmission.

These results imply that an inverse filter developed for one structure cannot be applied to another and still achieve good spectral reconstruction. The development of an inverse filter must account for these variations, and either develop processing algorithms insensitive to the variations (robust) or find a way of learning individual transfer functions for each machine rather than using a single inverse filter for a class of machines (adaptive).

6.11 MODELING RANDOMNESS IN TRANSFER FUNCTION MAGNITUDE

The statistics of transfer functions have long been of interest in room acoustics because of the need for frequency averaging to generate smooth response curves. An important parameter in these statistics is the modal overlap M, the ratio of modal bandwidth to modal spacing. When this parameter is large, as it often is in room response, the magnitude statistics become fairly simple.

Structures are different in the behavior of their transfer functions primarily because they tend to have low modal overlap. This means that the statistics of magnitude and phase of the transfer functions need to be reevaluated. We shall examine the magnitude statistics here and discuss phase statistics in the next chapter.

The easiest form of a transfer function to study is a ratio of polynomials:

$$H(\omega) = \frac{N(\omega)}{D(\omega)} = A\frac{(\omega - \omega_a)(\omega - \omega_b)\cdots}{(\omega - \omega_1)(\omega - \omega_2)\cdots}, \tag{6.6}$$

where $\omega_a, \omega_b, \ldots$ are the roots of $N(\omega)$ and the zeros of $H(\omega)$. The roots of $D(\omega)$, ω_1, $\omega_2, \ldots,$ are the poles of $H(\omega)$. If we assume $\exp(j\omega t)$ time dependence, then system damping will cause the poles of H to move off the real frequency axis into the upper half-plane, as shown in Figure 6.30.

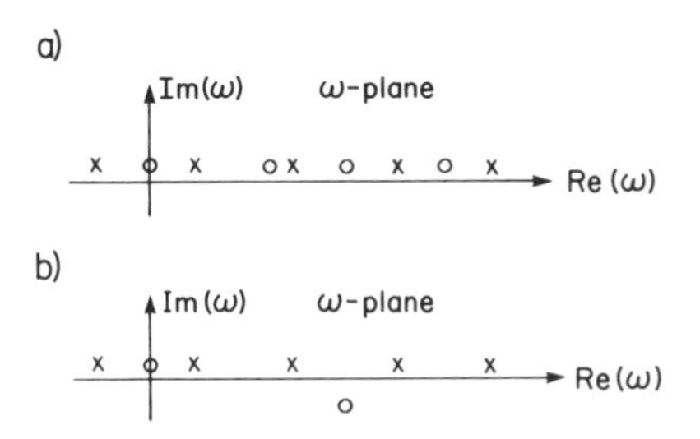

Figure 6.30 Pole–zero patterns in the complex frequency plane for an input function (a) and a transfer function (b).

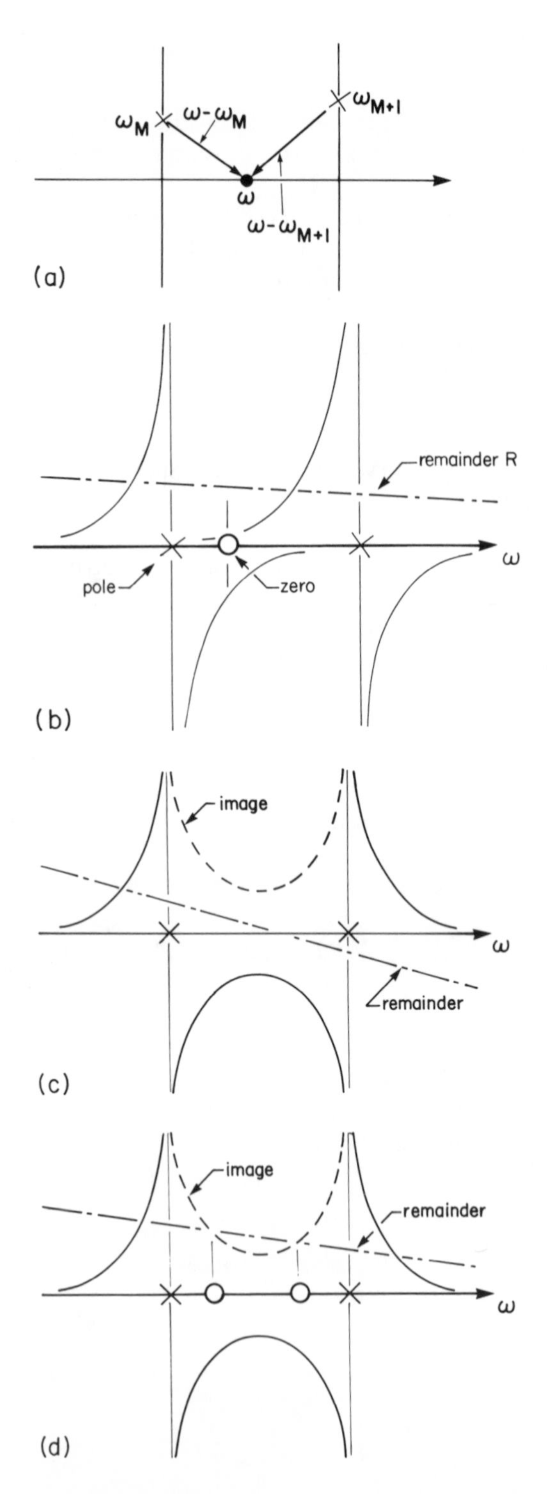

Figure 6.31 Phasors that determine the transfer function magnitude and phase in the frequency interval between two adjacent resonances (a) and the possibility of formation of zero(s) in this interval depending on the relative signs of residues (b, c, d).

Input functions, for which the excitation and observation locations are the same, have poles and zeros that alternate along the real frequency axis. With damping, both poles and zeros move into the upper half-plane, as shown in Figure 6.30(a). (Note the discussion in connection with Figure 3.11.) When the source and observer are at different locations, the poles and zeros do not alternate and zeros become less numerous than poles. With damping, some of the zeros may move into the lower half-plane, as shown in Figure 6.30(b).

Another way to represent the transfer functions of dynamical systems such as rooms and structures is by the modal expansion of Equation 3.56. In that equation, the ω_n's are the resonance frequencies or poles. The zeros of this expression are difficult to determine in general, but when the separation between modes is large compared to their bandwidth ($M \ll 1$), as in Figure 6.30, rules for their occurrence can be found.

Consider a test frequency between two resonances, as shown in Figure 6.31(a). The two phasors $\omega - \omega_M$ and $\omega - \omega_{M+1}$ will dominate the summation in Equation 3.56, with the rest of the terms producing a fairly slow "remainder" function:

$$H(\omega) \doteq \frac{B}{\omega^2 - \omega_M^2} + \frac{C}{\omega^2 - \omega_{M+1}^2} + \text{remainder}. \tag{6.7}$$

If the numerators of these terms have the same sign, the situation is as shown in Figure 6.31(b). The terms plus remainder must add to zero at some place between the poles. As a special case when $x_s = x_0$, we have an input function and the sign of all terms in Equation 3.56 is the same, leading to the required alternation of poles and zeros.

If the signs of successive terms in Equation 6.5 are opposite, then we have the situation in Figure 6.31(c) or (d), depending on the value of the remainder function. In Figure 6.31(c), no zero occurs, but if the remainder has the proper sign and magnitude as in Figure 6.31(d), a "double zero" may be produced. These double zeros are infrequent in transfer functions with low modal overlap (spacing large compared to the modal bandwidth).

The general picture, therefore, is a transfer function defined by a set of zeros and poles slightly offset from the real frequency axis. As frequency is increased, there is a peak in the magnitude of this function each time a pole is passed. If we express the magnitude in logarithmic units, there will also be a notch in the magnitude when a zero is passed as shown in Figure 6.32. It is useful, therefore, to think of the log magnitude as a sequence of pulses along the frequency axis, much like "shot noise."

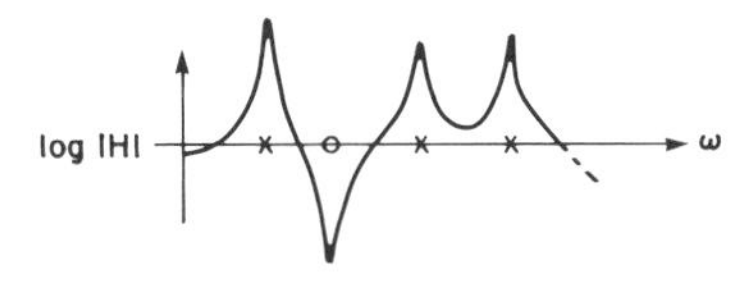

Figure 6.32 The transfer functions as determined by poles and zeros show the "pulse" nature of the magnitude curve.

When the modal overlap $M \equiv \pi f\eta/2\overline{\delta f}$ is small, the individual resonances may be assumed to determine the magnitude of response independently. When M is not small, the individual resonances interfere with each other, and the magnitude function becomes quite complicated. The extreme situations tend to have fairly simple statistics, but the intermediate situation of $M \sim 1$ is more complicated.

If we assume that the frequency spacings between resonances are Poisson distributed, the mean and variance of the squared magnitude of H as expressed by Equation 3.56 can be calculated. The mean is given by Equation 3.64 or, in the notation of this section,

$$m_{H^2} = (4f\overline{\delta f}\eta m_s^2)^{-1}, \tag{6.8}$$

where m_s is the system mass. The variance calculation is rather complicated, but the result is similar to that in Equation 4.38 and is

$$\frac{\sigma_{H^2}^2}{m_{H^2}^2} = 1 + \frac{1}{2\mathrm{M}} \frac{\langle \psi_M^4 \rangle^2}{\langle \psi_M^2 \rangle^4}, \tag{6.9}$$

so the exact result depends on mode shape. The value of the last factor on the right depends on the dimensionality of the system.

Since we know the mean and variance of the magnitude, we can use them to determine a probability density. The density chosen is the gamma density. If we let $\theta = |H|^2/m_{H^2}$, then the mean value of θ, $m_\theta = 1$, and

$$\phi(\theta) = \frac{\theta^{\mu-1} e^{-\theta}}{\Gamma(\mu)} \qquad (0 < \theta < \infty) \tag{6.10}$$

where $\mu = \sigma_\theta^2 = \sigma_{H^2}^2/m_{H^2}^2$ is given in Equation 6.9 and $\Gamma(\mu)$ is the gamma function. When $M \to \infty$, then $\mu = 1$ and we have an exponential distribution for θ, which has been shown to be exact.

If we express the magnitude in logarithmic terms (i.e., in dB), then we are interested in the statistics of $l = \ln\theta$. The probability density for l is

$$\Phi(l) = \phi(\theta)\frac{d\theta}{dl} = \frac{\exp(\mu l - e^l)}{\Gamma(\mu)} \qquad (-\infty < l < \infty). \tag{6.11}$$

This density function has very interesting properties. Its moment generating function (mgf) is

$$M_l(\beta) = \langle e^{\beta l} \rangle = \int_{-\infty}^{\infty} e^{\beta l}\,\Phi(l)\,dl = \frac{\Gamma(\beta+\mu)}{\Gamma(\mu)}, \tag{6.12}$$

and the cumulant generating function (cgf) is

$$K_l(\beta) = \ln M_l(\beta) = \ln\Gamma(\beta+\mu) - \ln\Gamma(\mu). \tag{6.13}$$

The mean value of l is the first cumulant:

$$m_l = K_l'(\beta)\big|_{\beta=0} = \frac{d}{d\mu}\ln\Gamma(\mu), \tag{6.14}$$

and the variance is the second cumulant:

$$\sigma_l^2 = K_l''(\beta)\big|_{\beta=0} = \frac{d^2}{d\mu^2}\ln\Gamma(\mu). \tag{6.15}$$

The logarithmic derivatives of the gamma function have been fairly extensively studied, and the results for these moments are

$$m_l = -0.577 - \frac{1}{\mu} + \sum_{n=1}^{\infty}\frac{\mu}{n(n+\mu)}, \tag{6.16}$$

$$\sigma_l^2 = \sum_{n=0}^{\infty}\frac{1}{(n+\mu^2)}, \tag{6.17}$$

and, therefore, both the mean and the standard deviation can be expressed in terms of the modal overlap M by using Equations 6.6 and 6.7. The result of the calculation of σ_l is shown in Figure 6.33.

It is of particular interest to compare the standard deviation of the level l as determined from Equation 6.17 to measured values for structural transfer functions. This is done in Figure 6.34. The measured standard deviation is for the diesel engine data of Figure 6.29, and comparison indicates that the modal overlap for the structure should be from 1 to 2.

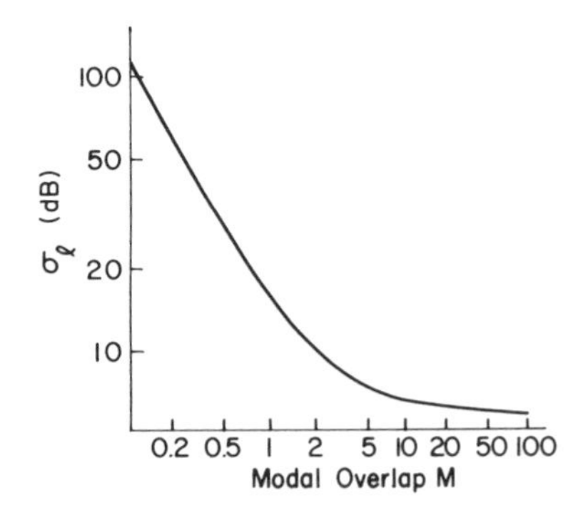

Figure 6.33 The gamma distribution for the transfer function magnitude leads to a standard deviation of level that depends on modal overlap.

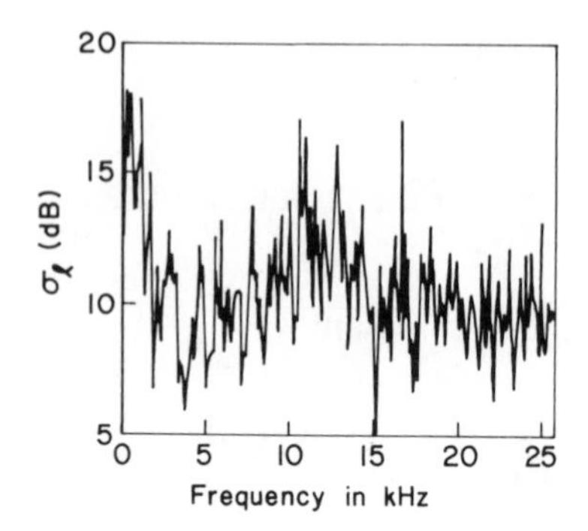

Figure 6.34 Measured standard deviation in level indicates a range of modal overlap of about 1 to 2.

CHAPTER 7

Diagnostics Using Signal Phase

7.1 INTRODUCTION

This chapter is concerned with recovering the temporal waveform of a source of vibration, and other aspects of diagnostics that use the phase of a signal. In the preceding section, we were able to ignore the phase because it does not affect the power spectrum or related measures. Features of the power spectrum as a signature were studied to determine the nature of a source event—for example, the impact between a valve and its seat. Now, we want to recover features of a time waveform so that we can have more detailed information about what is happening during machine operation.

The time waveform of interest is the cylinder pressure (Figure 2.46). The pressure–time curve of a cylinder determines the power output of the cylinder, and the rate of pressure rise in combustion for a naturally aspirated diesel engine indicates the combustion process in the cylinder. This signature is therefore a good diagnostic signal for engines and is directly measured by a pressure transducer during engine development.

7.2 CYLINDER PRESSURE WAVEFORM RECOVERY

The cylinder pressure waveform of Figure 2.46 can be obtained experimentally by drilling a hole in the head of a diesel engine and inserting a pressure transducer. Some measured waveforms at various load conditions are shown in Figure 7.1. These curves have no dc (time average) component because frequencies below 5 Hz are missing as a result of these data being tape recorded. One can see the increase in strength of the combustion pressure as the load on the engine rises and the amount of fuel injected to the cylinder is increased. This part of the waveform is of particular interest in diagnostics because the magnitude and the timing of this combustion pulse strongly depend on conditions within the combustion chamber.

The vibratory acceleration of the engine casing at one location on the engine skirt (call it "A") that results from combustion is shown in Figure 7.2. Only one pressure pulse is shown because only the pressure in cylinder 1 was directly

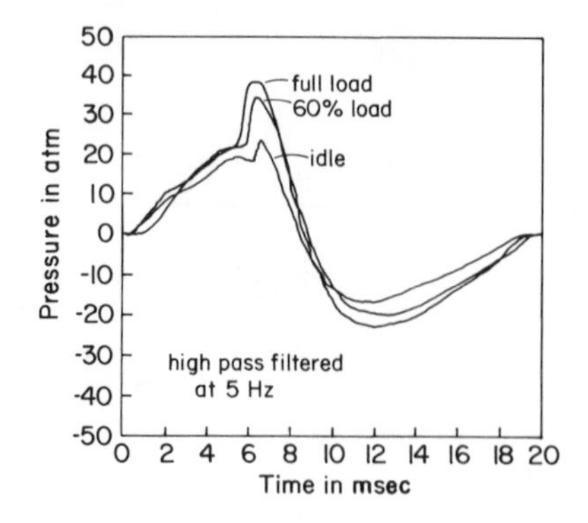

Figure 7.1 Directly measured cylinder pressure in a four-cylinder diesel engine for various load conditions. Time average (dc) component has been removed by high-pass filter.

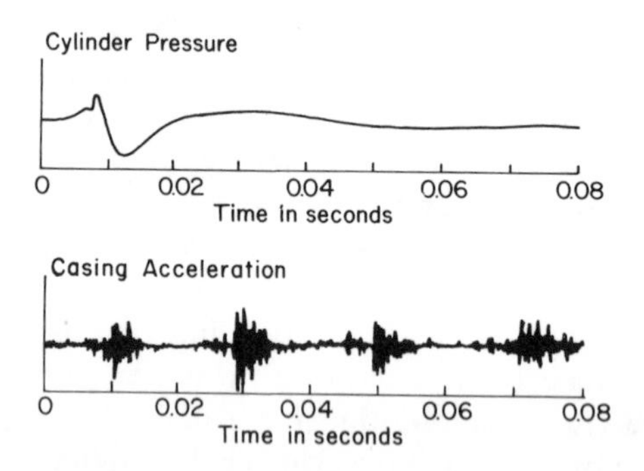

Figure 7.2 Directly measured cylinder pressure and engine casing vibration at a skirt location A. Note that vibration due to all four cylinders is about the same at this location.

measured, but the vibration record during an engine cycle shows four acceleration pulses due to combustion in each of the four cylinders. The acceleration waveform does not look much like the pressure waveform because of the filtering effect of the transfer function of the engine structure. The dispersion and the reverberation within the engine structure are responsible for this change in waveform. The combustion pressure rise lasts about 2 msec, whereas the resulting vibration lasts about 10 msec.

Figure 7.3 shows the vibration at a second location on the engine (call it "B"). We see that the vibration levels due to firing of various cylinders are not much different; indeed, it would be difficult from these traces to determine whether point A or point B is closer to any particular cylinder. This observation is consistent with the earlier discussion in connection with Figure 4.4. Thus, we cannot use the location of the accelerometer as a way to discriminate against various sources of vibration, as is

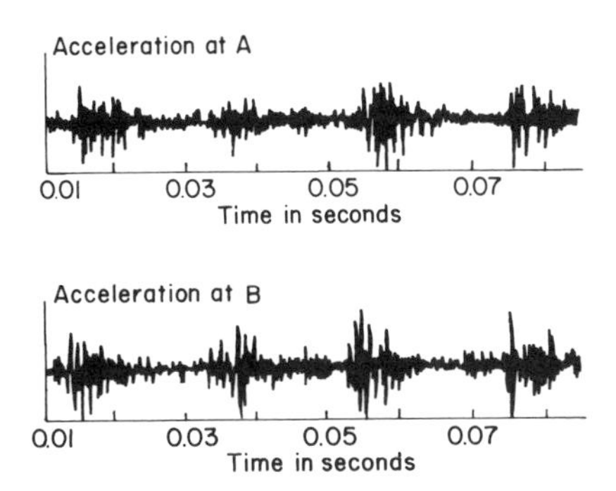

Figure 7.3 Comparison of simultaneous vibration records at two locations on the engine casing shows similar vibration levels.

often done for bearing vibrations. Bearing vibrations tend to be monitored at much higher frequencies, and an accelerometer or other gauge can be placed close to the bearing. In such cases, good discrimination against other noise sources is possible. We cannot discriminate by choice of accelerometer location in the present case, but fortunately the vibration due to one cylinder event does tend to die away before the next one occurs, as we can see from Figures 7.2 and 7.3.

The pressure waveforms in Figure 7.1 represent one-half revolution of the engine. The time feature of the pressure pulse due to combustion is about 2 msec, and the very rapid pressure rise occurs in less than 1 msec. The acceleration that results at point A on the engine for 60% load is shown in Figure 7.4. As noted earlier, this acceleration is spread out in time because of dispersion and reverberation in the engine structure. We also note that the characteristic "frequency" of this waveform is much greater than that in Figure 7.1. Counting peaks, we find that there are

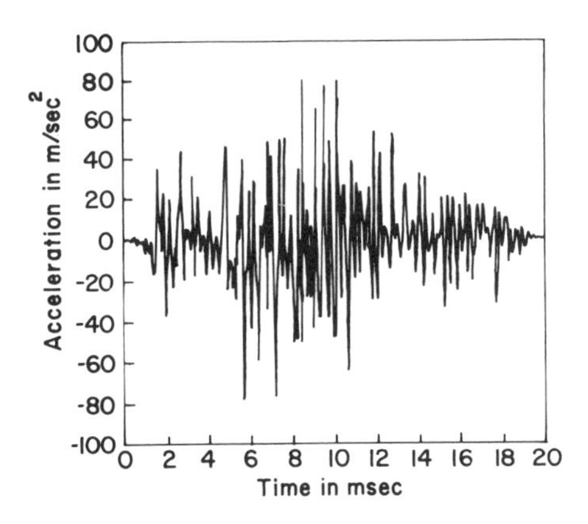

Figure 7.4 Acceleration pulse on engine casing due to combustion pressure pulse at 60% load. Complete engine cycle is 40 msec.

roughly $3\frac{1}{2}$ peaks/msec, implying a characteristic frequency for this signal of about 3 kHz. If we examine the spectrum in Figure 2.47, it is obvious that the vibration is due to combustion and will be greatly reduced when the engine is simply being motored.

The vibration signal that we can easily observe, shown in Figure 7.4, is quite complicated and difficult to use in a diagnostic system. We need to convert it to a waveform that gives us more information, like the pressure waveform shown in Figure 7.1. If we can recover Figure 7.1 from the data in Figure 7.4, then we can determine whether the engine is operating correctly—that is, whether there is a fault or some operating parameter such as amount of fuel injected or injection timing that should be altered to improve engine performance.

The high-frequency content of the acceleration signal is of particular interest in light of the combustion pressure spectrum in Figure 2.47. As noted in section 2.10, there is a large increase in the spectral strength from 250 Hz upward, due to the combustion pulse. We note from Figure 2.47 that the nature of the spectrum does not change as the load changes. The magnitude of the spectrum simply drops as the load is reduced.

The transfer function between cylinder pressure and acceleration is determined by taking the ratio of the spectrum of the acceleration signal in Figure 7.4 to the spectrum of the pressure in Figure 7.1. This may be done using a two-channel spectrum analyzer or a computer. The resulting transfer function is shown in Figure 7.5.

We can see in Figure A.30 that the log magnitude and phase of a signal are the sums of the logarithm of the transfer function and the logarithm of the input and

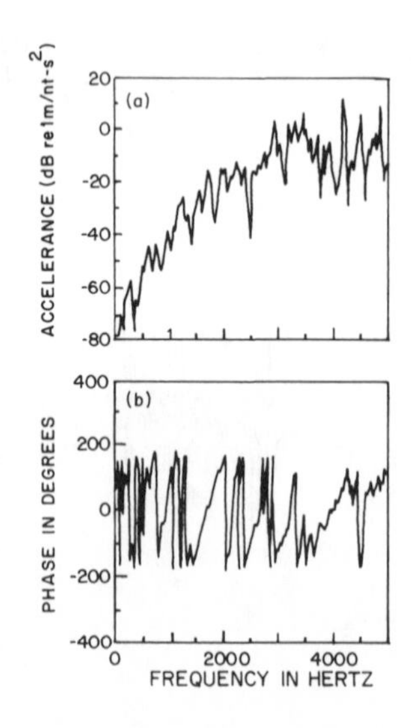

Figure 7.5 Magnitude and phase of transfer function from combustion force to casing acceleration. Note sawtooth appearance of phase curve at higher frequencies, indicative of phase progression.

therefore have additive properties. Suppose we place an inverse filter in series with the transfer function of the engine structure so that the overall system transfer function has uniform magnitude and constant phase. We thus need to generate an inverse filter that has exactly the negative of the log magnitude for gain and the negative of the phase curve as a phase component. Then the system output should reproduce the input exactly; mathematically, we have

$$Y_{out} \times H_{inv} = X_{in} \times H_{struct} \times H_{inv} = X_{in} \tag{7.1}$$

if $H_{inv} = 1/H_{struct}$. We can therefore use the data in Figure 7.5 to generate a transfer function for the inverse filter.

We noted in the discussion of valve and valve seat impact that simple inversion of a measured transfer function may not be a desirable procedure because the magnitude of the transfer function has some very sharp zeros. When these become peaks in the transfer function, there can be difficulties. A certain amount of smoothing in the data is therefore desirable. In Chapter 6, we used a simple running average of the transfer function, but other forms of smoothing, such as cepstral smoothing, may be more useful.

Because of our desire to retain phase information, we must now use the full complex cepstrum and not just the power cepstrum. If we wish to do any operation on the data in Figure 7.5 in which the phase is involved, then the discontinuities in the phase curve that result from the function being placed in the interval $(-\pi, \pi)$ represent a difficulty. The jumps in the phase curve are, in a sense, arbitrary because we can always add or subtract 2π to the phase and not change the function being represented. Nevertheless, the phase needs to be a continuous function of frequency if we wish to compute a phase cepstrum.

We see that the phase in Figure 7.5 develops a general sawtooth pattern toward the higher frequencies, due to the repeated subtractions of 2π from the data in its presentation. The process of removing these arbitrary jumps in the phase curve is called *phase unwrapping*; standard algorithms exist in signal processing to accomplish this. This phase data in Figure 7.5 are replotted in an unwrapped form in Figure 7.6. We see a rather large progression in phase delay as the frequency

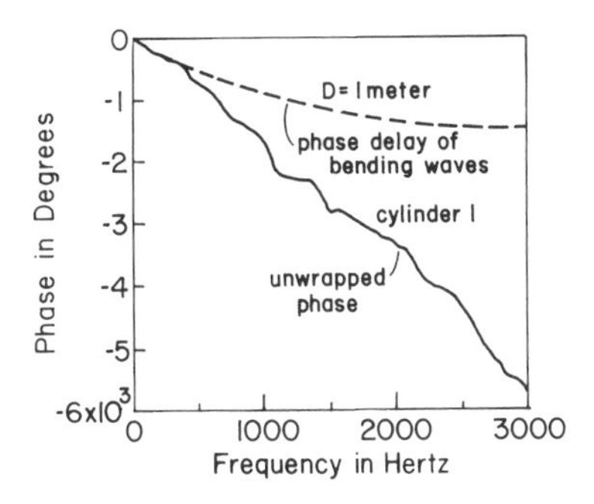

Figure 7.6 Unwrapped phase of diesel engine transfer function. The progressive phase exceeds the flexural wave propagation phase by a significant amount.

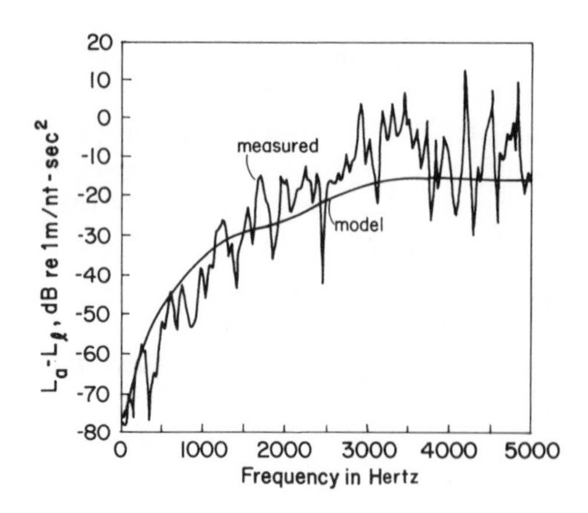

Figure 7.7 Smoothed magnitude of transfer function used to generate an inverse filter for recovery of cylinder pressure.

increases from 0 to 3000 Hz. The phase is now in the form of a continuous function that could, for example, be used to compute a phase cepstrum.

This phase curve, however, represents a phase delay significantly greater than the phase delay that we would expect from the slowest structural waves, which are the bending waves from the source to the measurement location. The phase factor for such a wave, as discussed in Chapter 3, is $e^{-jk_b r}$, where r is the distance between source and receiver, and $k_b = \sqrt{\omega/\kappa c_1}$ is the propagation constant for bending waves. In this engine, the distance between source and receiver is about 1 m; and if we assume that the wall thickness of the engine is about 1 cm, then the phase curve for bending waves is shown by the dashed line in the figure. It is clear that the measured phase delay is much greater than that due to the simple infinite system bending wave. We shall return to this point later in the chapter.

Since measured combustion pressure and casing acceleration have been used to generate the transfer function, then, if an inverse filter based on the data in Figure 7.5 is used to generate a system output, we will, of course, recover the source waveform without any distortion, as indicated in Equation 7.1. We can use the data, however, to show how sensitive the recovered waveform is to various forms of smoothing. For example, if we form the magnitude of the inverse filter by using a simple analytical model as in Figure 7.7, but determine its phase exactly from measured data and apply this inverse filter to the measured laboratory data to generate an input source waveform, we get the input reconstruction or estimate shown in Figure 7.8. The quality of the reconstructed waveform in this instance is good, and quite a bit of information could be read from the reconstructed waveform to determine the operation of the engine.

If, instead, we use the exact form of the log magnitude of the transfer function, but smooth the phase to generate an inverse filter, the reconstructed source waveform is as shown in Figure 7.9. Clearly, the reconstructed waveform in this instance is of no value in determining engine operating parameters. As we might

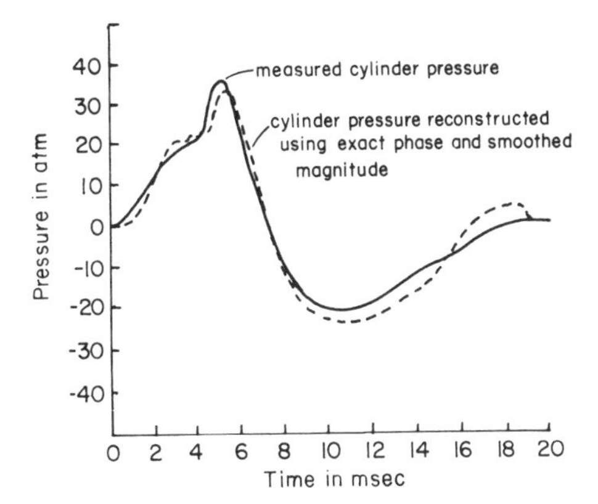

Figure 7.8 Comparison of directly measured and inverse filter recovered cylinder pressure waveform using exact phase inverse and smoothed magnitude.

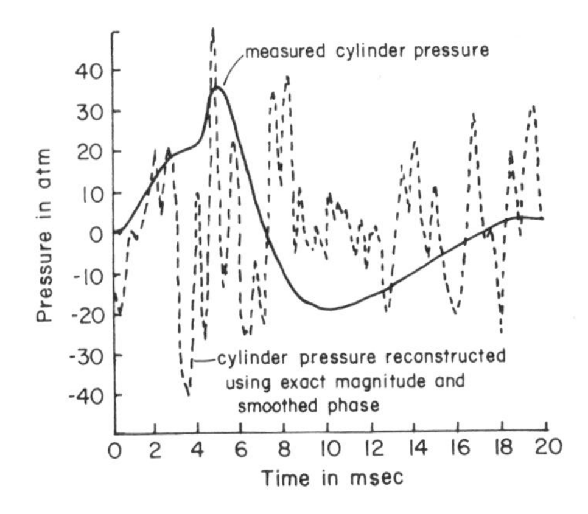

Figure 7.9 Comparison of directly measured and recovered cylinder pressure waveforms using an inverse filter with exact magnitude and smoothed phase.

have suspected, the phase is far more important than the magnitude in waveform reconstruction; if we are to successfully perform waveform reconstruction for diagnostics, we must have prccedures for making good estimates of transfer function phase or the phase cepstrum.

Another example of cylinder pressure recovery is shown in Figure 7.10. We used magnitude smoothing and exact phase to obtain a signature recovery using the transfer function from cylinder pressure in cylinder 1 to pick up location A, which is the "correct" inverse filter for vibration data sensed at that location. We have also applied an inverse filter to this location A data, but this time the signal is due to the firing of cylinder 4, and the inverse filter is therefore "incorrect." The signature recovery in this case is now useless for diagnostic interpretation.

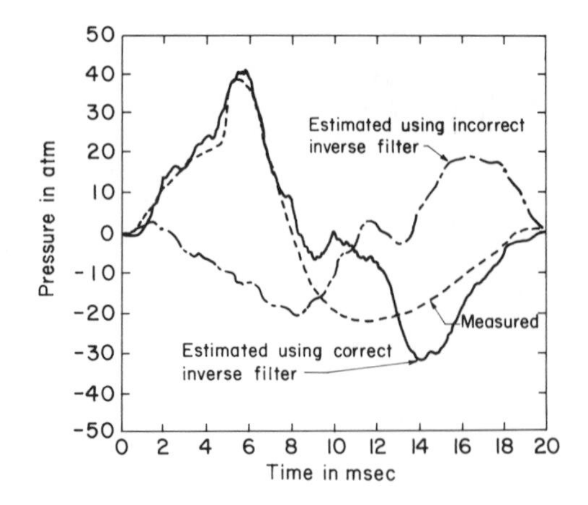

Figure 7.10 When the cylinder pressure (motoring plus combustion) is recovered, the waveform recovered is very dependent on whether the correct inverse filter is used.

7.3 CEPSTRAL ANALYSIS IN WAVEFORM RECOVERY

Figure A.30 shows how the input and the impulse response of a system become folded together so that the separate extraction of source and path information is very difficult in the signal time domain. We have also seen that the impulse response can be thought of either as a superposition of propagation paths (e.g., the combination of paths through the head and the piston–connecting rod–crankshaft) or as a combination of modal responses, as discussed in Chapters 3 and 4.

The diagnostics problem might be to determine the path transfer function, perhaps as a way of constructing an inverse filter, or to discover a fault in the structure. It is not always convenient to measure the transfer functions for every machine on which we wish to perform waveform analysis; therefore, some method of inferring the transfer function from response data is desirable. In applications of a diagnostic system to machines, however, we are more commonly interested in recovering source waveforms. The problem in this case, therefore, is to recover the input force or other source characterization, such as transmission error from output data.

Simple windowing in the frequency or time domain has been a widely used technique for trying to separate propagation paths in structures. Consider the situation in Figure 7.11 in which there are two paths of vibration transmission in a structure. Path a might be considered a low-frequency path, and path b a high-frequency path. Suppose this structure is excited by a periodic impulsive force, perhaps due to a faulty tooth in a gear train. It is very difficult to see the input waveform $l_{in}(t)$ in the output signal $O(t)$ because of the ringing and interference effects of these resonant paths.

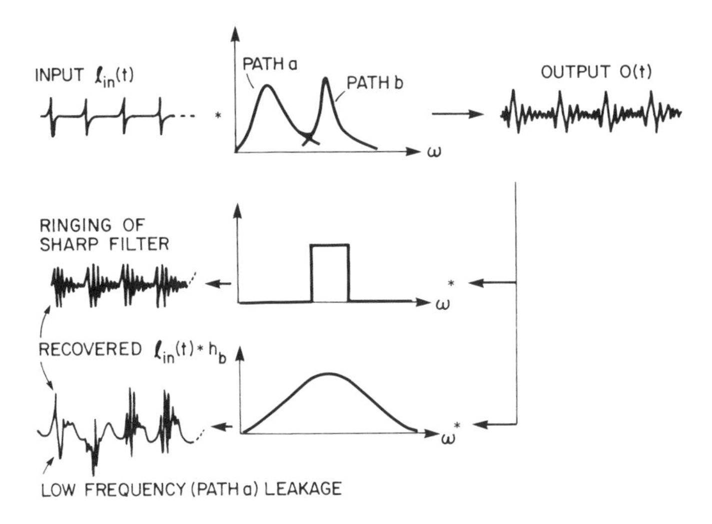

Figure 7.11 Attempts to separate a path by frequency windowing either result in time distortion by filter ringing or leakage from unwanted path.

If we want to separate the contribution of path b, we might consider applying a frequency window or filter to the output data. Because the frequencies of a and b are fairly close to each other, we need a filter with sharp skirts to separate them, but if we apply such a filter to the output signal the filter will ring and there will be a loss in time resolution and a distortion of the waveform. This ringing might in part be due to the properties of path b, but for a very sharp filter the ringing is also due to the filter. Thus, not only can we not recover the input waveform, but we cannot even recover the correct time behavior for path b.

If we try to solve the problem of ringing by making the filter on path b less sharp, then some of the low-frequency energy from path a will leak into the output signal; we then get a distorted waveform for the input, and we have again failed to recover the time waveform of the signal through path b.

Suppose, on the other hand, we have an input force applied to a transmission path that has the echo structure shown at the top of Figure 7.12. The convolution of the input and the impulse response will produce an output that for early arrivals will allow the source and path to be separated, but at later times the pulses overlap to produce a general smearing of the waveform. The early paths can be separated, therefore, by simple time windowing (or gating) as shown, and a good quality of input waveform recovery is obtained. If we try to window a later path, however, the window must be short enough to separate the different arrivals; but as the arrivals get very close together in time, the window becomes shorter than the input waveform, and, therefore, we obtain a very poor recovery of the input waveform.

The cepstrum can offer a possible solution to the problem of separating source and path characteristics. In other cases described in Chapter 6, we have shown how time windowing of the real cepstrum, separating it into low-time and high-time

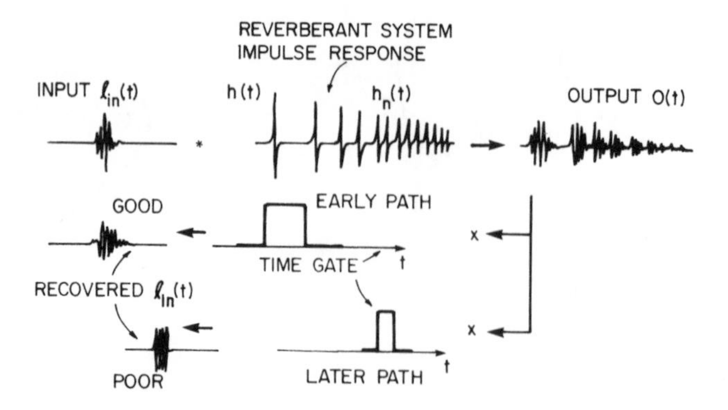

Figure 7.12 Attempts to separate paths by time windowing can be useful if time spreading by one path is small and if paths have significantly different delay times.

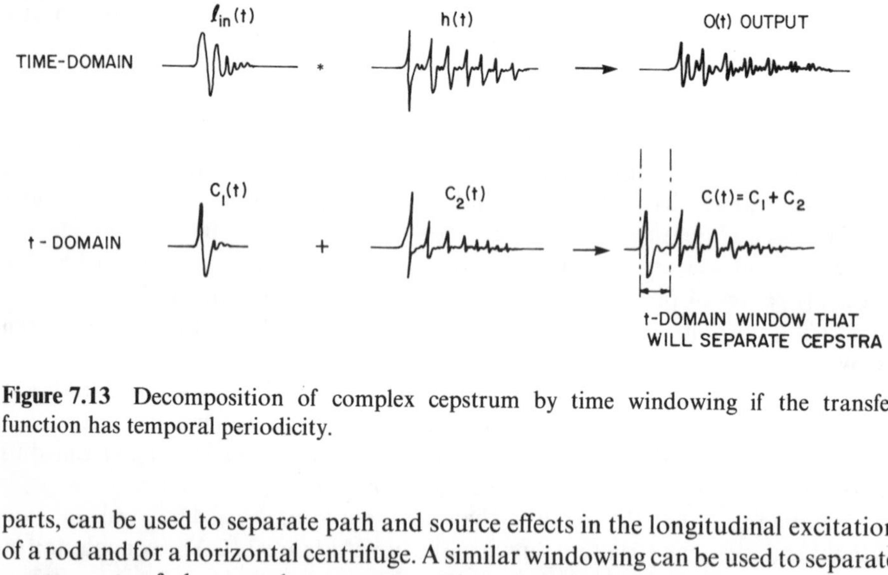

$\ell_{in}(t)$ h(t) O(t) OUTPUT

TIME-DOMAIN

$C_1(t)$ $C_2(t)$ $C(t) = C_1 + C_2$

t - DOMAIN

t-DOMAIN WINDOW THAT WILL SEPARATE CEPSTRA

Figure 7.13 Decomposition of complex cepstrum by time windowing if the transfer function has temporal periodicity.

parts, can be used to separate path and source effects in the longitudinal excitation of a rod and for a horizontal centrifuge. A similar windowing can be used to separate components of the complex cepstrum. This decomposition is illustrated in Figure 7.13, in which source and path cepstra are separated by time windowing of the cepstrum.

Suppose that we have windowed out what we believe to be the cepstrum due to the propagation path. Referring to Figure A.30, we note that the phase cepstrum is a Fourier (sine) transform of the phase, which is an odd function of frequency. Therefore, the phase cepstrum is an odd function of time, and similarly the magnitude cepstrum is an even function of time. If we take the total cepstrum of the windowed "path cepstrum," then by selecting out the even and odd parts of the resulting cepstrum, the phase and the log magnitude of the path can be determined.

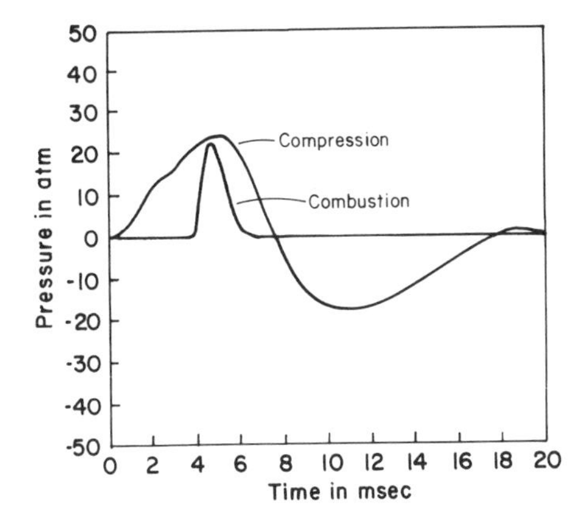

Figure 7.14 Decomposition of cylinder pressure into compression and combustion terms. Compression is expected to be a high-time component, and combustion a low-time component of the pressure cepstrum.

For now, however, we shall apply the cepstral decomposition idea to the source or input function. The decomposition we have in mind is separation of the compression and the combustion parts of the diesel engine cylinder pressure waveform. A graph of these components is shown in Figure 7.14. For diagnostic reasons, we might be interested in separating only the combustion part of the pressure and eliminating the compression pulse.

A flowchart for calculating the cepstrum is shown in Figure 7.15. A data file containing the time record is read by the program, and a Fourier transform is taken. If the complex cepstrum is to be computed, the phase must be unwrapped to make it a continuous function of frequency. The cepstrum is then computed and generally will then be windowed either to separate low- from high-time components or perhaps to separate out a sequence of cepstral peaks if either the input function or the transfer function is periodic in time. The selection of the shape of the window is something of an art, and there are no hard and fast rules at this stage for determining whether we should use an exponential, Hanning (lifted cosine), rectangular, or some other form of window.

Once the cepstrum has been windowed and the desired parts have been selected, we compute an inverse cepstrum to obtain the log magnitude and phase of the function. If we wish to get back to the original temporal waveform, then additional steps are required to exponentiate and invert the resulting Fourier transform to recover the original signal. Considering the steps required in this process, it is obvious why cepstral analysis as a useful technique depends on digital computation.

The cepstrum of the directly measured waveforms of cylinder pressure in Figure 7.1 is shown in Figure 7.16. The spectra of this waveform in Figure 2.47 shows that the combustion pressure component has a wide-band high-frequency power

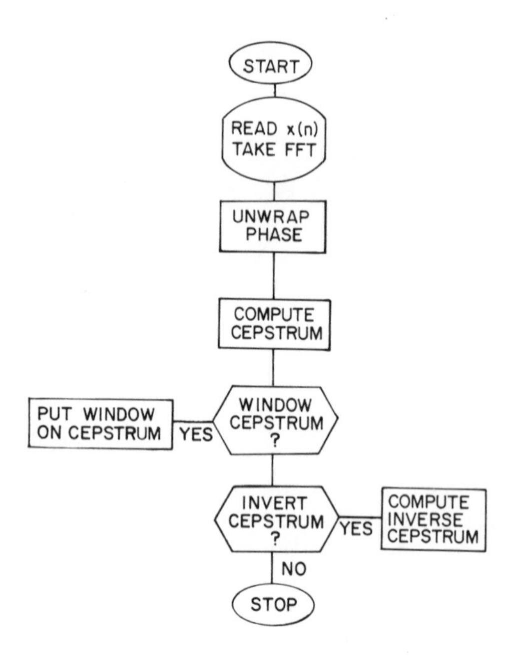

Figure 7.15 Flowchart of a program to compute complex cepstra and to provide windows for separation of components.

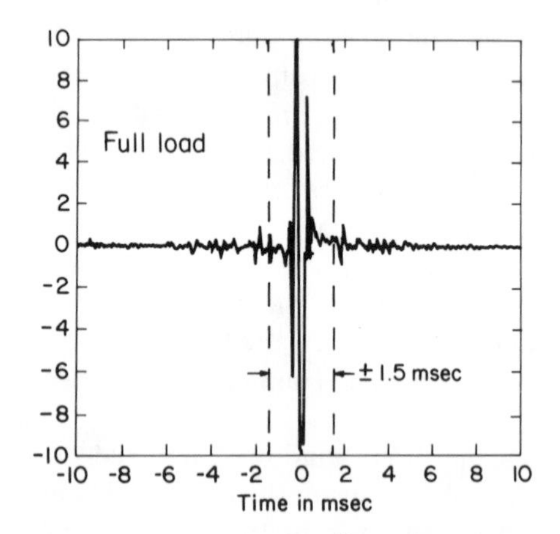

Figure 7.16 Complex cepstra for cylinder pressures of four-cylinder diesel engine at full load. The window is chosen to separate compression and combustion components.

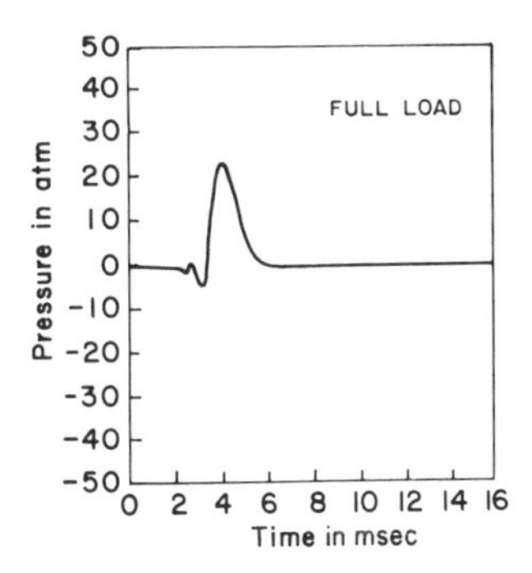

Figure 7.17 Cylinder pressure component corresponding to combustion pressure, recovered from windowed cepstrum. No inverse filter is used in this reconstruction.

spectrum, and we therefore expect that the low-time part of the cepstrum should be dominated by the combustion pressure pulse. Consequently, a simple low-time window was applied to the "full-load" cepstrum in Figure 7.16, and the time waveform recovered from this low-time cepstrum is shown in Figure 7.17. Aside from a slight transient at its beginning, this waveform shows the abrupt rise and rather slower decay that we would expect for the combustion component of the cylinder pressure.

We carried out a similar waveform recovery by using measured acceleration data. The transfer function from cylinder 1 to the observation location A was determined by mounting a pressure gauge in the cylinder and observing the

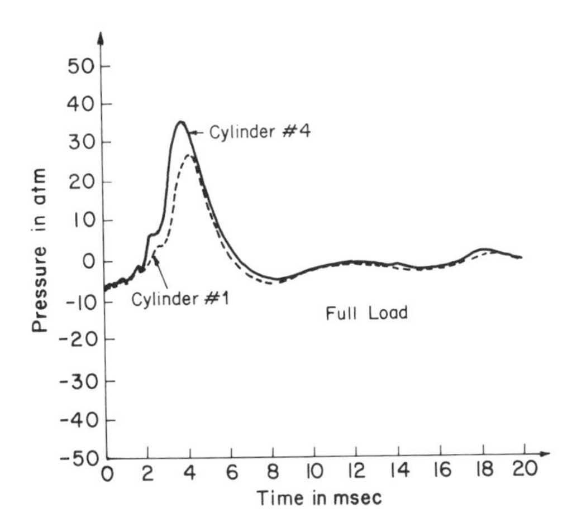

Figure 7.18 Recovered combustion pressure component from windowed cepstrum. This cepstrum is obtained as output of a filter that is inverse to the path from cylinder 1 but not cylinder 4.

vibration of the engine block. Let us now take vibration pulses that are due to the firing of cylinders 1 and 4 and apply this same inverse filter to both signals. We have shown in Figure 7.10 that the waveform recovered by this process for cylinder 1 will be fairly good, but for cylinder 4 the time waveform recovery is quite poor. However, if we now time window the cepstrum to extract the low-time components of each of those recovered cepstra and then compute the time–pressure waveform due to combustion pressure in these two cylinders, we obtain Figure 7.18. We see that the combustion pressure waveforms are very similar, and the recovery appears to be insensitive to the inverse filter used. This may be a clue as to how we may obtain an inverse filter that can be used for a variety of cylinders so that we need not use a separate inverse filter to recover pressure waveforms from each cylinder. It is too early to say whether this technique is useful, but the results in Figure 7.18 are encouraging.

7.4 PHASE CHARACTERISTICS OF STRUCTURAL TRANSFER FUNCTIONS

We have seen that the phase of transfer functions is of great importance when we wish to recover waveforms, and that the phase of the inverse filter should be the negative of the phase introduced by the transfer function of the structure. In this section, we discuss in more detail the properties of a structure that affect the transfer function phase and certain mathematical relationships that allow us to relate the phase to other measured quantities.

In the discussion concerning Figure 7.6, we compared the phase of the transfer function with the phase delay of a bending wave between source and receiver. This comparison was made because some experiments and theoretical calculations suggest that the expected or average phase of vibration in a finite structure should be equal to the propagating phase of the appropriate wave type in an unbounded structure. For platelike structures, this is a bending wave on an infinite plate.

A demonstration at one of the M.I.T. summer courses in acoustics by Leo Beranek consisted of measuring the vibration of a beam that has a sand termination at one end and is excited into flexural motion by a shaker at the other. The magnitude and phase of the vibration was read as the microphone was moved along the beam. The graph of the measured phase and amplitude is given in Figure 7.19. When there is very little damping, the standing-wave ratio (SWR) is quite high and the phase tends to remain constant until it suffers a jump of 180° at the nodes of vibration.

If the phase jumps of 180° are plotted as in the figure, they generate a staircaselike function. We can think of this as a form of phase unwrapping. As the damping at the far end of the beam is increased, the SWR is reduced and the traveling wave component of the vibration will increase relative to the standing-wave component. Now the phase changes more gradually, and as the damping is increased to the point where there is no reflection the phase becomes the straight line $k_b r$, where k_b is the bending wave number and r is the distance from the shaker to the

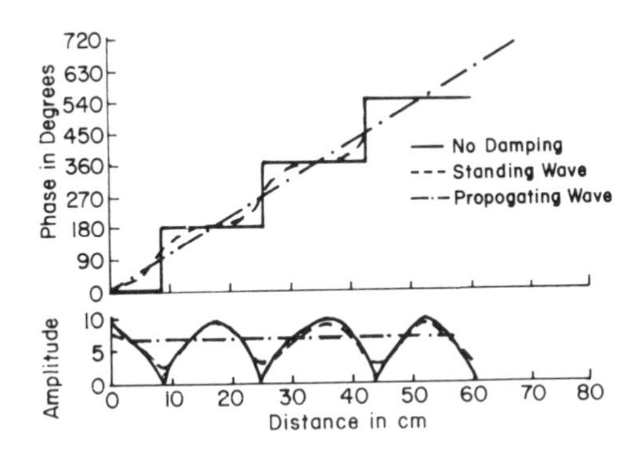

Figure 7.19 Demonstration that the progressive phase in a resonant bending beam converges to that for a propagating wave as damping is increased.

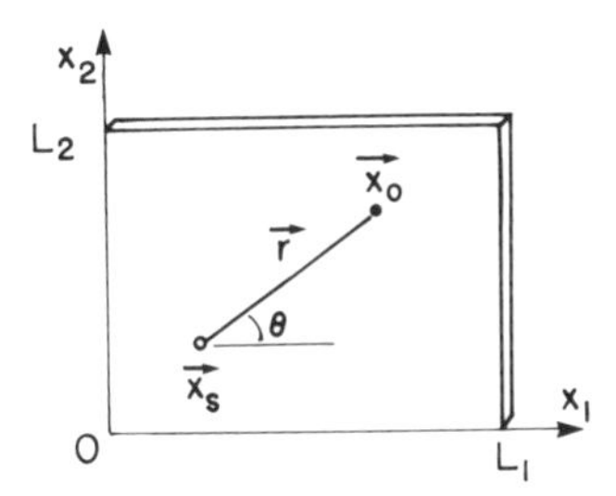

Figure 7.20 Two-dimensional rectangular system used to calculate the average response due to a point source at distance r from an observer.

microphone pickup. Therefore, even with very light damping if we plot the phase as the staircase we see that the free bending propagation represents an average or a trend line for the phase in this one-dimensional case.

A second example is the theoretical calculation of the phase of the average response of a two-dimensional system. Consider the two-dimensional acoustic space in Figure 7.20, which has dimensions L_1 by L_2. The response at the observation position $\mathbf{x}_0$, due to a source at the source point $\mathbf{x}_s$, is

$$\langle y(\mathbf{x}_0)\rangle = \text{const.} \sum_{m_1, m_2} \frac{\langle \psi_{m_1, m_2}(\mathbf{x}_s)\psi_{m_1, m_2}(\mathbf{x}_0)\rangle_{\mathbf{x}_s, \theta}}{k^2 - k_{m_1, m_2}^2}, \tag{7.2}$$

where the response is averaged over source location and the angle between source and observer, but the distance between source and observer is held fixed. If the mode shapes are of the form

$$\psi_M \sim \cos k_1 x_1 \cos k_2 x_2, \tag{7.3}$$

which are the correct mode shapes for an acoustical space where the allowed propagation constants k_1 and k_2 are

$$k_1 = \frac{m_1\pi}{L_1}, \quad k_2 = \frac{m_2\pi}{L_2}, \quad k_M = \sqrt{k_1^2 + k_2^2}, \tag{7.4}$$

then it is easy to calculate the average in Equation 7.2 to obtain

$$\langle \psi_M(\mathbf{x}_s)\psi_M(\mathbf{x}_0) \rangle_{\mathbf{x}_s,\theta} \propto J_0(k_M r), \tag{7.5}$$

where J_0 is the zero-order Bessel function. If the modes have some damping, it is possible to replace the summation in Equation 7.2 by an integral to obtain

$$\langle y \rangle_{\mathbf{x}_s,\theta} \propto \sum_M \frac{J_0(k_M r)}{k^2 - k_M^2} \propto H_0^{(1)}(kr), \tag{7.6}$$

where $H_0^{(1)}$ is the Hankel function and represents a cylindrical sound wave propagating away from the source point x_s. The Hankel function has the asymptotic form $e^{-jkr}/\sqrt{r}$, and, therefore, the phase of this average response is simply kr. Thus, the phase of the average response of a finite structure is kr, and if the wave is a bending wave then the phase is $k_b r$.

We see from these two examples that the phase of the average response should be kr, and yet when we measured the unwrapped phase of a transfer function in a finite structure, we found that the trend in the phase greatly exceeds kr. We resolve this dilemma in the following sections.

7.5 PHASE OF INPUT AND TRANSFER FUNCTIONS

The phase and magnitude of the transfer function are given in terms of its real and imaginary parts by

$$\phi_h = \tan^{-1}\left[\frac{H_i}{H_r}\right], \qquad |H| = \sqrt{H_r^2 + H_i^2}. \tag{7.7}$$

The tangent function is sketched in Figure 7.21, and the tangent varies from + to − infinity as the argument changes from $-\pi/2$ to $+\pi/2$. Indeed, any computer program or calculator will give the range of the arc tangent function from $-\pi/2$ to $+\pi/2$. We can extend the range of the phase presentation from $-\pi$ to $+\pi$ by noting the relative signs of H_i and H_r.

In Figure 7.22, H_r and H_i are positive in the first quadrant and negative in the fourth quadrant. In the third quadrant, H_i is positive and H_r is negative; in the second quadrant, H_r is positive and H_i is negative. Thus we can add the appropriate angle to that presented by the arctan function to generate a range in angle from $-\pi$ to $+\pi$ if we track the signs of the real and imaginary parts of H.

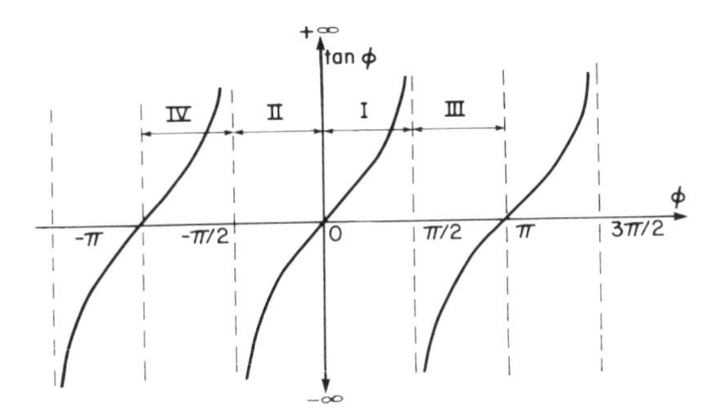

Figure 7.21 Tangent function covers range from $-\infty$ to $+\infty$ for the range of the argument from $-\pi/2$ to $+\pi/2$, quadrants I and II. Additional information must be supplied to assign the argument to quadrants III and IV.

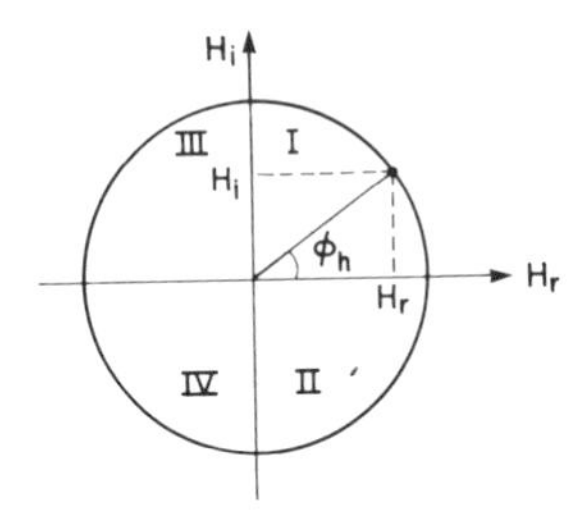

Figure 7.22 The phase can be assigned to quadrants III and IV if the real part of the transfer function, H_r, is negative.

There is a fundamental ambiguity in the phase of multiples of 2π, however, that cannot be removed by this process. Consider a wave e^{-jkx}, where k is some function of ω depending on the structure. If we plot the phase of this wave as a function of frequency (Figure 7.23), then the slope of this phase at any frequency is

$$\frac{d\phi}{d\omega} = x\frac{dk}{d\omega} = \frac{x}{c_g} \tag{7.8}$$

where c_g is the group speed. This ratio of separation to speed is called the *group delay*, and we see that a phase delay that increases with frequency produces a group delay proportional to the slope of the phase curve.

We have noted several times that to generate a phase cepstrum from the Fourier transfer of ϕ the phase must be a continuous function of frequency. Differentiating ϕ with respect to ω in Equation 7.7 gives

$$\frac{d\phi}{d\omega} = \frac{H_r\left(\dfrac{dH_i}{d\omega}\right) - H_i\left(\dfrac{dH_r}{d\omega}\right)}{H_r^2 + H_i^2}. \tag{7.9}$$

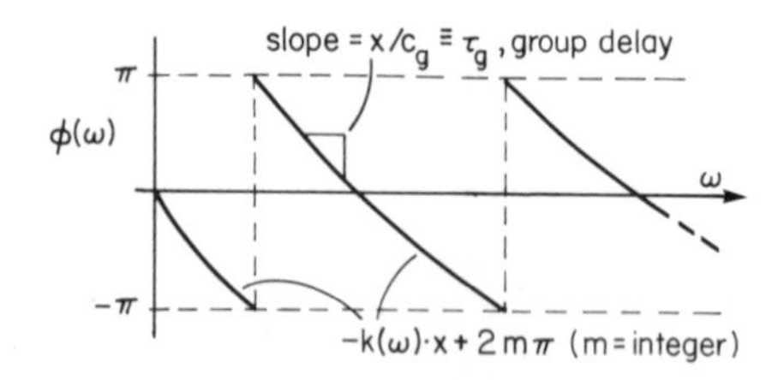

Figure 7.23 Propagating phase delay and its placement in the interval between $\pm\pi$. The slope of this curve is the group delay, the time taken for an energy packet to travel a distance x.

The calculation of $dH/d\omega$ in Equation 7.9 is performed by noting from Equation A.29 that

$$\frac{dH}{d\omega} = -j\int_0^\infty th(t)e^{-j\omega t}\,dt, \tag{7.10}$$

so the needed ingredients for determining ϕ are the Fourier transforms of $h(t)$ and $th(t)$. If we integrate Equation 7.9 with respect to frequency, we can generate a continuous function of frequency, which is the unwrapped phase.

There is generally too much error in the numerical integration to give a good absolute value of the phase, but integration over relatively short intervals can give an indication of the need to add or subtract 2π to generate phase as a continuous function. This is the basis of the Tribolet phase unwrapping algorithm.

7.6 POLES AND ZEROS OF TRANSFER FUNCTIONS

Consider a simple transfer function, the drive point mobility of a resonator:

$$H(\omega) = \frac{j\omega/K}{1 - \dfrac{\omega^2}{\omega_r^2} + \left(\dfrac{j\omega}{\omega_r}\right)\eta}. \tag{7.11}$$

If we express the numerator and denominator of Equation 7.11 as polynomials in ω, then the zeros (roots) of the polynomial in the numerator are the zeros of H and the zeros of the denominator are the poles of H. At the poles, the transfer function diverges, and these frequencies correspond to system resonances. When the system has damping, the poles are complex and can be plotted in the complex ω plane as shown in Figure 7.24.

According to Equation A.29, the impulse response is evaluated by integrating the transfer function along the real frequency axis. As indicated in Figure 7.24, this integral is converted into a contour integral by closing it over a large radius in the

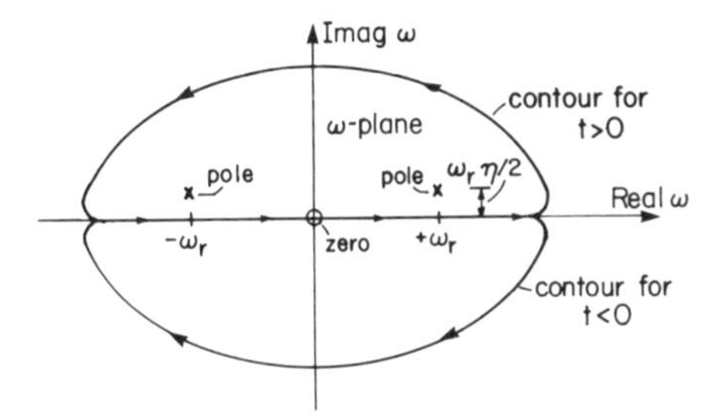

Figure 7.24 Closure of the contour integral for evaluation of the Fourier integral ensures causality and stability since poles must lie in the upper half of the complex ω plane.

upper half-plane for $t > 0$ and by a similar path in the lower half-plane for $t < 0$. Since only the upper contour encloses poles, the impulse response is nonzero for $t > 0$ and vanishes for $t < 0$, as required. Its value is

$$h(t) = \begin{cases} \dfrac{1}{\omega_r M} \sin \omega_r t \, e^{-\omega_r \eta t/2} & (t > 0), \\ 0 & (t < 0). \end{cases} \tag{7.12}$$

There are two consequences to the poles being in the upper half-plane: the impulse response vanishes for negative time, and the impulse response decays for large time, due to the damping. These two conditions are referred to as *causality* and *realizability*.

More complicated structures with many modes of vibration have a transfer function that can be expressed as the ratio of polynomials given in Equation 6.6. As noted in Chapter 6, when the numerator and denominator are factored into their roots, ω_a, ω_b, and so on, are the zeros of the transfer function and ω_1, ω_2, and so on, are the poles. If we plot the poles and zeros in the ω plane, as in Figures 6.30 and 7.25, the poles must be in the positive half-plane for the reasons described above. In this diagram, the factors in the numerator and denominator of Equation 6.6 become phasors pointing from the pole or zero to the frequency ω on the real axis.

As the test frequency ω goes by a pole, the phasor from the pole to the test frequency becomes very short and the magnitude of the transfer function increases,

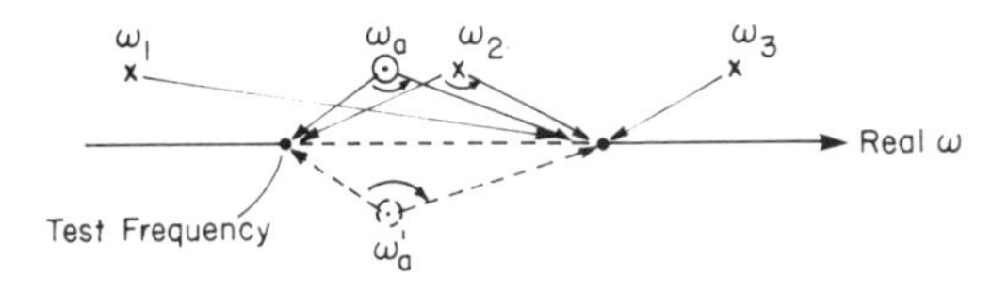

Figure 7.25 The rotation of phase factors in the numerator and denominator of a system function determines the system phase response.

producing a peak. By a similar argument, when the test frequency goes by a zero, the transfer function becomes very small. Also note that the phase angle of the phasor changes by $+\pi$ as the test frequency goes by a pole. Since this phasor is in the denominator of the transfer function, the transfer function phase changes by $-\pi$. A zero in the positive half-plane, by the same argument, produces a phase change of $+\pi$ in the transfer function, and a zero in the lower half-plane produces a phase change of $-\pi$. Thus by knowing where the poles and zeros are, we can estimate the change in the phase of the transfer function as the test frequency goes from the origin $\omega = 0$ to its final value.

Since the magnitude of H is determined by the lengths of the phasors, a zero that is a given distance from the real frequency axis will produce the same magnitude of transfer function whether that zero is in the upper or lower half of the ω plane. If it is in the upper half-plane, it will reduce the phase of H, but if it is in the lower half-plane, it will increase the phase of H. Since the poles must be in the upper half-plane, it is clear that for a given magnitude of transfer function we will have less phase delay if all the zeros are in the positive half-plane. For this reason, transfer functions with all their poles and zeros in the positive half-plane are called *minimum-phase* transfer functions.

7.7 DRIVE POINT SYSTEM FUNCTIONS

If we consider the drive point mobility of a structure

$$Y_{\mathrm{dp}} = \frac{V(\omega)}{L(\omega)}, \tag{7.13}$$

where V and L are Fourier transforms of velocity and force, the poles correspond to frequencies at which we can have a finite velocity for zero input force. In this case, the poles correspond to system resonances in which the drive point is free (no force). If we consider instead the drive point impedance

$$Z_{\mathrm{dp}} = \frac{L(\omega)}{V(\omega)} = \frac{1}{Y_{\mathrm{dp}}}, \tag{7.14}$$

the poles of this expression correspond to frequencies at which the force at the drive point may be finite for zero velocity. The poles in this case correspond to system resonances in which the drive point is held fixed. Since both of these conditions correspond to resonances of a real mechanical system, both the poles and zeros of the drive point mobility and impedance must be in the positive half-plane. Drive point impedances or mobilities are, therefore, minimum-phase functions.

For a transfer function, the poles of the transfer function still represent system resonances, since they represent finite velocity at some location for no driving force. The zeros, however, do not have such a simple interpretation because their values depend upon the excitation and observation locations. We cannot even be sure that the zeros are in the upper half-plane. To examine this question further, we consider some examples.

7.8 THE ONE-DIMENSIONAL ACOUSTICAL PIPE

Consider the acoustical pipe in Figure 7.26, which has a volume velocity source at location x_s and a microphone at x_0. The ratio of the pressure at the microphone to the volume velocity is the transfer acoustical impedance

$$\frac{p(x_0)}{U_s} = \frac{-j\rho_0 c}{A} \frac{\cos kx_s \cos k(L - x_0)}{\sin kL}, \tag{7.15}$$

where $k = \omega/c$ and system resonances occur when $k = m\pi/L$. Zeros occur when the distance between the source or observation points and the ends of the pipe are odd multiples of a quarter wavelength of sound, or

$$\begin{aligned} k &= \frac{m\pi}{L} && \text{(poles)}, \\ k &= \frac{(n + 1/2)\pi}{x_s} && \text{(zeros)}, \\ k &= \frac{(r + 1/2)\pi}{L - x_0} && \text{(zeros)}, \end{aligned} \tag{7.16}$$

where m, n, r are integers.

Since, as discussed in Section 3.3, the poles and zeros represent resonances of real physical systems, they, in this case, must all be in the positive half-plane. The difference between the number of poles and zeros that occur up to any frequency is then a direct indication of the average phase delay. The phase delay is $+\pi N_{\text{poles}}$ due to the poles and $-\pi N_{\text{zeros}}$ due to the zeros; the total phase is

$$\begin{aligned} \phi &= (N_{\text{poles}} - N_{\text{zeros}}) \cdot \pi = \left\{ \frac{k}{\pi/L} - \frac{k}{\pi/(L - x_0)} - \frac{k}{\pi/x_s} \right\} \pi \\ &= k(x_0 - x_s). \end{aligned} \tag{7.17}$$

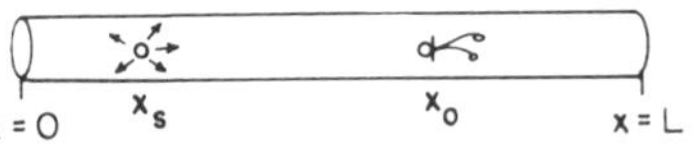

x = 0 x_s x_0 x = L

Figure 7.26 Acoustical pipe as example of one-dimensional system illustrating the phase trend equal to propagation phase.

We find that the average phase delay is the propagation phase, which is what we found in connection with the discussion of Figure 7.19.

The expression in Equation 7.15 for the transfer function is very special. In most mechanical systems, we cannot find such a simple analytical expression for the transfer function in which the poles and zeros are so easily identified. A more available expression is the modal expansion, for which this one-dimensional system is

$$\frac{p}{U_s} = \text{const.} \sum_m \frac{\cos(m\pi x_s/L)\cos(m\pi x_0/L)}{\omega^2 - \omega_m^2} \xrightarrow[x_s = 0,\, x_0 = 0]{} \sum_m \frac{1}{\omega^2 - \omega_m^2};$$

$$\xrightarrow[x_s = 0,\, x_0 = L]{} \sum_m \frac{(-1)^m}{\omega^2 - \omega_m^2}. \tag{7.18}$$

We note the form that the equation takes for a drive point impedance at $x = 0$ and for a transfer impedance between one end of the tube and the other.

From Equation 7.16, we see that the zeros become infinitely far apart when the source is at the origin and the receiver is at $x_0 = L$. Therefore, the transfer function in this case has no zeros. On the other hand, when the source and receiver are at the origin, the zeros of the input mobility are the zeros of the cosine function, which interlace and alternate with the zeros of the sine function. Note the similarity between this situation and that in Figure 3.11.

We saw in Chapter 6 that zeros alternate with poles when we have an input function (impedance or mobility). This corresponds to one situation in Equation 7.18. On the other hand, when the source and observation points are at opposite ends of the pipe, the residues alternate in sign and the situation is as in Figure 6.31(c), and there are no zeros, in agreement with Equation 7.16.

In the discussion in Chapter 6, it did not matter whether the zeros were above or below the real frequency axis insofar as magnitude was concerned, but it does matter for the phase. From Equation 6.7, we can show that for Figure 6.31(b), the zero is located at

$$\omega = \bar{\omega} + \frac{\delta(B - A)}{B + A} \tag{7.19}$$

where $\omega_M = \bar{\omega} - \delta$ and $\omega_{M+1} = \bar{\omega} + \delta$ and $\bar{\omega} = (\omega_M + \omega_{M+1})/2$.

This zero lies between ω_M and ω_{M+1} if A, B have the same sign. On the other hand, if $A = -B$, and the remainder function is R, then zeros occur at

$$\omega = \bar{\omega} \pm \delta\sqrt{1 - 2A/R\delta}. \tag{7.20}$$

When $2A/R\delta > 1$ (small remainder), we have the situation in Figure 6.31(b) and there are no zeros between the two poles; but if $2A/R\delta < 1$, we have a double zero, as shown in Figure 6.31(d). Note, however, that in Equations 7.19 and 7.20 the imaginary part of the zero is the same as that of the pole; therefore, for small damping at least, the poles and zeros are all above the real frequency axis. This means we can use simple pole and zero counting to estimate the phase, as in Equation 7.14, for both input and transfer functions for lightly damped structures (small modal overlap).

7.9 PHASE IN TWO-DIMENSIONAL SYSTEMS

In light of this discussion, consider the flat, rectangular room in Figure 7.20, which has mode shapes given by Equations 7.3 and 7.4. Two-dimensional systems such as this room are of much greater interest as models for machine structures, because such structures tend to be two dimensional. The model expansion of the transfer acoustic impedance in this case is

$$\frac{p}{U_s} \propto \sum_M \frac{\psi_M(\mathbf{x}_s)\psi_M(\mathbf{x}_0)}{\omega^2 - \omega_M^2} \xrightarrow[\left[\substack{\mathbf{x}_s = (0,0) \\ \mathbf{x}_0 = (L_1, L_2)}\right]]{} \sum_M \frac{1}{\omega^2 - \omega_M^2} \xrightarrow[\left[\substack{\mathbf{x}_s = (0,0) \\ \mathbf{x}_s = (0,0)}\right]]{} \sum_M \frac{(-1)^{m_1+m_2}}{\omega^2 - \omega_M^2}. \tag{7.21}$$

Again, we show forms for the input impedance at the origin and a transfer impedance from one corner of the room to the other.

When the source and receiver are at the origin, the terms in the expansion all have the same sign, so there is a zero between every adjacent pair of poles; poles and zeros alternate. We get the expected form for an input impedance, and there is no phase buildup. For the transfer function, if the source point is at the origin, the first modal factor becomes 1; if the observation point is at the opposite corner, L_1, L_2, then the second modal factor becomes $(-1)^{m_1+m_2}$.

The question is whether this factor changes sign as we go from one resonance frequency to the next. In the one-dimensional case, it was evident that it did; but in the two-dimensional case, adjacent succeeding resonances are determined by the modal lattice, as shown in Figure 7.27. The constant-frequency locus in this figure is a circle, and as the frequency increases and the circle increases its radius its intersection with the lattice points determines the sequence of resonance frequencies. These frequencies can jump from one mode number pair to another erratically. If we assume that $m_1 + m_2$ has a 50–50 chance of changing its parity (i.e., its evenness or oddness from one resonance frequency to the next), then half the time we would expect a zero to occur between two adjacent resonances and half the time we would not expect one to occur. With this assumption, we estimate the expected number of zeros to be just one half the number of poles, so the phase trend is

$$\bar{\phi} = -(N_p - N_z)\pi = -\tfrac{1}{2}\pi N_p. \tag{7.22}$$

The number of poles is the number of resonances up to some frequency. This number can be evaluated based on our discussions in Chapter 3 on modal density and mode count.

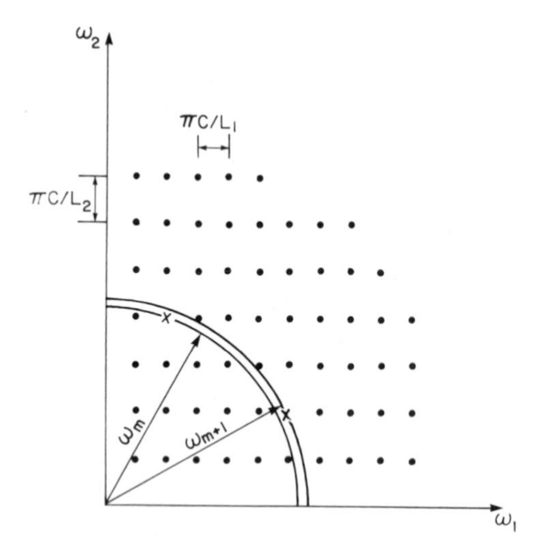

Figure 7.27 Wave number lattice for a two-dimensional room showing how the sequence of resonance frequencies is generated and how the modal index sum $m_1 + m_2$ tends to have a random parity.

Let us consider Equation 7.22 for a two-dimensional plate, where the mode count and the corresponding expected phase up to frequency ω are

$$\frac{1}{2}N_p\pi = \frac{\sqrt{3}\pi Af}{2hc_1} = \phi, \tag{7.23}$$

$$N_p = \frac{\omega A}{4\pi\kappa c_1}. \tag{7.24}$$

We apply this formula to the unwrapped phase for the engine structure used to obtain the data in Figures 7.5 and 7.6. At 1000 Hz, the phase shift is about 1500°, which gives an area-to-thickness ratio of

$$\frac{A}{h} = \frac{c_1\phi^\circ}{90\sqrt{3f}} = 167 \text{ ft}. \tag{7.25}$$

If we assume an average wall thickness of 1/2 in, then the effective plate area is 6.7 ft^2, which is reasonable for this engine structure. We note that the expected phase from Equation 7.24 should increase linearly with frequency; the unwrapped phase in Figure 7.6 does show such a trend.

Table 7.1 summarizes the estimation of transfer functions. The transfer function is represented by its magnitude and phase. For waveform reconstruction, where we do not need a detailed magnitude of the transfer function, we can use statistical energy analysis, which gives the transfer function magnitude in terms of

Table 7.1 Summary: Transfer Function from x_s to x_0

$$Y(\omega) = \text{const.} \sum_M \frac{\psi_M(x_s)\psi_M(x_0)}{\omega^2 - \omega_M^2}$$

$$= |Y(\omega)| e^{j\phi(\omega)}.$$

For waveform reconstruction, we may estimate magnitude by SEA:

$$|Y(\omega)|^2 = \frac{|V_0|^2}{|F_s|^2} = \frac{\pi}{2} \frac{n(\omega)}{M^2 \omega\eta} = \frac{G}{\omega\eta M}.$$

Phase up to frequency ω is determined by

$$\phi = -\pi(N_{\text{poles}} - N_{\text{zeros}}).$$

N_{poles} determined by SEA.
N_{zeros} determined by changes in sign of $\psi_M(x_s)\psi_M(x_0)$.
If probability of sign change is 1/2, then

$$N_{\text{zeros}} = \tfrac{1}{2} N_{\text{poles}} \quad \text{and} \quad \phi(\omega) = -\tfrac{1}{2}\pi N_{\text{poles}}.$$

the input drive mobility and the structural damping. We can estimate the phase up to some frequency by counting the difference between the number of poles and the number of zeros. The number of poles is determined by the mode count, and the number of zeros is determined by the changes in sign of the mode shape product at the source and observation locations. When there is some separation between source and receiver, we enter the statistical phase region in which the number of zeros is one half the number of poles, and the phase estimate is $-\pi/2$ times the number of poles.

We note that the accuracy of Equation 7.22 or 7.17 depends on the absence of double zeros and, therefore, on a small remainder function. In acoustical terms, this means we have either a small "direct field," which is usually the case for structures, or we are not affected by "rigid body" modes. Both of these factors may produce sizable remainder functions.

7.10 AN EXPERIMENTAL STUDY OF PHASE SHIFT

We have already seen that in one-dimensional systems the phase trend is equal to the propagation phase, and phase data taken on an engine structure showed a phase trend significantly greater than the expected propagation phase. To study this problem in more detail, we performed the experiment in Figure 7.28. A plate approximately 30 cm × 40 cm and about 0.1 mm thick was excited by a small shaker. The plate had strips of damping material applied to its surface and the

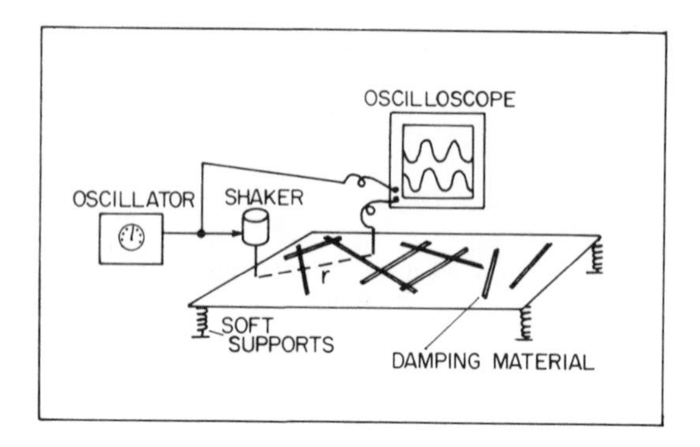

Figure 7.28 Measurement of phase delay versus range and frequency for a thin damped plate. Excitation is by a shaker, and the sensor is a fiber-optic pickup.

vibration was picked up at a distance *r* from the shaker by a noncontacting fiber-optic sensor. The relative phase of the excitation and the vibratory response of the plate were measured on an oscilloscope.

For each separation *r*, we start at a very low frequency where the source and response are in phase and the whole plate vibrates as a rigid body. As the frequency is gradually increased, the phase at the receiving point falls behind that of the excitation point, and we keep track of the total phase lag accumulated up to some maximum frequency. This process is then repeated for various separations *r*, and the accumulated phase as a function of range is graphed in Figure 7.29.

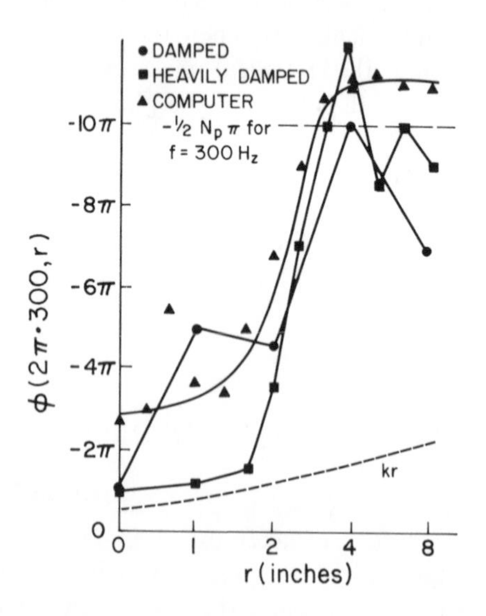

Figure 7.29 Result of phase measurement of thin plate showing phase delay greatly in excess of propagation phase.

There is one mode in this plate about every 15 Hz, so up to 300 Hz approximately 20 modes are encountered. Thus, the computed statistical or reverberant phase is 10π, which is shown as a dashed line in Figure 7.29. The data for accumulated phase for the plate when lightly damped and heavily damped are graphed in this figure. It is clear that the phase curves substantially exceed the propagation phase kr and that this transition occurs for $kr \sim 3$. A computer simulation is also included in the graph, in which the phase trend is calculated by following changes in the sign of adjacent numerators in the modal expansion of response. The agreement between the computer simulation and the data is encouraging.

7.11 PHASE VARIABILITY OF STRUCTURAL TRANSFER FUNCTIONS

In Chapter 6, we showed data on the mean and standard deviation of the magnitude of engine transfer functions. In Figure 7.30, the unwrapped phase data for this same population of structures is graphed. The average phase trend in this data is generally like that in Figure 7.6, but the degree of phase variability is remarkable. At 10 kHz, the standard deviation of unwrapped phase is nearly 20 rad (it only takes π to reverse the sign of a signal). In light of our earlier discussion of the sensitivity of waveform recovery to phase, this variability is of concern.

To understand how we should model this randomness in the phase, we again consider the pole–zero diagram in Figure 7.25. As the test frequency ω moves along the real axis, the phase of H changes by $+\pi$ when a pole is passed and by $-\pi$ when a zero is passed. If the damping is small and the poles and zeros are very close to the real axis, these phase changes will be very abrupt, much like steps in a two-dimensional phase-versus-frequency plane. It seems appropriate, therefore, to model this process as a two-dimensional random walk, as in Figure 7.31.

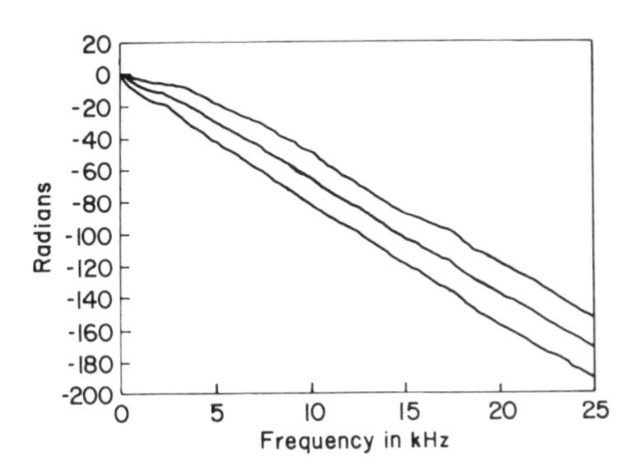

Figure 7.30 Mean and mean $\pm\ \sigma$ for the phase of transfer function of population of diesel engines.

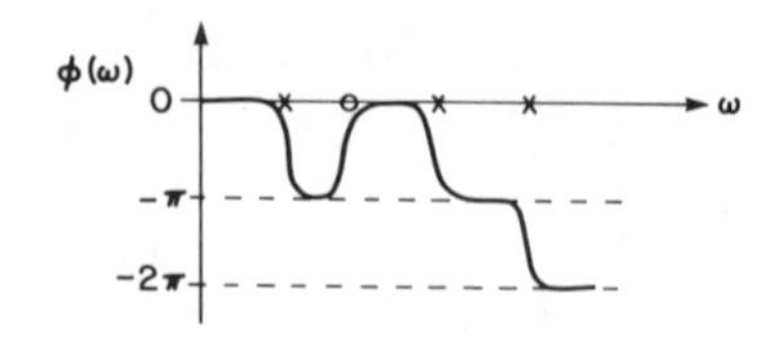

Figure 7.31 The phase transitions due to poles and zeros can be thought of as a random walk in the phase–frequency plane.

Let us first consider the case of equal steps along the frequency axis, although we realize that the poles and zeros are not equally separated in frequency. Let the probability of a pole at each step be P, that of a zero be R, and the probability of neither be Q; hence, $P + Q + R = 1$. For each step, therefore, the average change in phase is

$$m_\xi = \pi(P - R) \tag{7.26}$$

and the variance is

$$\sigma_\xi^2 = \pi^2(R + P) - m_\xi^2 = \pi^2\{R + P - (P - R)^2\}. \tag{7.27}$$

Therefore, after N independent steps, the average phase is $m_\phi = Nm_\xi$ and $\sigma_\phi^2 = N\sigma_\xi^2$, which leads to a ratio of the variance to mean given by

$$\frac{\sigma_\phi^2}{m_\phi} = \left\{\frac{R + P}{P - R} - (P - R)\right\}. \tag{7.28}$$

If we are in the reverberant field of the structure, then the probability that the sign of adjacent residues will be different is 0.5, as discussed in section 7.9. The probability of a zero is, therefore, half the probability of a pole, or $P = 2R$. If we substitute this relation into Equation 7.28, we get

$$\frac{\sigma_\phi^2}{m_\phi} = \pi(3 - R). \tag{7.29}$$

Since the largest possible value of R is $1/3$ and R will generally be much smaller than this, we obtain an estimate of the variance of the phase to be

$$\frac{\sigma_\phi^2}{m_\phi} \doteq 3\pi. \tag{7.30}$$

Experimental data on the phase variance of the transfer functions of these diesel engines extracted from Figure 7.30 is shown in Figure 7.32. We see that

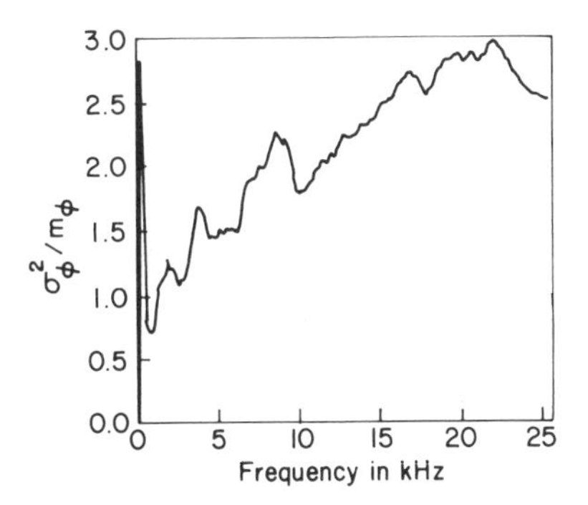

Figure 7.32 Measured ratio of variance to mean in phase indicates that the theoretical limit of 3 is nearly achieved, but lower values are also obtained.

Equation 7.30 is a good prediction of the upper bound of the variance, but much of the data has a smaller variance than this. If the frequency resolution of the spectrum analyzer is such that it cannot separate poles and adjacent zeros, then a zero in the same frequency band as a pole will "cancel" that pole insofar as its phase response is concerned. We can then model the phase as a random walk with a phase step of π for each uncanceled pole. If the probability of such a pole at each step is P, then the mean value for a step is

$$m_\xi = P\pi \tag{7.31}$$

and the variance for one step is

$$\begin{aligned} \sigma_\xi^2 &= P\pi^2 - P^2\pi^2 \\ &= P\pi^2(1-P). \end{aligned} \tag{7.32}$$

Then, for N steps, $m_\phi = NP\pi$ and $\sigma_\phi^2 = NP\pi^2(1-P)$, so

$$\frac{\sigma_\phi^2}{m_\phi} = \pi(1-P). \tag{7.33}$$

For this model, the ratio of variance to mean is a direct measure of the probability of an uncanceled pole, and this ratio has a maximum value of π when $P \to 0$. It is clear that the variance in phase is an excellent indicator of the appropriate model. The results show that the analyzer is combining poles and zeros in its resolution bandwidth for much of the data. When the poles and zeros are separated enough for the analyzer to resolve them, the ratio of phase variance to mean increases substantially.

7.12 NON-MINIMUM-PHASE SYSTEMS AND CEPSTRAL ANALYSIS

We have indicated that lightly damped structures should be minimum phase. As we shall see, this makes it possible to relate the phase and log magnitude by the Hilbert transform, which simplifies problems of transfer function determination. The development of these ideas is greatly assisted by considering the complex cepstrum of the system response function H.

Consider the transfer function pole–zero pattern in Figure 7.33. This transfer function has its poles and zeros in the positive half of the ω plane and it also has some zeros in the lower half-plane. As discussed earlier, such a system is not minimum phase because it has more phase shift for a given frequency amplitude spectrum than another system would have if the zeros in the lower half-plane were reflected across the real ω axis into the upper half-plane.

When we compute the log transfer function to evaluate the complex cepstrum of the system, the poles of the transfer function become poles of the log, but so do the zeros, because $\log(0) \to -\infty$. From our previous discussion of the relationship between the location of the poles and the behavior of the time transform, we see that these poles in the lower half-plane make the cepstrum nonzero for negative time. The poles of $\log H$ are shown in Figure 7.34. The zeros are not shown, but they will occur whenever $H = 1$. The non-minimum-phase transfer function has a nonzero cepstrum for negative time and is shown in Figure 7.35.

Because the system output is the product of the system function $H(\omega)$ times the applied force transform, the output of the system may be non-minimum phase, either because the transfer function is non-minimum phase or because the source is non-minimum phase. Of course, it is possible that both the source and the transfer function might have non-minimum-phase components.

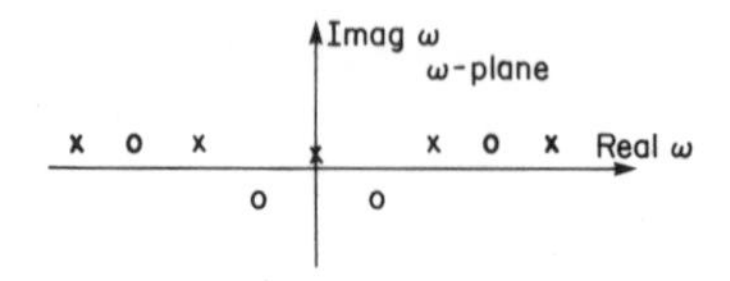

Figure 7.33 Pole–zero pattern of transfer function that is not minimum phase because of zeros in lower half-plane.

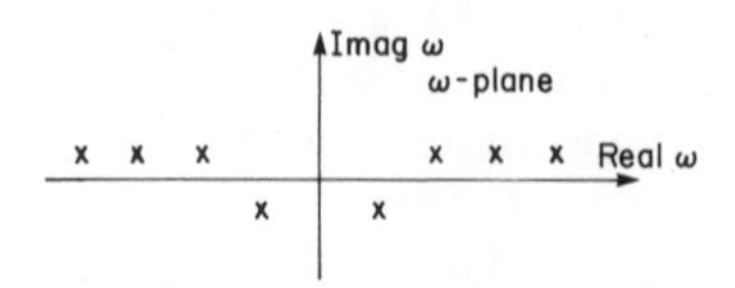

Figure 7.34 Pole pattern of log of transfer function. Zeros are not shown. Non-minimum-phase zeros become poles that generate nonzero cepstrum for $t < 0$.

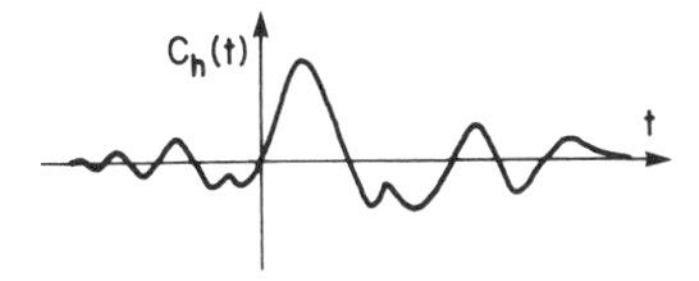

Figure 7.35 Complex cepstrum of a non-minimum-phase system that has poles in the negative half-plane leading to a nonzero cepstrum for $t < 0$.

Suppose we have a minimum-phase system so that the cepstrum vanishes for negative time and is nonzero for positive time. We see from Section A.13 that the complex cepstrum is composed of an even part, which is the real or power cepstrum, and a second part, which is the phase cepstrum. These two components are even and odd in time, respectively, and that is true whether the system is minimum phase or not. A minimum-phase transfer function, therefore, satisfies the Hilbert transform relation between real and imaginary parts of the log transfer function, or from Equation A.62,

$$\begin{aligned} \phi_{\min}(\omega) &= \frac{1}{\pi}\int_0^\infty dt \sin \omega t \int_0^\infty d\omega' \cos \omega' t \ln|H|, \\ \ln|H| &= \frac{1}{\pi}\int_0^\infty dt \cos \omega t \int_0^\infty d\omega' \sin \omega' t \, \phi_{\min}(\omega'), \end{aligned} \tag{7.34}$$

where $\phi_{\min}$ is the minimum-phase part of the transfer function phase and is determined by the log magnitude of the transfer function. If we have measured the power spectrum of the log magnitude of the transfer function, then we can determine $\phi_{\min}$ by Equation 7.34. This does not tell us if there are other components of the phase due to pure time delays or all-pass phase components.

We can always represent a transfer function as a product of a minimum-phase transfer function and an all-pass transfer function:

$$H(\omega) = H_{\min \phi} \times H_{\text{all-pass}}. \tag{7.35}$$

Suppose we have the transfer function expressed in terms of its pole–zero pattern, as in the upper part of Figure 7.36, and this transfer function has some zeros in the lower half-plane making it non-minimum phase. We can express this as the transfer function shown in the middle of the figure, in which the zero has been reflected across the real frequency axis and, therefore, represents a minimum-phase transfer function times a transfer function that has poles and zeros occurring only in pairs reflected across the real frequency axis from each other and where the poles cancel out the zeros introduced into the minimum-phase transfer function. The transfer function represented by the pole–zero pattern in the lower part of the figure is an all-pass transfer function because the magnitude of this frequency response is always

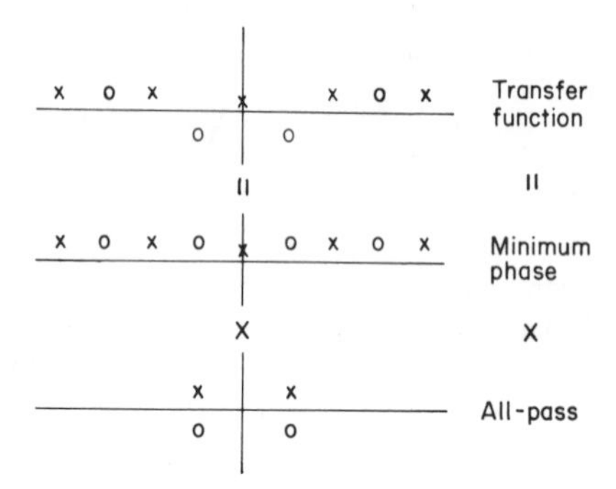

Figure 7.36 Decomposition of transfer function into all-pass and minimum-phase components.

unity. Whenever the test frequency goes between the pole–zero pair, it picks up a phase delay of 2π, however, so that $H_{\text{all-pass}}$ has a phase delay that adds to $\phi_{\min}$ to produce the total system phase shift.

The cepstrum of the all-pass transfer function is purely odd and creates the nonzero part of the complete cepstrum for $t < 0$. Thus, we can extract the all-pass portion of the system function by subtracting this part of the cepstrum and leaving only the minimum-phase part of the system function.

CHAPTER 8

Advanced Topics in Diagnostics

8.1 INTRODUCTION

Our discussions in the last two chapters have dealt with procedures that are either in regular use in diagnostics or represent procedures that could be implemented fairly easily. This chapter discusses other aspects of diagnostic procedures likely to be important in future machine monitoring systems.

To set the context for the discussion, consider the elements of a diagnostic system shown in Figure 8.1. We have discussed many of the issues of the processing of diagnostic signatures in order to develop signatures that are simple enough to reveal faults or operating parameters. These signatures are then examined in a comparator, which classifies the signal into a "faulty" or "safe" condition. We shall discuss this act of classification later in this chapter under the topic of decision analysis.

In most machines, many events occur that produce vibratory signals that may overlap in both frequency spectrum and in time. Therefore, information about the spatial location of the faults needs to be used to separate the sources. To provide this separation, we need to use multiple sensors. We shall discuss procedures by which multiple sensors can be used to separate simultaneous sources in the following section.

8.2 WAVEFORM RECOVERY FOR SOURCES THAT OPERATE AT THE SAME TIME

The structure involved in this study is a two-dimensional plate with two force gauges and four accelerometers as shown in Figure 8.2. The plate rests on a layer of foam

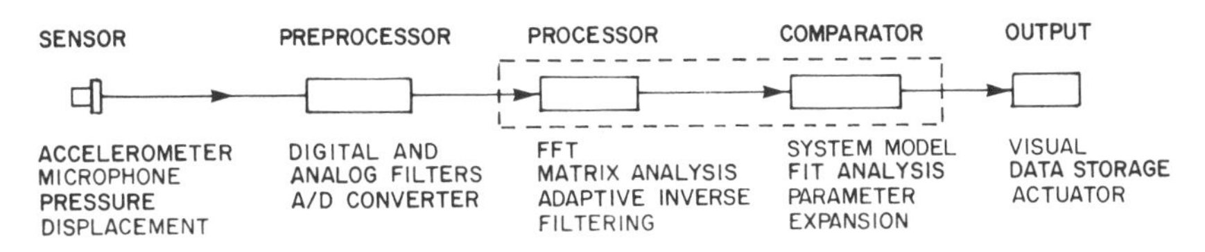

Figure 8.1 Schematic diagram of the components of a diagnostic system.

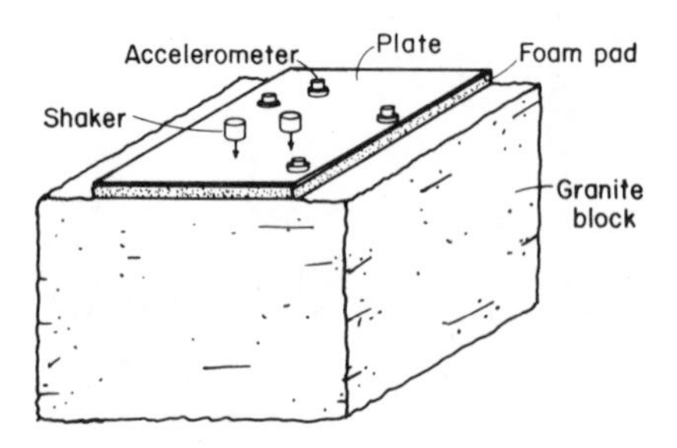

Figure 8.2 Diagram of thin plate sitting on foam pad used for experiments on simultaneous source recovery.

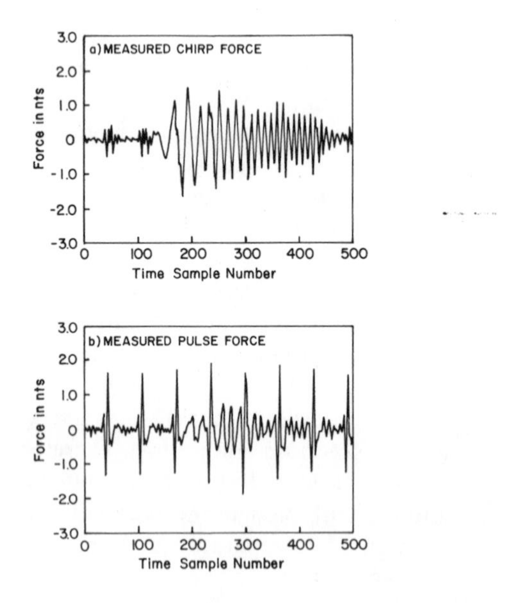

Figure 8.3 Directly measured forces applied to plate at the two shaker locations. Note there is some transmission from each excitation waveform to the other shaker.

that, in turn, sits on a granite block. The locations of the force gauges and accelerometers are shown in the figure. Signals with different waveforms are applied to the two force gauges. One signal is a very fast sine sweep, a "chirp," and the second is a sequence of pulses. These excitations overlap in time.

Time records of the force waveforms measured by the force gauges are shown in Figure 8.3. Small amounts of one excitation appear at the other shaker because the bending wave generated by one shaker reflects off of the other shaker, which produces a force at that shaker. Therefore, we regard Figure 8.3 as the actual force–time waveform at the source locations that is to be recovered from the vibration signals.

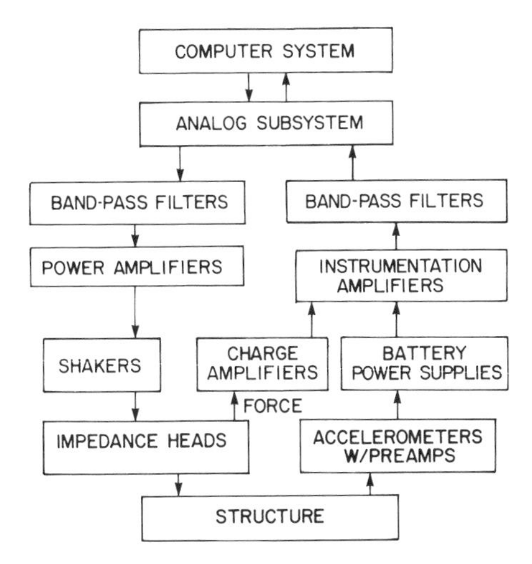

Figure 8.4 Computer-based system for generating excitation waveform and processing of signals received by transducers.

A block diagram of the signal processing setup is shown in Figure 8.4. The computer generates the analog signals through a digital-to-analog (D/A) converter and the structure is excited by the shakers that drive it through the impedance heads. The acceleration and force signals are picked up and passed back to the computer through appropriate signal-conditioning amplifiers and the analog-to-digital (A/D) converter. Thus the computer generates the source waveforms and also does the signal processing on the data that is recorded.

This form of processing has an advantage that is not obvious. Since the discrete Fourier transform (DFT) is a periodic function, then, if the excitation is periodic with the same period, there is no loss of coherence, because all frequency components have a basic period equal to that of the DFT. In systems where the response time is windowed and the structure has reverberation, it is possible for signals that were generated during a time not included in the excitation time window to appear in the windowed time period, causing a loss in coherence. When the entire process is based on the DFT cycle, coherence is not lost because of time leakage.

The analysis considers a general structure that has n inputs and m outputs and an impulse response matrix that couples the inputs to outputs as shown in Table 8.1. In the example we are considering, there are two inputs and four outputs, so $n = 2$ and $m = 4$; thus there are eight impulse responses. If we express the excitation and response by the Z-transform, we get the matrix equation in the lower part of Table 8.1 and an $m \times n$ matrix represents the system transfer function.

One of the directly measured transfer functions in the 4×2 matrix is shown in Figure 8.5. The log magnitude function has a character similar to other transfer functions we have looked at except that it has fewer modes and, therefore, less fluctuation over the frequency range of measurement. The phase curve shown in the

Table 8.1 Representation of the Input–Output Relations in a Multi-input–Multi-output System in the Time and Z-Transform Domains

$$
\overset{n \text{ inputs}}{\begin{bmatrix} x_1(t) \\ x_2(t) \\ \vdots \\ x_j(t) \\ \vdots \\ x_n(t) \end{bmatrix}} \Rightarrow \begin{array}{c} \text{Linear} \\ \text{time-invariant} \\ \text{structure} \\ \boxed{h_{ij}(t)} \end{array} \Rightarrow \overset{m \text{ outputs}}{\begin{bmatrix} y_1(t) \\ y_2(t) \\ \vdots \\ y_i(t) \\ \vdots \\ y_m(t) \end{bmatrix}}
$$

(a) Multi-input–multi-output system in the time domain. If $m > n$, the system is overdetermined.

$$
\begin{bmatrix} H_{11}(z) & \cdots & H_{1n}(z) \\ \vdots & & \vdots \\ H_{i1}(z) & \ddots & \\ \vdots & & \\ H_{m1}(z) & \cdots & H_{mn}(z) \end{bmatrix} \begin{bmatrix} X_1(z) \\ \vdots \\ X_n(z) \end{bmatrix} = \begin{bmatrix} Y_1(z) \\ \vdots \\ Y_m(z) \end{bmatrix}
$$

(b) Multi-input–multi-output system in the Z-transform.

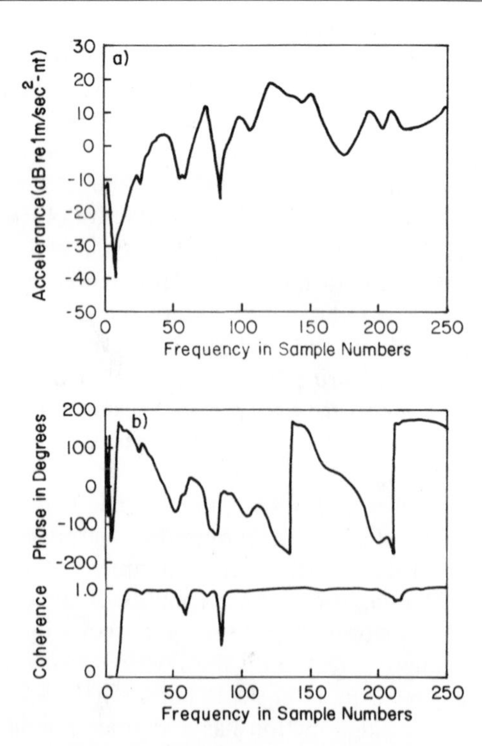

Figure 8.5 Magnitude and phase of the directly measured (1, 1) element of the transfer function matrix. Phase unwrapping is not necessary in this analysis.

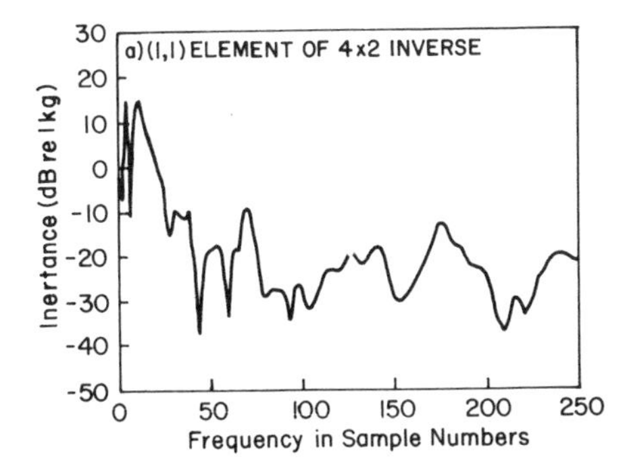

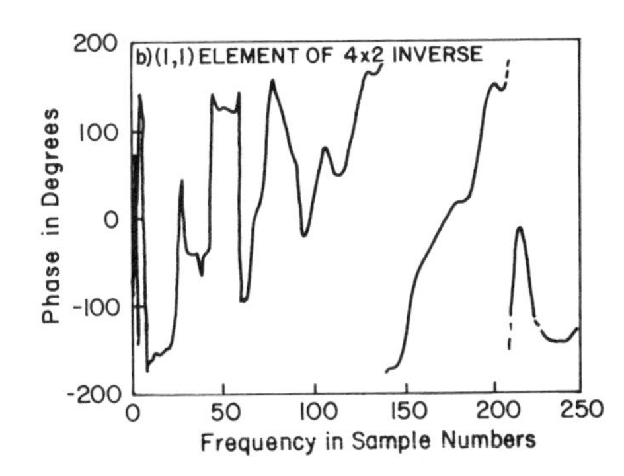

Figure 8.6 Phase and magnitude of the (1, 1) element of the matrix inverse of the transfer function matrix. This is not the simple inverse of the (1, 1) element.

lower part of the figure has not been unwrapped, but phase unwrapping is not necessary if we are using the phase directly in an inverse filter. Phase unwrapping is only necessary when we calculate the cepstrum, for which phase as a continuous function of frequency is required.

The matrix $H_{ij}(z)$ is inverted by an algebraic technique known as singular value decomposition (SVD). The inverse filters, therefore, are functions of frequency, and individual elements of the inverse filter can be inverted to construct impulse responses. As an example, the magnitude and phase of the frequency spectrum of the (1, 1) component of the inverse matrix are shown in Figure 8.6.

If one has a single-input–single-output system, then the inverse filter has a log magnitude that is simply the negative of the transfer function. We note by comparing Figures 8.5 and 8.6 that the inverse of the (1, 1) element of the system does not follow such a simple rule. There is a general similarity in the phase and magnitude of the inverse element with the negative of the forward transfer function element, but the inversion is found by matrix inversion rather than by simply taking the reciprocal of the individual matrix element. We notice that the phase of the inverse component does increase with frequency as we anticipated in our earlier discussion.

A sample impulse response of the 4×2 inverse is shown in Figure 8.7. This impulse response has nonzero response values for negative time in accordance with our discussion of the required phase properties of an inverse filter in Appendix A.

The vibration that results at an accelerometer due to simultaneous action of the two sources shown in Figure 8.2 is shown in Figure 8.8. It is clear that the two sources produce reverberation and a smearing of the individual sources as well as mixing of the two sources together in a nearly unrecognizable form.

It is theoretically possible to reconstruct the two source waveforms from a pair of accelerometers by straightforwardly generating a 2×2 transfer function matrix that has a 2×2 inverse. If that is done and the inverse filter matrix is applied to the data from the two accelerometers, the reconstructed waveforms of the source are as

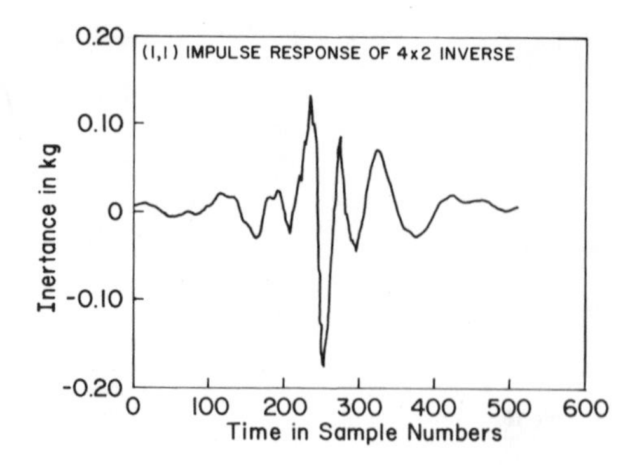

Figure 8.7 The impulse response of the inverse (1, 1) element is noncausal, which we expect from the properties of a filter that has a negative phase delay characteristic.

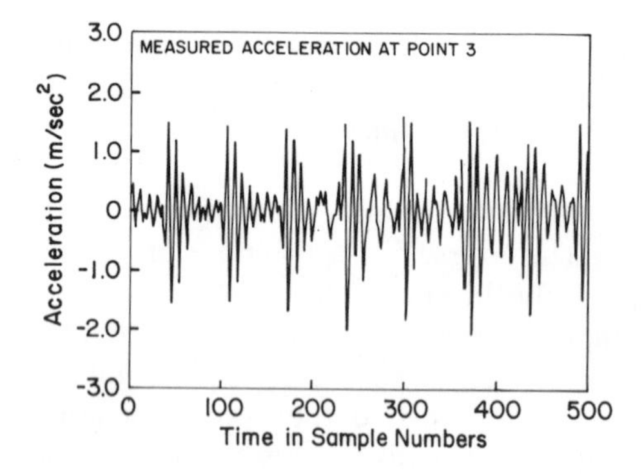

Figure 8.8 A typical accelerometer output in this experiment. The responses due to the two shakers are mixed in such a way that their separate visual identification is not possible.

shown in Figure 8.9. The waveform reconstruction is somewhat successful but not very satisfactory. One can see a kind of ringing in the reconstructed waveforms that is not in the original. The ringing frequency is at about 100 Hz, or at sample 64 in the frequency space.

Examining the eigenvalues of the 2×2 inverse matrix s_1 and s_2 plotted in Figure 8.10, we see that the second eigenvalue in this frequency range is getting close to the error in the reconstruction of the transfer function. This error is related to the coherence measured during the direct measurement of the transfer functions. Normally if the error exceeds one of the eigenvalues s_1 or s_2, the processing program causes that eigenvalue to be ignored in the construction of the inverse filter. The error in this frequency range is artificially increased to exceed s_2, causing that eigenvalue to be ignored in this frequency range. With this editing of the error, the 2×2 reconstructed waveforms are as shown in Figure 8.11. We see that now the

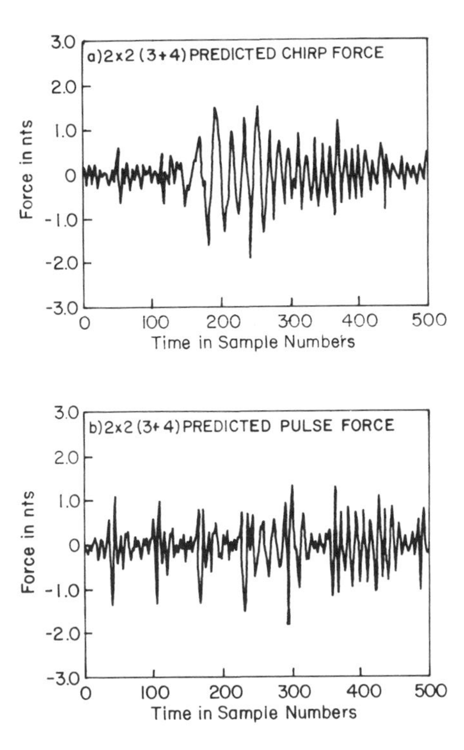

Figure 8.9 A pair of recovered excitation waveforms using a 2 × 2 matrix and its inverse. The extra ringing in the inferred output arises from a false peak in the inverse filter.

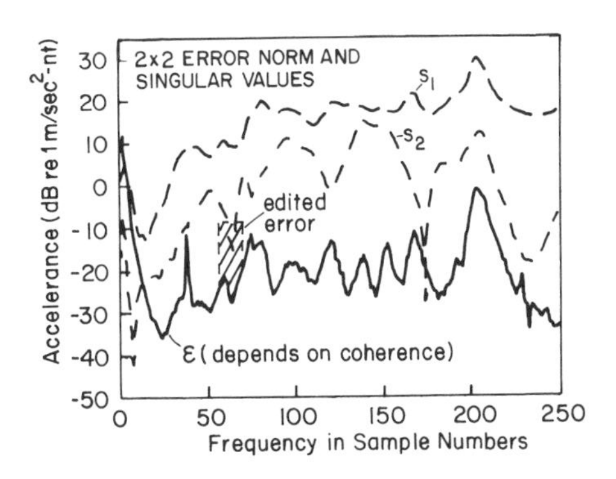

Figure 8.10 Comparison of the eigenvalues of the inverse filter with error in data determined from coherence. The proximity of the s_2 value to the error of sample 64 produces the 100-Hz ringing.

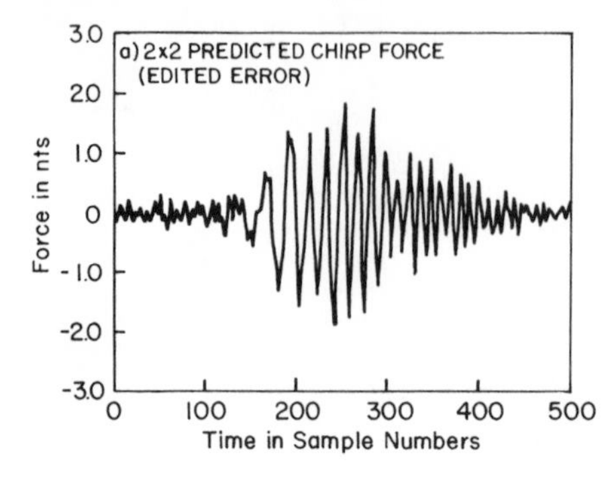

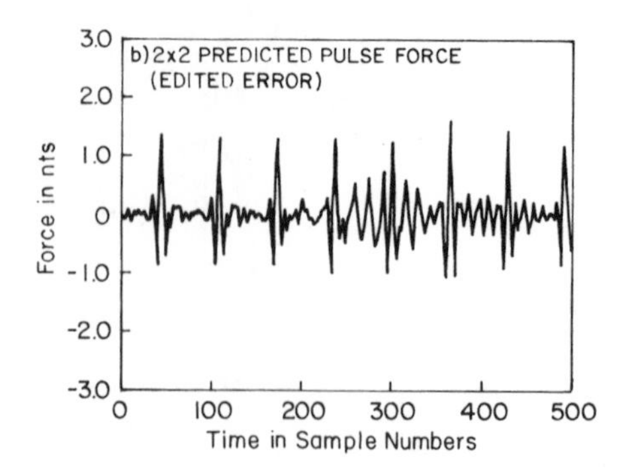

Figure 8.11 Recovered excitation waveforms when error is edited (changed) so that the eigenvalue s_2 is ignored in the frequency range near sample 64 (100 Hz).

ringing at 100 Hz has been eliminated, and the quality of the reconstructed waveform is quite good.

These results suggest how we might design diagnostic systems using waveform recovery when a number of sources operate simultaneously in a machine. Much remains to be done in achieving such a system however. For example, a knowledge of the matrix of a transfer function assumed to be available for the system and measured in the experiment just described may not be available. An adaptive system able to learn the transfer functions might need to be developed.

8.3 DIAGNOSTIC SYSTEM DESIGN AND APPLICATION

In Chapter 7, we saw that cepstral windowing of received signals allowed recovery of the combustion component of cylinder pressure, even when an incorrect transfer function is used. This is an example of a "robust" processor; that is, the system is able to ignore differences of phase and magnitude of the transfer function and the inverse filter by smoothing of both parts. Low-time windowing of the cepstrum induces short-range smoothing in the frequency domain that appears to be helpful.

The uncertainty with a robust processor is that in the process of reducing sensitivity to unwanted variations, the sensitivity to variations that are of interest may be reduced or eliminated. Therefore the alternative of an adaptive processor may be attractive. An adaptive processor is able to ignore the effects of the unknown or uncertain transfer function by separating the source and path portions of the received signal and examining each for the effects of a fault.

We also saw in Appendix A that cepstral windowing can be used to separate path from source effects. A prototype adaptive diagnostic system based on this principle and shown in Figure 8.12 has been developed by Cambridge Collaborative, Inc., of Cambridge, Massachusetts. This system, based on a concept of flexibility, is capable of dealing with a large variety of sensors and has a substantial

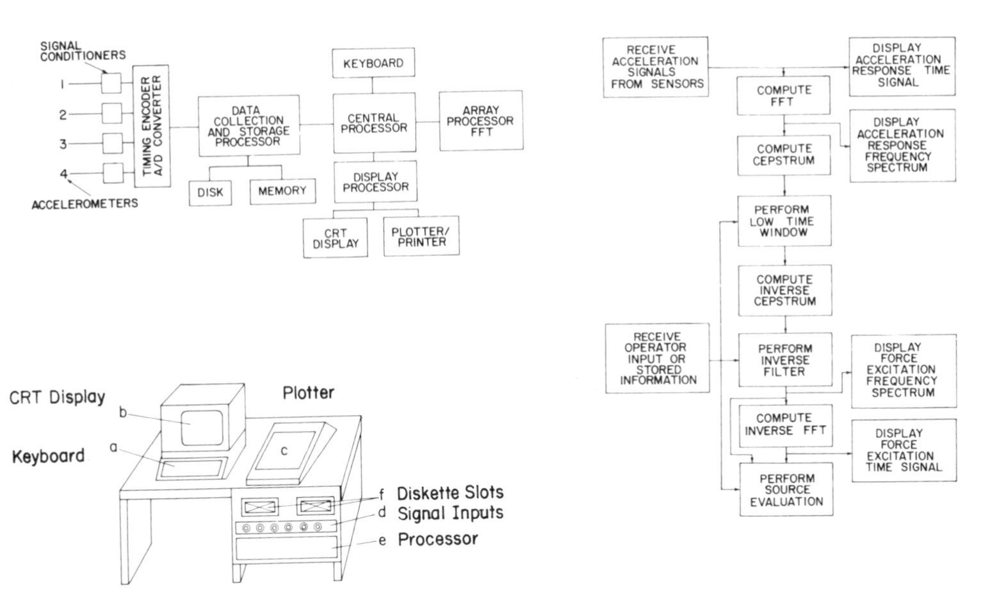

Figure 8.12 An advanced diagnostic system capable of signature analysis of transient forces in a machine that may overlap each other in time requires multiple sensors and rather sophisticated signal processing procedures.

library of special programs to do various kinds of signal processing. The specialization of the system to particular processes or product diagnostics results from the preparation of executive programs that select the appropriate combination of signal processing programs and also from a specific hardware design of the sensor front end of the system to bring in data to the processor that is appropriate to that process.

The principal considerations in this design are that the elaborate systems described here are fairly expensive to develop and that flexibility in a design allows the development costs to be distributed over a variety of applications. The number of potential systems that can be sold to diagnose any process may be too limited to cover the development costs unless the basic software library can service a variety of applications.

The diagnostic system in Figure 8.12 shows several sensors connected to the machine, a high-speed A/D converter, and a set of low-speed A/D converters for slowly varying signals that pass data directly to a storage processor and to disk memory. Sensor locations and mountings for diagnostics tend to be very closely specified. Current systems also use one transducer per machine element, typically at bearing caps or shaft journals. More advanced systems like that in Figure 8.12 will be required to have sensors placed at greater distances from the source and to use several sensors, but fewer than the number of sources diagnosed. This requires multidimensional signal processing for recovery of the source waveforms or spectra. These signals are then recovered and processed by a central processor using an array processor for very fast, nearly real-time analysis. The outputs can be in the form of CRT display or plotter/printer, and data can also be transmitted to a central data bank via a local network or a standard IEEE 448 or RS-232 interface.

One possible application for such a system is the testing of diesel engines. During production, component assemblies such as a head with valves and cam assemblies can be tested by rotating the cams and allowing the valves to impact the head. The spectral analysis procedures described in Chapter 6 can be applied to determine whether the valves are operating properly and are properly sealed.

Another application is at the end of the production line, at which point the engine is completed and tested. Recovery of cylinder pressure waveform will allow us to calculate the operation of injectors and other components as well as determine the horsepower per cylinder developed by the engine. The system in Figure 8.13 can also be used to test machines in service bay situations. Cummins Engine Co. has a system called Compucheck, which measures flywheel acceleration, among other quantities, for an unloaded engine. Flywheel acceleration can allow us to determine the impulse produced by each cylinder provided that there are not too many

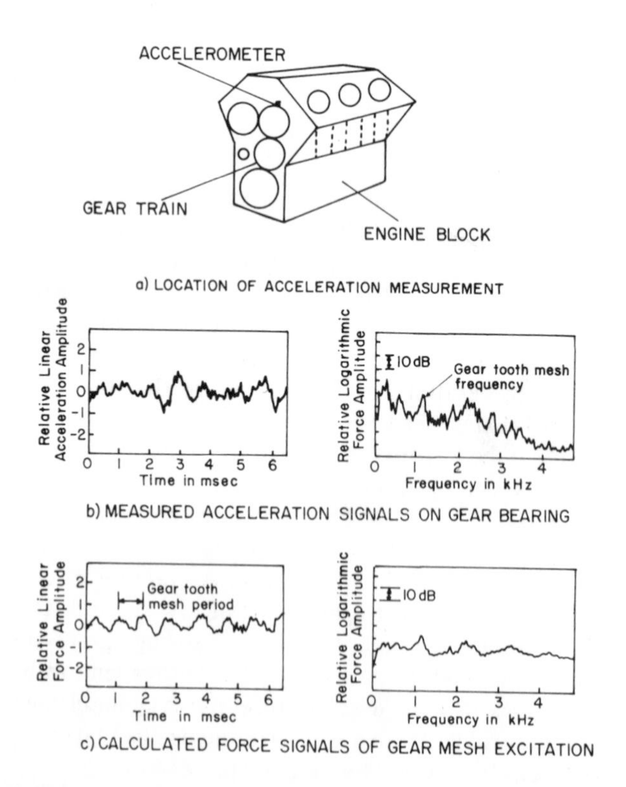

Figure 8.13 Application of diagnostic system to recover a gear noise signature from a diesel engine. In this case, transfer function data is not required a priori, but is extracted from the data, along with the source waveform.

cylinders. A 12-cylinder engine may fire too frequently to allow individual acceleration records to be determined, but a 6-cylinder engine produces fewer pulses per second and can be evaluated. Also, the flywheel acceleration technique does not allow the engine to be highly loaded. A system based on recovery of cylinder pressure from vibration data would not have such limitations, and it offers a possibility for a significant advance in service bay testing.

Another application is in-service monitoring and control of diesel engines. Some diesel engines in earth moving equipment are so large and have so many cylinders that even if one cylinder is inoperative because of a burned valve or some other malfunction, the operator may not sense the change in horsepower. The change in sound may not be noticed because the environment is so noisy. A system that can detect the individual operation of each cylinder by vibration analysis would be very valuable. Remember, however, that the in-service environment is quite hostile, and any system designed to operate there has to have the same degree of ruggedness that any other component or an engine must have. It is clearly possible to develop such sensor arrays and cabling, and such like, but it is not a trivial matter and will require a great deal of engineering effort.

Another application area is papermaking. In this case, as in other areas of manufacturing, there are two sets of problems. The first has to do with the quality of the production process itself. This includes the machinery, motors, bearings, flow-handling equipment, binding equipment, and so on. At present, many of the motors and rollers in a paper mill have vibration monitoring equipment. The high-level diagnostics system of Figure 8.12 could allow us to use fewer sensors and also give greater information about the nature of the developing machinery faults.

The second issue has to do with the quality of the product itself. Paper manufacturers suspect that problems in product quality, such as fiber density and water content, may be related to vibrations or pulsations in the pulp flow into the header, vibration of the moving screen, and other effects. In this case, the high-level diagnostic system would allow us to develop correlations between various parameters as well as to monitor the quality of the paper.

From these examples, it is clear that the diagnostic system has to deal with a variety of inputs other than sound or vibration. Some other important variables are flow rates, temperatures, optical transparency, radiographic absorption, chemical content, resistivity, forces, and torques. Each of these measurement fields has its own array of sensors, such as accelerometers, microphones, thermocouples, and optical sensors, and the associated signal processing and amplification equipment.

The flexibility in the design results from the programmability of the system in terms of gain, bandwidth, sampling rate, and other parameters of the input system. These parameters depend on the nature of the sensor, the time rates of data to be gathered, and the requirements for an executive program that is able to assemble the processing software into the proper packages to do the analyses of input data. A sample arrangement of processing programs for waveform recovery from acceleration signals is also shown in Figure 8.12. The various steps involved in cepstral processing of the data, including Fourier transforming, windowing, inverse filtering, and presentation of the information, are indicated in the figure. This arrangement of programs is selected by the executive program for each computational component needed for that particular processor. Other signal processing program components that might be in such a package include signal averaging, Hilbert transforms, and threshold detection.

Figure 8.13 shows application of the system of Figure 8.12. In this case, a single accelerometer is measuring the vibration near the bearing of a cam shaft, and the vibration waveform in the upper left corner is shown. The power spectrum of this vibration is shown in the upper right corner of the figure, and although we can see the gear tooth mesh frequency and some of its harmonics, many other noise sources are present.

The cepstrum of this spectrum (including phase) is evaluated, and the peaks in the cepstrum at the tooth meshing period and its multiples are extracted from the spectrum and, in turn, are used to compute a new spectrum of vibration, which is shown in the lower right corner of the figure. Now the peaks in the gear mesh are much clearer, and when the time waveform in the lower left corner is reconstructed, the periodicity and the fluctuations due to individual gear meshes are much clearer than they were originally. Thus, cepstral windowing can provide useful source data and a much clearer waveform of the gearing-mesh-induced vibration, uncontaminated by structural propagation effects.

8.4 SYSTEM IDENTIFICATION USING ORTHONORMAL FUNCTIONS

We shall discuss two procedures in this section. The first is a fairly old one, developed by Y. W. Lee and Norbert Wiener, that considers the functional representation of the system impulse response $h(t)$ shown in Figure 8.14. The impulse response is expanded in a sequence of orthonormal functions, the Laguerre functions $l_n(t)$. The first four Laguerre functions are sketched in the figure. The formal

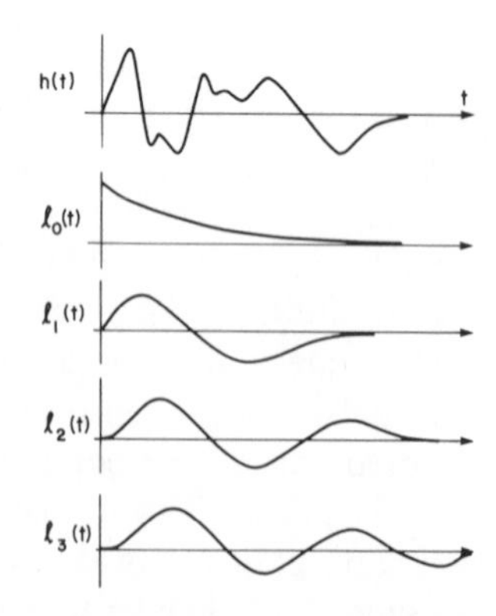

Figure 8.14 Decomposition of a system impulse response into a series of Laguerre functions. The decomposition leads directly to a system representation.

expansion is

$$h(t) = \sum_{n=0}^{\infty} A_n l_n(t). \tag{8.1}$$

Because of the orthogonality properties of the Laguerre functions, the coefficients in this expansion can be evaluated from

$$A_n = \int_0^{\infty} h(t) l_n(t)\, dt. \tag{8.2}$$

Functional representations for the Laguerre functions are:

$$\begin{aligned} l_0(t) &= \sqrt{2p}\, e^{-pt}, \\ l_1(t) &= \sqrt{2p}(2pt - 1)e^{-pt}, \\ &\vdots \\ l_m(t) &= L_m(2pt)e^{-pt}, \end{aligned} \tag{8.3}$$

where the quantities $L_m(2pt)$ are the standard Laguerre polynomials. The Fourier transform of the Laguerre functions is

$$L_m(\omega) = \frac{\sqrt{2p}(p - j\omega)^n}{2\pi(p + j\omega)^{n+1}}, \tag{8.4}$$

which have zeros at $p = j\omega$ and poles at $p = -j\omega$.

The interesting thing about the Laguerre functions is that they represent the impulse responses of a set of lattice networks (Figure 8.15). Thus, if the coefficients A_n in the expansion of the system impulse response can be evaluated by integration, the equivalent circuit for the system and its inverse filter can be constructed in a straightforward way.

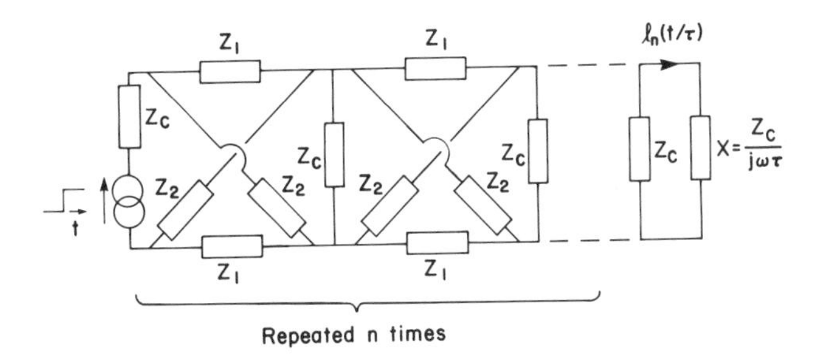

Figure 8.15 This lattice network generates a Laguerre function output and can be used to synthesize an arbitrary impulse response.

Laguerre functions have been widely used to represent the step or impulse response of systems with relatively few degrees of freedom. The system may have nonlinear characteristics that can be represented by parameter adjustment. For example, the step response of a human operator to a change in visual or auditory input has often been represented by Laguerre functions. A human operator is then modeled by a Laguerre network. The Laguerre functions, however, have deficiencies in the presentation of continuous systems, particularly if they are used to represent the cepstrum of a transfer function; that transfer function must be minimum phase, since Laguerre functions represent the signal only for $t > 0$.

Another problem of the Laguerre functions is that they represent systems in terms of a sequence of models, each of which has relatively few degrees of freedom. The lowest Laguerre network has one degree of freedom, the next has two, the next three, and so on. Since a structure over a frequency range of interest may have literally hundreds of modes, it is obvious that to get the proper order in the frequency spectrum or the complexity in the impulse response, it will require a large number of terms in the Laguerre expansion. It will also be difficult to correlate properties of the impulse response or derived network properties with the structure, since a force flow mobility analogy structure cannot have a multitude of ungrounded capacitor masses. Thus, the network cannot physically represent the mechanical system even though it may mimic the system in terms of impulse response.

The Hermite functions do not have the problems just cited with respect to the representation of transfer functions, since they exist for positive and negative values of the argument. The lowest-order Hermite function is the Gaussian, and the higher-order Hermite functions are the successively higher derivatives of the Gaussian, as shown in Figure 8.16. They also have the advantage that odd-order Hermite functions are odd functions, and even-order Hermite functions are even functions. Therefore, we can expand the even functions of time or frequency in even-order Hermite functions, and odd functions such as phase or phase cepstrum in the odd-order Hermite functions:

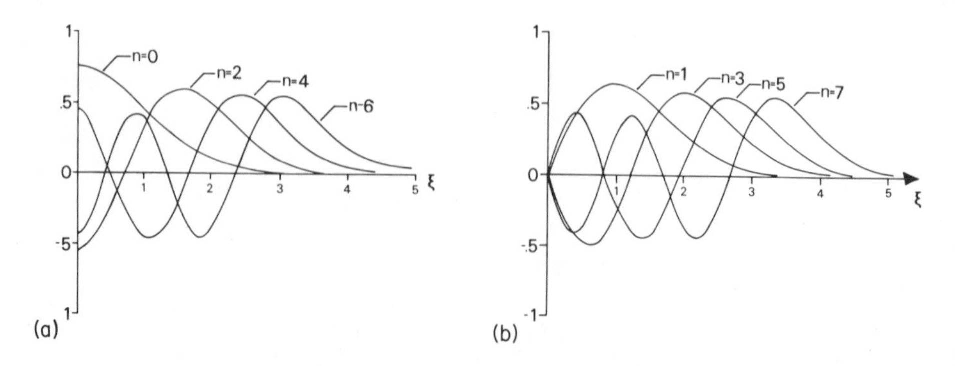

Figure 8.16 Even and odd Hermite functions $h_n(\xi)$ used for expansion of the log magnitude and phase, respectively, of transfer function. They are also used for expression of cepstra.

$$\hat{M}(\omega)e^{-(\omega/\omega_w)^2/2} = \sum_{m=0}^{\infty} C_{2m}h_{2m}\left(\frac{\omega}{\omega_w}\right),$$
$$\phi(\omega)e^{-(\omega/\omega_w)^2/2} = \sum_{m=0}^{\infty} C_{2m+1}h_{2m+1}\left(\frac{\omega}{\omega_w}\right). \quad (8.5)$$

The Hermite functions have another interesting property in that they are their own Fourier transform, and, therefore, the magnitude and phase cepstra are also expandable in these functions and the coefficients of expansion are closely related to those of the frequency functions:

$$\tilde{C}_{\hat{M}}(t) = \sum_{m=0}^{\infty} (-)^m C_{2m}h_{2m}(\omega_w),$$
$$\tilde{C}_{\phi}(t) = \sum_{m=0}^{\infty} (-)^m C_{2m+1}h_{2m+1}(\omega_w t). \quad (8.6)$$

The Hermite functions are products of the Hermite polynomial and the Gaussian weighting functions. The lowest-order Hermite polynomial H_0 is 1, and the next-order function is $H_1(x) = 2x$. If we think of these as the log magnitude and phase of a transfer function, then they represent the magnitude and phase of a transmission line. Thus, the lowest two Hermite polynomials represent a continuous system, in contrast to the lower-order Laguerre functions, which represent lumped systems with very few degrees of freedom.

From the definition of complex cepstrum,

$$\hat{Y} = \hat{M} + j\phi \underset{\text{F. T.}}{\longleftrightarrow} C_y(t) = C_{\hat{M}}(t) + C_{\phi}(t). \quad (8.7)$$

If we deal with a minimum-phase system, then the cepstrum vanishes for negative time, so the magnitude cepstrum must equal the phase cepstrum for $t > 0$, and the magnitude cepstrum must equal the negative of the phase cepstrum for $t < 0$. If we expand the log magnitude function in terms of the Hermite functions, then we can express the magnitude cepstrum in terms of Hermite functions also. If we wish, we can also express the phase cepstrum in terms of a Hermite expansion and determine the coefficients in the expansion of the phase cepstrum from the conditions in Equation 8.8:

$$\begin{aligned} \tilde{C}_{\hat{M}} &= \tilde{C}_{\phi} \qquad \text{for } t > 0, \\ &= -\tilde{C}_{\phi} \qquad \text{for } t < 0. \end{aligned} \quad (8.8)$$

On the other hand, if we have expanded the magnitude cepstrum in terms of Hermite functions and if we know the phase must be the Hilbert transform of the magnitude cepstrum, then we may wish to expand the phase in terms of the Hilbert transforms of the Hermite functions. These latter functions are called Beranek functions and are graphed in Figure 8.17. The expansion of the phase in terms of even-order Beranek function is

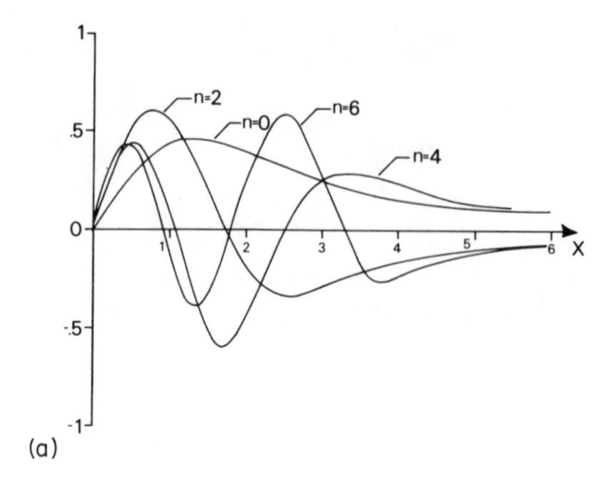

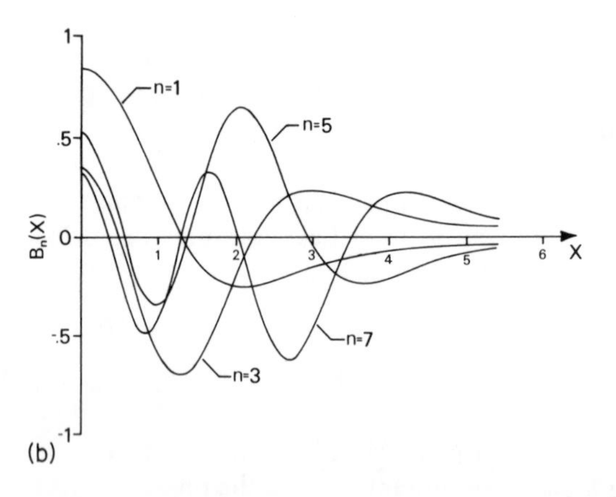

Figure 8.17 Even and odd Beranek functions $B_n(x)$, which are the Hilbert transforms of the Hermite functions. These are used for expansions of minimum-phase system functions.

$$\phi(\omega) = e^{+(\omega/\omega_w)^2/2} \sum_{m=0}^{\infty} C_{2m} B_{2m}\left(\frac{\omega}{\omega_w}\right). \tag{8.9}$$

On the other hand, if we have expanded the phase in terms of Hermite functions, then the log magnitude is expressed in terms of odd-order Beranek functions according to

$$\hat{M}(\omega) = e^{+(\omega/\omega_w)^2/2} \sum_{m=0}^{\infty} C_{2m+1} B_{2m+1}\left(\frac{\omega}{\omega_w}\right). \tag{8.10}$$

The Beranek functions are defined by the integrals

$$
\begin{aligned}
B_{2m}(\xi) &= \frac{2}{\pi}(-)^{m+1}\int_0^\infty d\tau \sin \xi\tau h_{2m}(\tau) \qquad \text{(odd in } \xi\text{)}, \\
B_{2m+1}(\xi) &= \frac{2}{\pi}(-)^{m+1}\int_0^\infty d\tau \cos \xi\tau h_{2m+1}(\tau) \qquad \text{(even in } \xi\text{)},
\end{aligned} \tag{8.11}
$$

which are the Hilbert transforms of the Hermite functions.

First, comparing Figures 8.16(b) and 8.17(a), we see that the odd Hermite functions have been converted to even Beranek functions, which is what we expect to happen with the Hilbert transform. Also notice that the Beranek function decays more slowly at large values of the argument than the Hermite function does because it is, in effect, the Fourier transform of a function that has a cusp at the origin. Comparing Figures 8.16(a) and 8.17(b), we see that even-order Hermite functions are turned into odd Beranek functions, and again the Beranek function decays slowly at large values of the argument because it is, in effect, the Fourier transform of a function that has a jump at the origin.

It is still too early to say how useful these expansions will be in system representation, but research is proceeding and so far the results look encouraging.

APPENDIX A

Some Relevant Mathematics

A.1 INTRODUCTION

In this appendix, we describe some mathematical concepts and methods that are important in the discussions of the main text. The principal topic of the appendix is Fourier analysis in its various forms of importance to signal processing. But we also describe the related Z-transforms, important to discrete processes, and the Hilbert transform, which relates the real and imaginary parts of some frequency response and time domain functions.

A.2 COMPLEX VARIABLES

Complex variables may be considered as number pairs on a two-dimensional plane, as shown in Figure A.1. The number pair (a, b) has a projection a on the horizontal (real) axis and a projection b on the vertical (imaginary) axis.

In this context, there is nothing less "real" about the imaginary axis than the real axis. In addition to this number pair, we define a rotation operation j that rotates the vector from the origin to the point (a, b) counterclockwise by 90° or $\pi/2$ rad to the point (a', b'). As a consequence, j operating on $(1, 0)$ produces $(0, 1)$, and a subsequent operation by j converts the number pair $(0, 1)$ to $(-1, 0)$. Therefore j operating twice (or j^2) is the same as multiplying by -1. For this reason, the quantity j^2 can be represented as -1, and the operator is j often written as $\sqrt{-1}$.

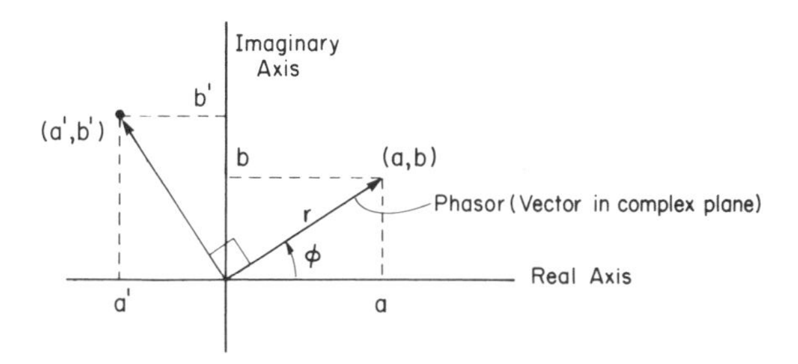

Figure A.1 Representation of a number pair as a vector or phasor in the complex plane.

But such a definition has very little meaning because there is a basic sign ambiguity in any square root, and, in any case, this representation gives j an appearance of unreality that it does not deserve.

Noting that the expansion of e^x is

$$e^x = 1 + x + \frac{x^2}{2!} + \cdots, \tag{A.1}$$

then, if we substitute $j\theta$ for x, we get

$$\begin{aligned} e^{j\theta} &= 1 + j\theta - \frac{\theta^2}{2} - j\frac{\theta^3}{3!} + \frac{\theta^4}{4!} + \cdots \\ &= \cos\theta + j\sin\theta. \end{aligned} \tag{A.2}$$

Thus $e^{j\theta}$ is the number pair $(\cos\theta, \sin\theta)$. Clearly then we can write the number pair (a, b) as

$$(a, b) = re^{j\phi} \tag{A.3}$$

where $r = \sqrt{a^2 + b^2}$ and $\phi = \tan^{-1}(b/a)$.

Thus operating on (a, b) with $e^{j\phi}$ produces

$$e^{j\theta}(a, b) = re^{j\phi}e^{j\theta} = re^{j(\phi+\theta)}. \tag{A.4}$$

The operator $e^{j\theta}$ is, therefore, a counterclockwise rotator by an angle θ.

Let us now write for the pair (a, b) the complex number z. The modulus (magnitude) of z is

$$|z| = r = \sqrt{a^2 + b^2}, \tag{A.5}$$

whereas its angle or argument is

$$\phi = \text{Arg}(z) = \tan^{-1}\left(\frac{b}{a}\right). \tag{A.6}$$

The complex conjugate of z, denoted z^*, is found by reflecting the phasor across the real axis:

$$z^* = (a, -b), \tag{A.7}$$

and, therefore, the product of z and its conjugate z^* is $r^2 = a^2 + b^2$.

Now suppose we want to represent the cosine wave in Figure A.2, which has an amplitude A and period T. The formula is

$$a(t) = A\cos(\omega t + \phi), \qquad \omega = \frac{2\pi}{T}. \tag{A.8}$$

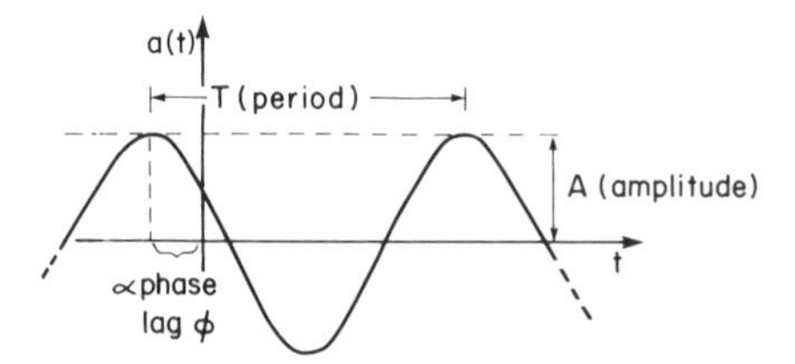

Figure A.2 Cosine wave of amplitude A and period T to be represented by a rotating phasor in the complex plane.

We now introduce the operators $\text{Re}\{z\}$, which sets the imaginary part of the complex number equal to zero, and $\text{Im}\{z\}$, which sets the real part of the complex number equal to zero. Consequently,

$$\text{Re}\{z\} = (a, 0) = a, \qquad \text{Im}\{z\} = (0, b) = b. \tag{A.9}$$

We can then represent $a(t)$ by

$$a(t) = \text{Re}\{Ae^{j(\omega t + \phi)}\} = \text{Re}\{A_c e^{j\omega t}\}. \tag{A.10}$$

We have introduced the complex amplitude A_c, which contains the phase of the time waveform as an offset of the rotating vector $e^{j\omega t}$ from the real axis at $t = 0$, as seen in Figure A.3. The product $Ae^{j\phi}$ is called the complex amplitude of $a(t)$ and is a "phasor."

Another way of writing Equation A.10 is

$$a(t) = \frac{A_c e^{j\omega t} + A_c^* e^{-j\omega t}}{2}; \tag{A.11}$$

the second term corresponds to a phasor rotating in the clockwise direction. We can think of this as the introduction of a negative frequency in order to produce a real

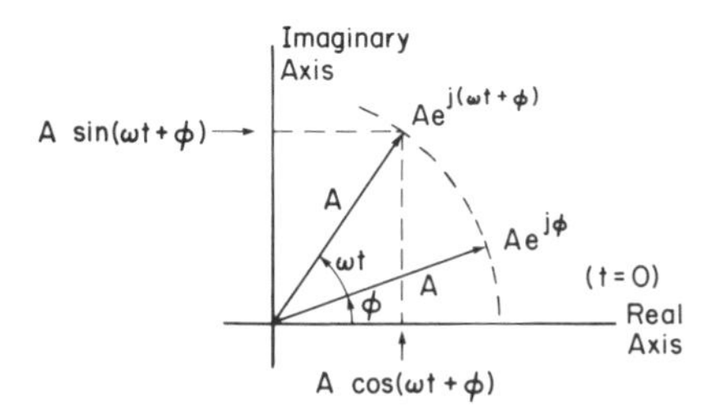

Figure A.3 Rotating phasor representation of cosine wave with phase shift ϕ.

signal $a(t)$. This requires that the amplitude of the negative frequency be the complex conjugate of the amplitude of the positive frequency term.

We can represent a time signal using positive frequencies only, but that requires that we allow the signal to be complex. Such complex time signals consisting of positive frequencies only are called "analytic." We shall have more to say about this in the discussion of Hilbert transforms.

Referring to Figure A.2, we clearly see that the average of $a(t)$ over any cycle or integral number of periods must vanish:

$$\langle a(t) \rangle_t = \frac{1}{T} \int_0^T a(t)\, dt = 0. \tag{A.12}$$

If we represent the sinusoid by $z = A_c e^{j\omega t}$, then by the same rule the average of z must also vanish. Suppose, however, we wish to average z^2. Then $z^2 = A_c^2 e^{2j\omega t}$, and the average of z^2 is 0 by the same rule. However, we notice that if z represents the sinusoid in Figure A.2, the average of the square of the sinusoid will not be zero but $A^2/2$, so the substitution of a complex quantity for a real one appears to be unable to provide a correct value of this average.

To explore this further, let us suppose that a load $l(t)$ and a velocity $v(t)$ are represented in terms of the sinusoids

$$v = A\cos(\omega t + \phi), \qquad l = B\cos(\omega t + \psi). \tag{A.13}$$

The time average of this product is

$$\begin{aligned} \langle lv \rangle_t &= AB\langle \cos(\omega t + \phi)\cos(\omega t + \psi) \rangle_t \\ &= \frac{AB}{2}\langle \cos(2\omega t + \phi + \psi) + \cos(\phi - \psi) \rangle_t \\ &= \frac{AB}{2}\cos(\phi - \psi). \end{aligned} \tag{A.14}$$

We need a rule that will allow us to use complex amplitudes to calculate time averages of products as in Equation A.14 . To accomplish this, we express the force and velocity as complex variables z_1 and z_2, where

$$z_1 = A^{j\omega t}, \qquad z_2 = Be^{j\omega t}, \tag{A.15}$$

where A and B are complex phasors given by $A = |A|e^{j\phi}$ and $B = |B|e^{j\psi}$. Then

$$v = \frac{(z_1 + z_1^*)}{2}, \qquad l = \frac{(z_2 + z_2^*)}{2}. \tag{A.16}$$

Since l and v are real quantities, we can use Equation 2.13 to compute the time average of their product:

$$\langle lv \rangle_t = \frac{\langle z_1 z_2 + z_1^* z_2^* + z_1 z_2^* + z_1^* z_2 \rangle_t}{4}$$

$$= \frac{|A||B|}{2} \cos(\phi - \psi). \tag{A.17}$$

The first two terms on the right side of Equation 2.16 vanish because they average a sinusoid over a cycle of vibration, but the third and fourth terms are twice the real part of the product of $z_1 z_2^*$. The general rule for calculating a time average of a product of two sinusoids when expressed as complex variables is, therefore,

$$\langle a(t)b(t) \rangle_t = \tfrac{1}{2}\mathrm{Re}\{AB^*\}. \tag{A.18}$$

A.3 PERIODIC FUNCTIONS: FOURIER SERIES

The function $e^{j\omega t}$ has a period $T = 2\pi/\omega$. Indeed, any function $e^{jn\omega t}$ is a periodic function with the same time period as long as n is an integer. Suppose the arbitrary function in Figure A.4 has this period. We can specify a periodic function by

$$l(t) = l(t + nT). \tag{A.19}$$

Let us define the "fundamental" radian frequency by $2\pi/T \equiv \omega_1$. The superposition of these periodic sinusoids may now be represented by

$$l(t) = \sum_{n=-\infty}^{\infty} L_n e^{jn\omega_1 t}. \tag{A.20}$$

Negative frequencies are included because we know they will be needed if $l(t)$ is to be real. The summation in Equation A.20, called a *Fourier series*, may be used to represent the periodic function $l(t)$ under a wide range of circumstances. We will not be concerned here with the exceptions, and we assume that any periodic function we wish to represent may be expressed by a Fourier series. The complex amplitude phasors L_n are found by multiplying both sides of the equation by $e^{-jm\omega_1 t}$ and

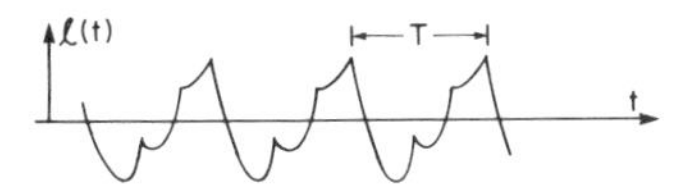

Figure A.4 Periodic function of period T. This function has the property that $l(t + nT) = l(t)$ where n is any integer. It has finite power but infinite energy.

integrating over a period T. This gives

$$\int_0^T e^{-jm\omega_1 t} l(t)\,dt = \sum_{n=-\infty}^{\infty} \int_0^T e^{j\omega_1 t(n-m)}\,dt, \tag{A.21}$$

and since the integral on the right side vanishes except for $n = m$ we get an expression for the amplitude L_m:

$$L_m = \frac{1}{T}\int_0^T e^{-jm\omega_1 t} l(t)\,dt = \langle e^{-jm\omega_1 t} l(t)\rangle_t. \tag{A.22}$$

Equation A.22 provides a rule for calculating the Fourier amplitudes or, more generally, a Fourier transformation, of the continuous function $l(t)$ to a discrete set of values L_m, as indicated by Figure A.5. It is the periodicity of $l(t)$ that causes the transformation to be discrete. If we had started with the Fourier series, we could have derived the periodic function $l(t)$ from it, and we would then regard the transformation as being from a discrete set of values in the frequency domain to a continuous set of values periodic in T.

Later in this appendix we will examine transformations that take us from a discrete representation in the time domain to a continuous and periodic representation in the frequency domain. These are the so-called Z-transforms used in digital signal processing. We will also be concerned with transformations that take us from continuous domains in time to continuous domains in frequency, the Fourier transform, and a discrete time to discrete frequency representation, which is the discrete Fourier transform (DFT).

To continue our discussion, suppose that $l(t)$ is constrained to be a real function, as it generally will be for physical quantities such as force, acceleration, and so on. This puts certain restraints on the values of L_m. Let us consider the two terms $m = \pm p$ and $L_{-p}e^{-jp\omega_1 t} + L_p e^{+jp\omega_1 t}$, and require this combination to be real. Since $e^{\pm jp\omega_1 t}$ are complex conjugates, for this combination to be real requires that the coefficient phasors be complex conjugate; that is,

$$L_{-p} = L_p^*. \tag{A.23}$$

If we also require that $l(t) = l(-t)$—that is, $l(t)$ is even—considering the same two terms, we now have

$$L_p^* e^{-jp\omega_1 t} + L_p e^{+jp\omega_1 t} = L_p^* e^{+jp\omega_1 t} + L_p e^{-jp\omega_1 t}, \tag{A.24}$$

Figure A.5 Fourier transformation converts a function defined continuously over a finite interval of time to a sequence defined for discrete values of radian frequency ω.

which by inspection requires that $L_p = L_p^*$. The only way that L_p and L_{-p} can both equal their own complex conjugates is that they equal each other and are real; therefore, combining the two terms in Equation A.24 gives $2L_p \cos p\omega_1 t$. Thus the complex exponential expansion of Equation 2.19 becomes a cosine series for real even functions. Note that L_p is also an even real function of its index; that is, $L_{-p} = L_p$. Similarly, if we have an odd function of time, $l(t) = -l(-t)$, the same argument will show that Equation A.20 becomes a series of functions $\sin p\omega_1 t$, the Fourier sine series. The requirement on the amplitudes is that $L_{-p} = -L_p$, and Equation A.23 then tells us that L_p must be purely imaginary.

As an example, consider the periodic square wave in Figure A.6, which, as drawn, is an odd function of time. From Equation A.22, we compute the complex amplitudes L_m by the formula

$$L_m = \frac{1}{T}\int_{-T/2}^{T/2} l(t)e^{-jm\omega_1 t}\,dt = -\frac{2jA}{T}(1-\cos m\omega_1 t)_0^{T/2} = -\frac{jA}{\pi m}[1-(-)^m], \qquad \text{(A.25)}$$

$$\text{where } [1-(-)^m] = \begin{cases} 0 & \text{for even values of } m, \\ 2 & \text{for odd values of } m. \end{cases}$$

We note that the Fourier amplitudes in Equation A.25 decrease by $1/f$ as f increases since they depend on $1/m$. This rate of dropoff is related to the fact that the function $l(t)$ has a step in value at the origin. For functions that rise linearly at the origin with time (i.e., have a discontinuity in slope), we find that the high-frequency dropoff goes as $1/f^2$ instead. We see this in examples of source spectra discussed in Chapter 2. If we calculate the Fourier series of a triangular wave, for example, we would find that the amplitudes vary as $1/f^2$.

If we think of the Fourier series as a specific example of the Fourier transformation from periodic and continuous waveforms to discrete functions, the properties of such transformations may be described as shown in Table A.1. The first important property shown in the table is uniqueness; that is, there is only one Fourier transform for any function. The second important property is Parseval's theorem, which, in effect, states that a calculation of power in the time domain is the same as the calculation of power in the frequency domain and is found by a summation of the squares of the magnitudes of the Fourier components.

Superposition states that if we have two periodic functions L and M and we add them with arbitrary coefficients, then the transform of this combination is the combination of the Fourier series amplitudes, L_n and M_n, combined with the same

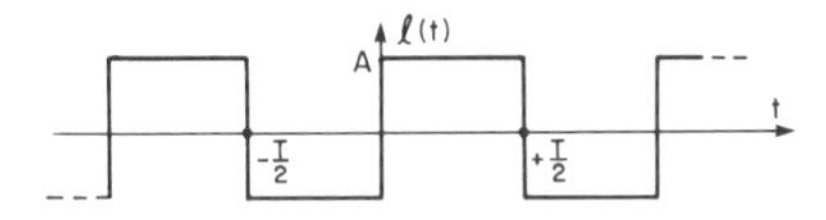

Figure A.6 Periodic square wave of amplitude A. This wave has a frequency spectrum that drops off $1/f$ at high frequencies.

Table A.1 Properties of Fourier Series Transforms (Periodic Cont. $\leftrightarrow$ Discrete)

Uniqueness $l(t) \leftrightarrow L_n$

Parseval's theorem $\frac{1}{T}\int_0^T l^2(t)\,dt = \sum_{n=-\infty}^{\infty} |L_n|^2$

Superposition $\alpha l(t) + \beta m(t) \leftrightarrow \alpha L_n + \beta M_n$

Differentiation $\left(\frac{d}{dt}\right)^m l(t) \leftrightarrow (jn\omega_1)^m L_n$

Delay $l(t - t_0) \leftrightarrow L_n e^{-jn\omega_1 t_0}$

Modulation $l(t)e^{j\Omega t} \leftrightarrow L(n\omega_1 - \Omega) = L_{n-n'} \quad \left(n' = \frac{\Omega}{\omega_1}\right)$

Multiplication $l(t) \cdot m(t) \leftrightarrow \sum_{s=-\infty}^{\infty} L_s M_{n-s}$ (convolution in freq. domain)

Convolution $l(t) * m(t) = \frac{1}{T}\int_0^T l(\lambda)m(t - \lambda)\,d\lambda \leftrightarrow L_n M_n$

coefficients. Also note that the mth time derivative is found by multiplying the Fourier amplitudes by the mth power of the frequency $jn\omega_1$. Another important property is the effect of a time delay in the waveform by the amount t_0, which transfers to a multiplication of the Fourier amplitude by the factor $e^{-jn\omega_1 t_0}$. This tells us that a linear frequency phase shift of the Fourier amplitudes is equivalent to a simple time delay, an observation that is important in our discussions of diagnostics.

The modulation theorem says that if the waveform $l(t)$ is multiplied by any sinusoid periodic in T, then the Fourier amplitudes are shifted by an index n' that is the ratio of the modulating frequency to the fundamental frequency. A closely related property is multiplication of two periodic waveforms $l(t)$ and $m(t)$. The Fourier series of the product is a discrete convolution of the Fourier amplitudes. This reduces to the modulation theorem in the event that there is only one frequency component in the waveform $m(t)$. This multiplication theorem has its inverse in the convolution theorem, which states that the convolution in the time domain is equivalent to a multiplication of the Fourier amplitudes in the frequency domain. All of the properties in Table A.1 have their counterparts for Fourier transforms in the other domain transformations mentioned above.

A.4 NONPERIODIC FUNCTIONS: FOURIER INTEGRAL TRANSFORMS

Let us now consider a pulse $l(t)$ that is repeated at period T, but we will now allow the shape of the pulse to remain unchanged as the period gets very large. Referring to

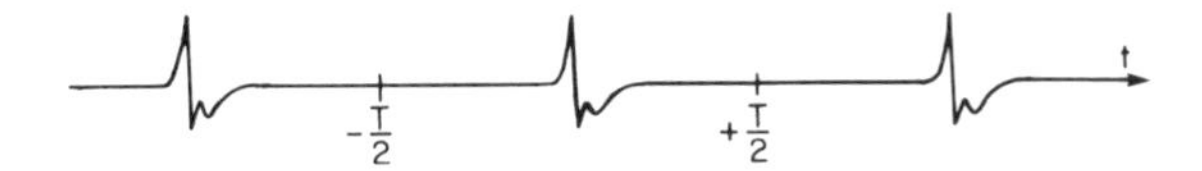

Figure A.7 Periodic signal in which pulse waveform is preserved as period and time between pulses becomes large.

Figure A.7, we can express this function by

$$l(t) = \sum_{n=-\infty}^{\infty} L_n e^{j\omega_n t}, \tag{A.26}$$

where we have used ω_n instead of $n\omega_1$, the nth frequency in the frequency spectrum. Then the Fourier amplitudes are

$$L'_n = T \cdot L_n = \int_{-T/2}^{+T/2} l(t) e^{-j\omega_n t}\, dt, \tag{A.27}$$

where L'_n is such that it will not change very much as T gets very large. As T gets large, the spacing between the frequencies ω_1 will get very small, so in any frequency interval $\Delta\omega$ the number of frequency components Δn is $\Delta\omega/\omega_1$, as shown in Figure A.8.

If we collect the terms in Equation A.26 into groups of size Δn, assuming that L_n does not change very much from one line to the next, we can change the summation over n into one over frequency intervals $\Delta\omega$:

$$\begin{aligned} l(t) &= \sum_{\substack{\text{groups} \\ \text{of lines}}} L_n e^{j\omega_n t} \Delta n \\ &= \frac{T}{2\pi} \sum_{\text{groups}} L_n e^{j\omega_n t} \Delta\omega \xrightarrow[T \to \infty]{} \frac{1}{2\pi} \int_{-\infty}^{\infty} \mathscr{L}(\omega) e^{j\omega t}\, d\omega, \end{aligned} \tag{A.28}$$

where we have used $\mathscr{L}(\omega)$ as the limiting form of L'_n as $T \to \infty$.

The Fourier transform pair between the frequency and time domains is therefore

$$\begin{aligned} \mathscr{L}(\omega) &= \int_{-\infty}^{\infty} l(t) e^{-j\omega t}\, dt, \\ l(t) &= \frac{1}{2\pi} \int_{-\infty}^{\infty} \mathscr{L}(\omega) e^{+j\omega t}\, d\omega. \end{aligned} \tag{A.29}$$

Note that $\mathscr{L}(0)$ is the area under the pulse $l(t)$. This transformation represents the conversion from the continuous infinite time domain to the continuous infinite frequency domain, as shown in Figure A.9. The properties of the Fourier integral

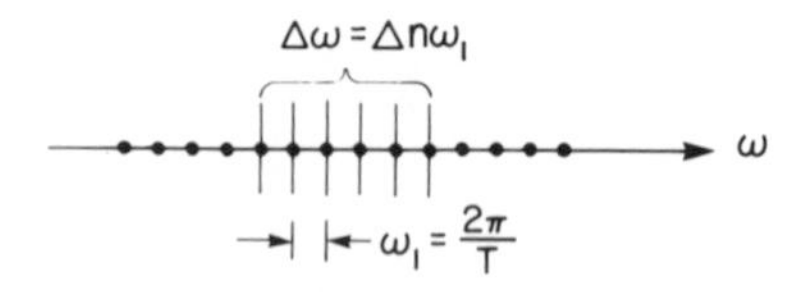

Figure A.8 Convergence of discrete frequency sequence to a continuous function as the time between pulses gets large.

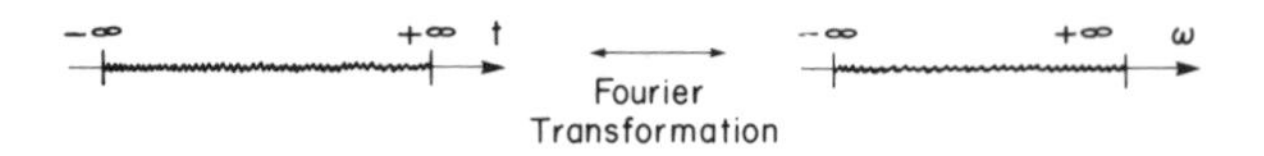

Figure A.9 Fourier integral transformation converts a function defined continuously over an infinite time domain to a function defined continuously over an infinite frequency domain.

transform are similar to those in Table A.1, but the difference in domains leads to slightly different representations of some of the properties. Parseval's theorem now takes the form as shown in the first line of Table A.2. The left side of this equation may be interpreted as the energy in the pulse, so the quantity $|\mathscr{L}|^2/2\pi$ is defined as the energy spectrum of the pulse.

In addition, since the functions are continuous, we can use continuous scaling of them to indicate what happens as we scale the time by some number a. The effect on the Fourier transform is shown in the second line of Table A.2. The modulation and multiplication formulas are now as shown, and since the domains are continuous, these properties are essentially reversible between the frequency and time domains. Also if $l(t)$ is real, we have $\mathscr{L}(\omega) = \mathscr{L}^*(-\omega)$, and if $l(t)$ is even, $\mathscr{L}(\omega)$ is even and real. Similarly if $l(t)$ is odd, $\mathscr{L}(\omega)$ is a purely imaginary odd function. Some of these properties are important in diagnostic and waveform analysis.

Table A.2 Properties of Fourier Integral Transforms

Parseval's theorem	$\int_{-\infty}^{\infty} l^2(t)\,dt = \frac{1}{2\pi}\int_{-\infty}^{\infty} d\omega\,|\mathscr{L}|^2$
Scaling	$l\left(\frac{t}{a}\right) \leftrightarrow a\mathscr{L}(\omega a)$
Modulation	$l(t)e^{j\omega t} \leftrightarrow \mathscr{L}(\omega - \Omega)$
Multiplication	$l(t)m(t) \leftrightarrow \frac{1}{2\pi}\int_{-\infty}^{\infty} \mathscr{L}(\Omega)M(\Omega - \omega)\,d\Omega$

A.5 RELATION BETWEEN FOURIER SERIES AND FOURIER INTEGRAL TRANSFORMS

We have seen that the Fourier series is a transformation between a function that is periodic and continuous in time to one that is discrete in frequency. The Fourier integral is a transformation between functions that are continuous over the infinite interval in time to functions that are continuous over the infinite interval in frequency. We now explore a very useful interpretation of the Fourier series, which shows its relation to the integral transform. Let us consider the function $r(t)$ shown in Figure A.10, which is a sequence of impulses. An impulse function is a pulse with finite area but extremely narrow in time.

We now expand $r(t)$ in a Fourier series:

$$r(t) = \sum_{n=-\infty}^{\infty} R_n e^{j\omega_n t} = \frac{1}{T} \sum_{n=-\infty}^{\infty} e^{j\omega_n t}, \qquad \text{(A.30)}$$

where we have evaluated the Fourier amplitudes R_n by

$$R_n = \frac{1}{T} \int_{-T/2}^{T/2} r(t) e^{-j\omega_n t}\, dt = \frac{1}{T}. \qquad \text{(A.31)}$$

In this case, the transform is independent of frequency.

Now consider the single pulse $p(t)$ in Figure A.11. It can be turned into a periodic function $q(t)$ by convolving $p(t)$ with $r(t)$; that is,

$$q(t) = \int_{-\infty}^{\infty} p(\lambda) \sum_{m=-\infty}^{\infty} \delta(\lambda - mT - t)\, d\lambda = \sum_{m=-\infty}^{\infty} p(t + mT) \qquad \text{(A.32)}$$

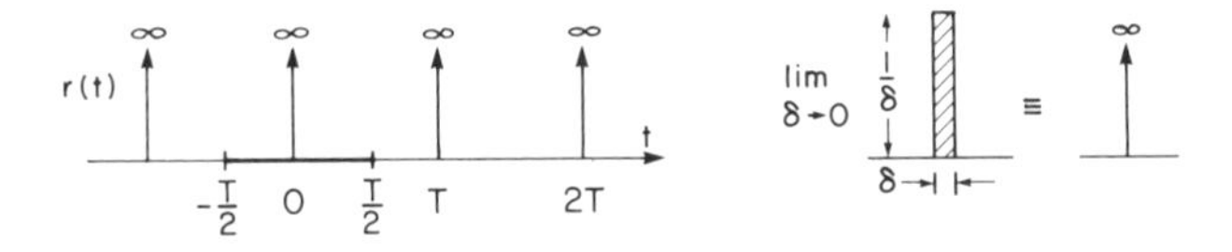

Figure A.10 Periodic sequence of impulses used to convert single pulse to a periodic function of convolution.

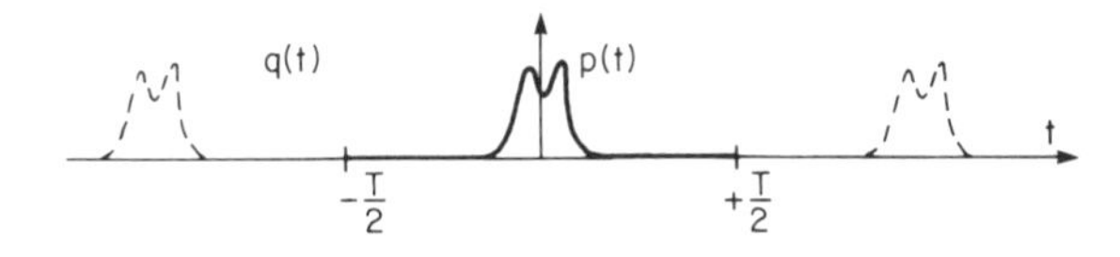

Figure A.11 Conversion of single pulse to a periodic function as a result of convolution with periodic sequence of impulses.

according to the definition of the delta function. Since $q(t)$ is a convolution, we use the convolution property of the Fourier transform to say that the transform of the periodic function $q(t)$ is

$$Q_n = \frac{1}{T}\int_{-T/2}^{T/2} q(t)e^{-j\omega_n t}\,dt = \frac{1}{T}\int_{-T/2}^{T/2} p(t)e^{-j\omega_n t}\,dt. \tag{A.33}$$

Therefore, the Fourier amplitudes of the periodic function Q_n are given by $1/T$ times the Fourier integral transform of the function $p(t)$; that is, $\mathscr{P}(\omega_n)$. Since $\mathscr{P}(\omega)$ is a continuous function, it is clear from this formula that the discrete Fourier series amplitudes are samples of this continuous function at the frequencies ω_n. Thus, when $p(t)$ is made into a periodic function, its transform becomes a line spectrum and the values of the spectral lines are given by samples of the continuous function $\mathscr{P}(\omega)$ at the discrete frequencies ω_n, as shown in Figure A.12.

We now apply Parseval's theorem to the periodic function $q(t)$; that is,

$$\frac{1}{T}\int q^2\,dt = \sum_n |Q_n|^2 = \frac{1}{\omega_1 T^2}|\mathscr{P}(\omega_n)|^2\omega_1$$

$$= \frac{1}{T}\sum_n \frac{1}{2\pi}|\mathscr{P}(\omega_n)|^2\omega_1. \tag{A.34}$$

We note that $|\mathscr{P}(\omega_n)|^2/2\pi$ is, from earlier discussion, the energy spectrum of the individual pulse. Also, the product of this energy spectrum and the frequency spacing ω_1 is the energy contained in the interval between the discrete line frequencies of the Fourier series. Since $1/T$ is the rate of occurrence of pulses in a periodic function with period T, the product of the rate times the energy per pulse is interpreted as the power in the periodic sequence $q(t)$ within the frequency interval ω_1.

Thus, the interpretation of Equation A.34 is that the power in any frequency interval is the energy in that same frequency interval times the rate of occurrence of pulses. Consequently, if the pulses do not occur in strict periodic sequence, but occur randomly in time, we can replace $1/T$, the periodic rate of occurrence, by an average rate of occurrence for random pulses ν. The power spectrum of a signal due to pulses

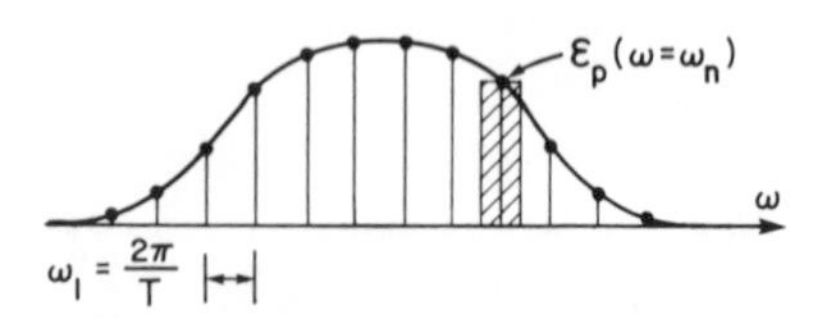

Figure A.12 Line spectrum of $q(t)$ interpreted as a periodic sampling of the single-pulse energy spectrum at frequencies separated by ω_1.

that occur randomly in time is

$$\text{power spectrum} = \nu \frac{1}{2\pi} |\mathscr{P}(\omega)|^2. \tag{A.35}$$

If there is jitter in the occurrence of pulses, then the line spectrum of the Fourier series will not be created, but a continuous power spectrum will occur instead. We can think of the line spectrum of a Fourier series being smeared out over the frequency interval ω_1 into a continuous power spectrum. In general, as long as the analysis bandwidth is greater than ω_1, it will not be possible to distinguish between the power spectrum due to a random sequence and the power spectrum due to a periodic sequence. Also, we may note that the jitter in a nearly periodic sequence of pulses due to machine operation, for example, may cause line spectra to become smeared out. But the basic relationship between the energy spectrum of the single event and the power spectrum of a sequence of pulses is valid whether we are considering a strictly periodic sequence or one that has jitter, which should, therefore, be considered to be a random sequence.

A.6 ANALYSIS OF SEQUENCES

Consider the analog signal $x_a(t)$ sampled at intervals Δt to form a sequence $x(n)$ as in Figure A.13. Sequences are generally derived from the sampling of analog time functions, but once we have the set of numbers we remember that there is no such thing as time in the computer. The processing done by the computer is directed at the sequence and not at an analog function of time. We should think of this sequence or the processing undertaken not as an approximation to an analog system but as a set of mathematical quantities and operations with their own validity and set of rules.

There are special sequences of particular importance in digital processing. Some are shown in Figure A.14. The left sequence (a) is 0 except for the value $n = 1$, for which it is 1. This is the unit sequence labeled $\delta(n)$. The next sequence is the unit step (b), which is unity for $n \geq 0$ and 0 for negative values of n. Finally, there is the

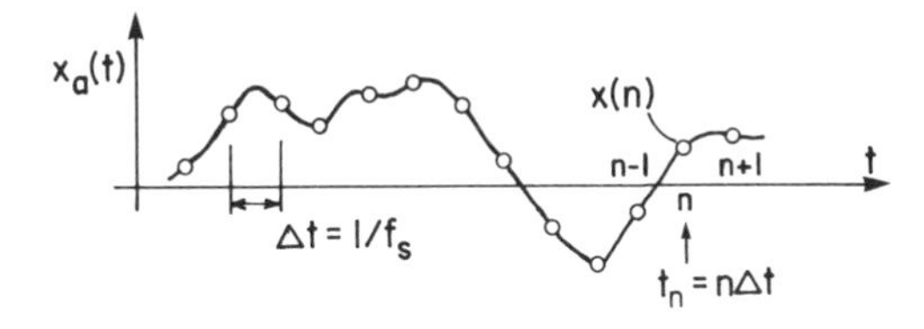

Figure A.13 Illustration of how an analog signal $x_a(t)$ is converted to the sequence $x(n)$ by a sampling process.

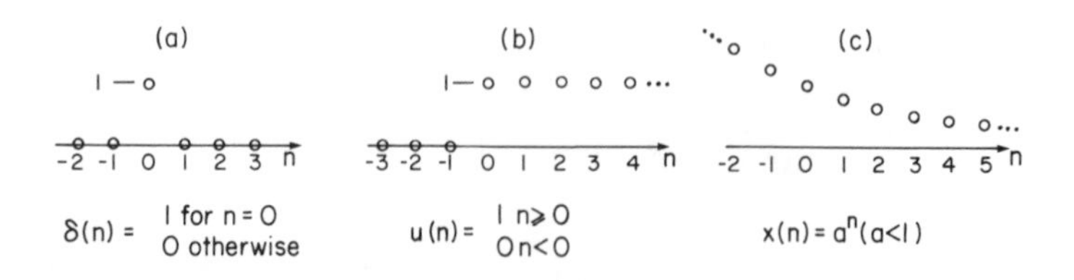

Figure A.14 Three important fundamental sequences: (a) unit; (b) step; and (c) exponential.

exponential sequence (c) for $a < 1$. From the definition of the unit sequence, we can write

$$x(n) = \sum_{k=-\infty}^{\infty} x(k)\delta(n-k), \tag{A.36}$$

since the only nonzero value of the unit sequence is for $n = k$.

In the digital world, the system (transfer) function that simulates a physical system can be written $h(n - k)$ because the output of the system should depend only on previous values of the input. Accordingly, the output of a system that has a transfer function h (Figure A.15) is

$$y(n) = \sum_{k=-\infty}^{\infty} x(k)h(n-k). \tag{A.37}$$

As an example, consider a system that has the transfer function

$$h(n) = \begin{cases} a^n & \text{for } n \geq 0 \\ 0 & \text{for } n < 0 \end{cases} \quad (a < 1), \tag{A.38}$$

which is graphed in Figure A.16(b). Suppose this system is excited by the rectangular pulse sequence in Figure A.16(a). The mathematical representation of the output is found by substituting Equation A.38 into A.37 to obtain

$$y(n) = \sum_{k=0}^{\substack{N-1\,(N \leq n) \\ n\,(0 \leq n < N)}} a^{n-k}. \tag{A.39}$$

Computing the output of this system is quite straightforward, and it is shown in Figure A.17. This output is similar to that of a continuous-time system, but it

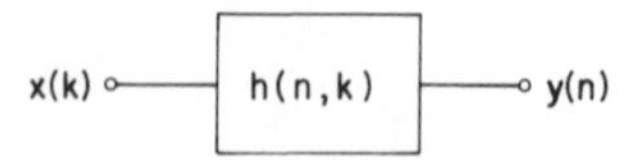

Figure A.15 An input sequence $x(k)$ is converted to the output sequence $y(n)$ by a transfer operator $h(n, k)$. Causality and invariance conditions on the system place constraints on h.

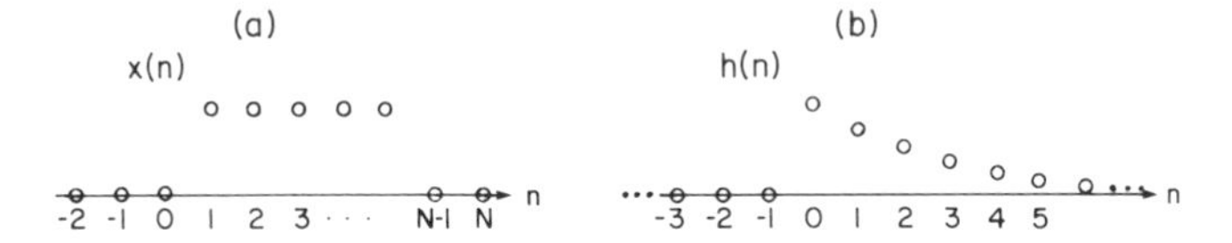

Figure A.16 Input unit pulse and exponential decay system response function used to excite discrete processor.

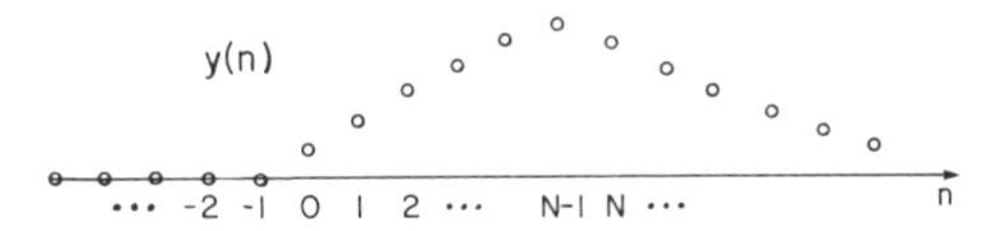

Figure A.17 System response sequence for the rectangular pulse input to discrete system.

should be emphasized that the calculations are not approximations to a continuous system but are precise in their own right.

A.7 FREQUENCY ANALYSIS

Suppose we consider the special sequence $x(n) = e^{j\omega n}$. Since n in an integer, any increase in the parameter ω by 2π will cause the sequence to come back to the same value. As we discussed in Section A.3, a function defined at discrete points in one domain is periodic and continuous in the other. Therefore, we define the range of ω to be $(-\pi, \pi)$. If we now let $x(n)$ be the input to the system in Figure A.15, the output is

$$y(n) = \sum_{k=-\infty}^{\infty} h(k)e^{j\omega(n-k)} = e^{j\omega n} \sum_{k=-\infty}^{\infty} h(k)e^{-j\omega k}$$

$$= e^{j\omega n} H(e^{j\omega}), \tag{A.40}$$

where the summation over k in the second term is, by definition, the Fourier transform of the sequence $h(k)$. Therefore,

$$H(e^{j\omega}) = \sum_{k=-\infty}^{\infty} h(k)e^{-j\omega k}. \tag{A.41}$$

The inverse transform is constructed as in Equation A.22, so the periodic function H, which is continuous over the interval $(-\pi, \pi)$, has the discrete transform

$$h(n) = \frac{1}{2\pi} \int_{-\pi}^{\pi} H(e^{j\omega})e^{+j\omega n}\, d\omega. \tag{A.42}$$

Note that $H(e^{j\omega})$ is the Fourier transform of a sequence, and it is a continuous function of frequency. It is not the function computed by a spectrum analyzer, which is the discrete Fourier transform (DFT). We shall see that the DFT is a sampled version of H.

Let us suppose that $x(n)$ is a real sequence, and it has a transform $X(e^{j\omega})$ that has real and imaginary parts:

$$X(e^{j\omega}) = \text{Re}[X] + j\,\text{Imag}[X]. \tag{A.43}$$

As we noted earlier in our discussion of Fourier transforms, the real part of the transform is an even function of ω, and the imaginary part is an odd function of ω.

A.8 SAMPLING FORMULAS

We can generate relationships between the transform of a continuous analog function $x_a(t)$ and the sample sequence $x(n)$ by assuming that the sequence $x(n)$ is arrived at by sampling the continuous analog time function, as shown in Figure A.13. Therefore,

$$x_a(n\Delta t) = x(n). \tag{A.44}$$

Substituting in Equations A.29 gives us the transform pair

$$x_a(t) = \frac{1}{2\pi}\int_{-\infty}^{\infty} X_a(j\Omega)e^{j\Omega n\Delta t}\,d\Omega, \tag{A.45a}$$

$$X_a(j\Omega) = \int_{-\infty}^{\infty} x_a(t)e^{-j\Omega t}\,dt. \tag{A.45b}$$

The frequency Ω is the analog frequency, and we are interested in seeing how it relates to the variable ω, the transform variable for discrete sequences. To see this, we use Equation A.44, and we break up the integral in Equation A.45a into frequency segments of length $2\pi/\Delta t$ to get

$$x(n) = x_a(n\Delta t) = \frac{1}{2\pi}\sum_{r=-\infty}^{\infty}\int_{(2r-1)\pi/\Delta t}^{(2r+1)\pi/\Delta t} X_a(j\Omega)e^{j\Omega n\Delta t}\,d\Omega. \tag{A.46}$$

If we set $\Omega = \omega/\Delta t$, then reversing the order of integration and summation gives

$$x(n) = \frac{1}{2\pi}\int_{-\pi}^{\pi}\frac{1}{\Delta t}\sum_{r=-\infty}^{\infty} X_a\left(\frac{j\omega}{\Delta t} + j\frac{2\pi r}{\Delta t}\right)d\omega. \tag{A.47}$$

Therefore, by identification from Equation A.42,

$$X(e^{j\omega}) = \frac{1}{\Delta t}\sum_{r=-\infty}^{\infty} X_a\left(\frac{j\omega}{\Delta t} + j\frac{2\pi r}{\Delta t}\right). \tag{A.48}$$

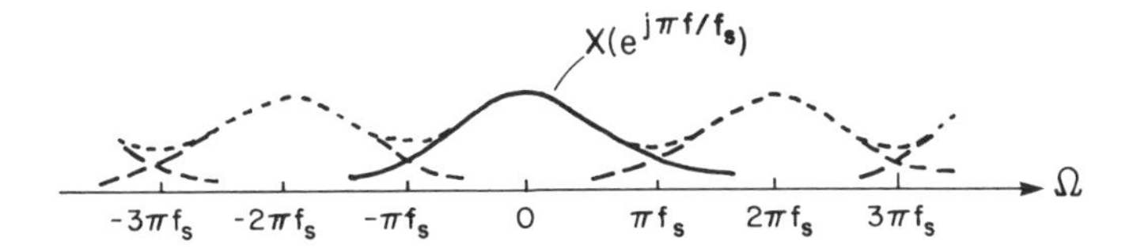

Figure A.18 Aliasing of the analog spectrum when the analog spectrum exceeds the Nyquist frequency $f_s/2$.

We see, therefore, that the transform over ω defined for the range from $-\pi$ to $+\pi$ is given by a summation of analog spectra shifted by an amount 2π times the sampling frequency $f_s = 1/\Delta t$. If the analog spectrum does not exceed the band $(-\pi f_s, +\pi f_s)$, then there is no ambiguity. But, if it does exceed this band, then there will be a leakage from one frequency interval to the other, as seen in Figure A.18. It is the purpose of antialiasing filters to eliminate this leakage. Such filters are "windows," and we speak of the spectrum generated thereby as a "windowed" spectrum.

A.9 ANTIALIASING IN THE FREQUENCY DOMAIN

A typical A/D processing system is shown in Figure A.19. A sensor (accelerometer) picks up a vibration signal from the machine and passes it through a preamplifier and a filter to eliminate various noises or allow us to concentrate on parts of the spectrum of interest. The signal may be directly monitored by an oscilloscope. Before passing into the digital part of the processing, we must use an antialiasing filter.

The conversion of analog signals into digital signals involves two steps. The first step is to turn the continuous-time analog waveform into a sequence of sampled values of the waveform. This process is illustrated in Figure A.20. We show a low-frequency waveform being sampled and producing the set of sampled values $x(n)$. These samples are separated by a time Δt_s, and the sampling rate f_s is the reciprocal of this time. For example, if Δt_s is 1 msec, then the sampling rate is 1000 samples/sec.

We have also sketched a second, higher-frequency sine wave that will produce the same set of sampled values that the low-frequency wave does. This process,

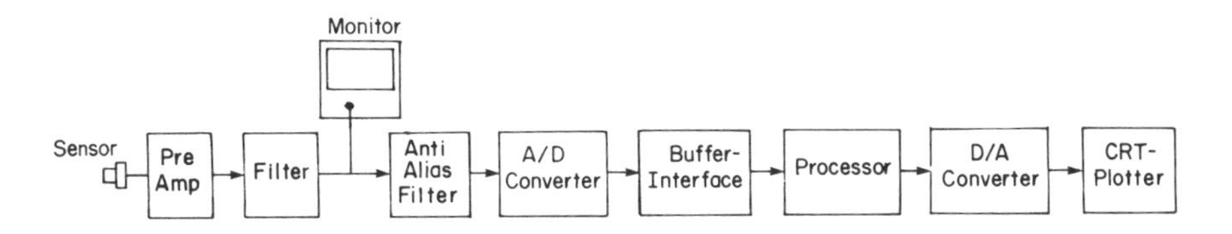

Figure A.19 Typical instrumentation lineup for digital processing of data. The transition points from analog to digital and vice versa will vary from one system to another.

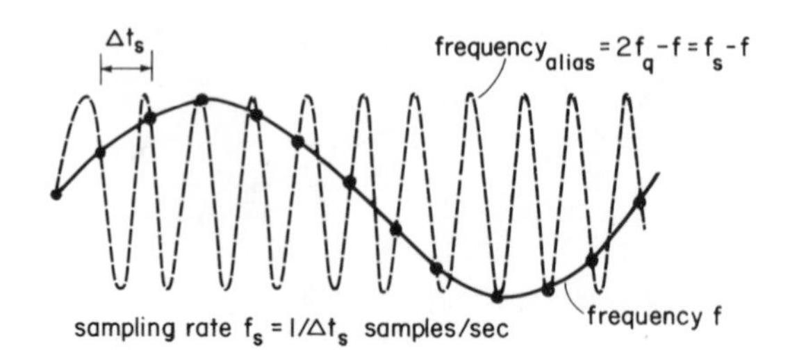

Figure A.20 Sampling of a continuous sine wave produces a sample sequence that may not be unique because a sine wave at a different frequency can produce the same sequence.

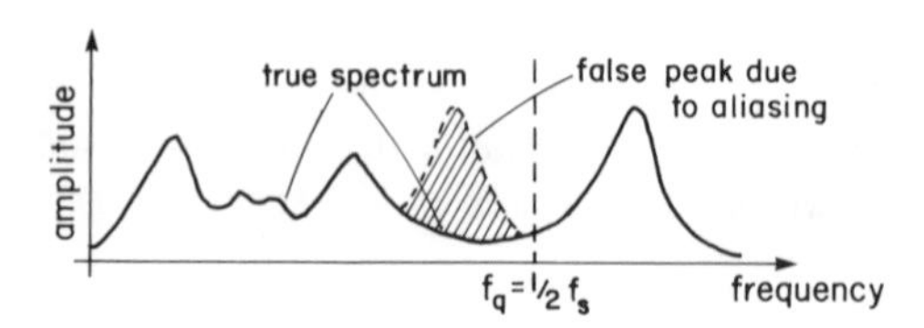

Figure A.21 Aliasing will cause spectral energy above the Nyquist frequency to appear below that frequency unless an antialiasing filter is used.

whereby two waves at different frequencies can produce the same set of sample values, is the aliasing described above. The frequency of the higher-frequency second sine wave f_{alias} is related to the lower-frequency f by the relation $f_{alias} = f_s - f$.

If we define a *Nyquist frequency* f_q, which is one half the sampling frequency, the alias frequency can be regarded as reflected across the Nyquist frequency to produce a false indication of a spectral line in the frequency range below the Nyquist frequency. This ambiguity is also indicated in Figure A.18.

As an example, if the sampling rate is 1000 samples/sec, then f_q is 500 Hz; therefore, a high-frequency signal of 900 Hz will be sampled as shown in Figure A.20 to be a signal of frequency 100 Hz. This aliasing uncertainty is illustrated in Figure A.21. If we have a continuous spectrum as shown in the figure, then any part of the spectrum greater than f_q will be reflected into the range below f_q as an apparent part of the spectrum up to f_q if these frequencies are not eliminated by an antialiasing filter.

Aliasing represents an error in determining the spectrum and has to be removed by eliminating frequencies from the input to the analyzer or the processor that are greater than f_q. We eliminate the undesired frequency components from the input to the processor by an antialiasing filter, which is designed to cut off contributions above the Nyquist frequency. Sketches of a passband behavior of typical antialiasing filters are shown in Figure A.22.

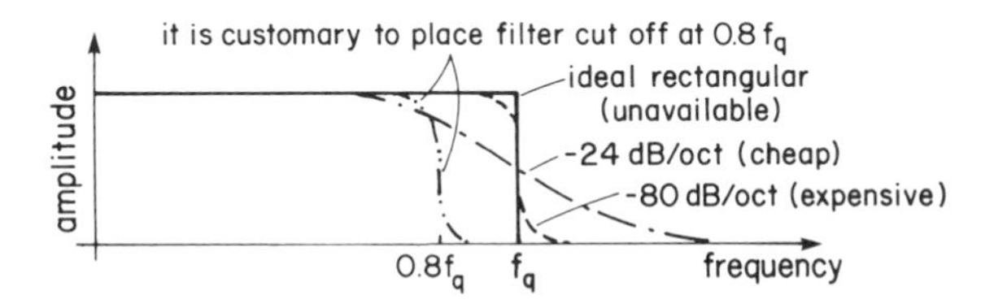

Figure A.22 Frequency response characteristics of an antialiasing filter are a compromise between desired performance and cost.

A.10 Z-TRANSFORM

The Fourier transform for sequences is defined by Equations A.41 and A.42. Since the range of ω is the interval $(-\pi, \pi)$, we can think of it as a unit circle in a complex plane, which we call the z plane and which is shown in Figure A.23. If we substitute the variable z for $e^{j\omega}$, allowing us to define the transform in terms of any point on the z plane, then the Z-transform is

$$Z[x] = X(z) = \Sigma\, x(n)z^{-n}. \tag{A.49}$$

For example, consider the Z-transform of the one-sided exponential

$$x(n) = a^n u(n) \qquad (n \geq 0), \tag{A.50}$$

which from Equation A.49 is

$$X(z) = \sum_{n=-\infty}^{\infty} a^n u(n) z^{-n} = \frac{1}{1 - az^{-1}} = \frac{z}{z - a}, \tag{A.51}$$

which is valid as long as the quantity $|az^{-1}| \leq 1$ or $|z| > a$.

The Z-transform converges in this case for the region outside the circle of radius a. If $a < 1$, it converges on the unit circle and a Fourier transform exists; but

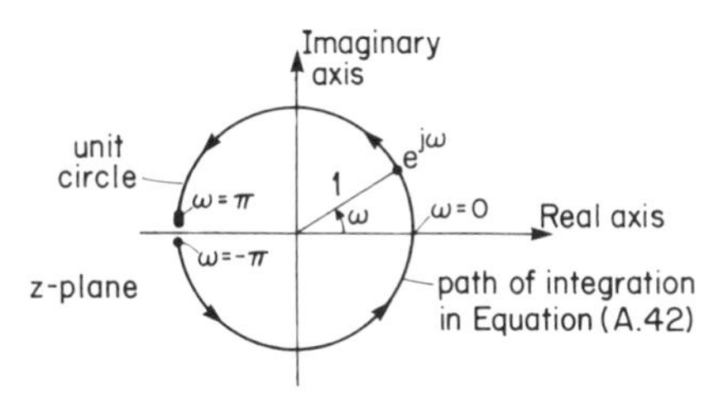

Figure A.23 The Fourier transformation may be considered to be an integration around the unit circle in the z plane.

if $a > 1$, it converges in a region outside the unit circle and a Fourier transform does not exist. Thus, the relationship between the Z-transform and the Fourier transform for discrete sequences is very much like that between the Fourier integral and the Laplace transform, in which certain functions have a Laplace transform but may not have a Fourier integral transform because of convergence problems. The region of convergence of the Z-transform for the sequence in Equation A.50 is shown in Figure A.24.

If the sequence is finite in length and nonzero for values $n_1 \leq n \leq n_2$, then its Z-transform is

$$X(z) = \sum_{n=n_1}^{n_2} x(n)z^{-n}. \tag{A.52}$$

This sum will converge for any finite value of z, except possibly for $z = 0$, so that the Z-transform exists for any finite sequence.

The sequence described by Equation A.50 is called a *right-sided sequence.* Any right-sided sequence will converge if it decays at least as rapidly as β^n for $n<0$, and $X(z)$ will exist for $|z| > \beta$. If one has a right-sided sequence with a finite number of values for $n < 0$, then it can be combined into a finite sequence for negative n and a right-sided sequence, and the region of convergence will be determined by the behavior of the sequence for $n > 0$.

If we have a double-sided sequence as in Figure A.25, which grows exponentially as α^n for negative values and decays as β^n for positive values, then this

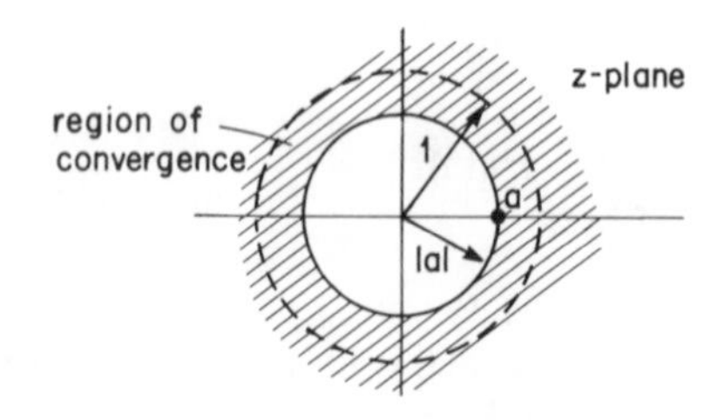

Figure A.24 Region of convergence for the transform of $a^n u(n)$. If $a<1$, this region includes the unit circle and the Fourier transform exists.

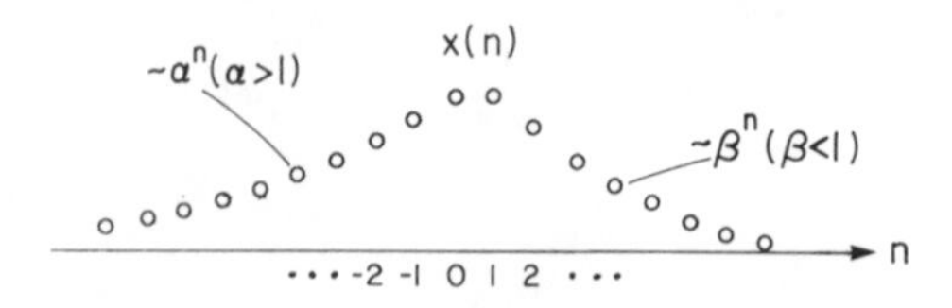

Figure A.25 Two-sided sequence that grows and decays exponentially. This is similar to an analog pulse.

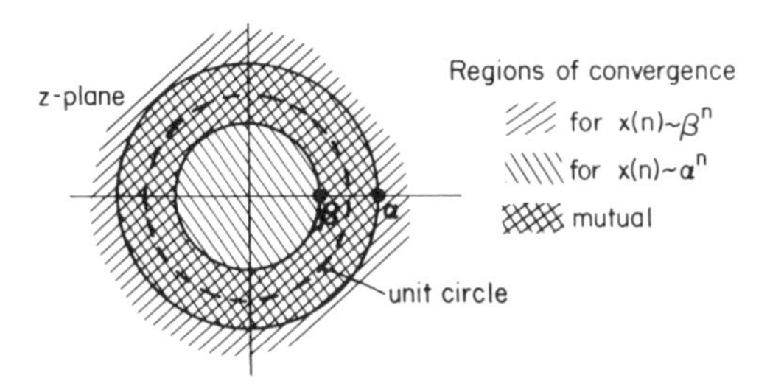

Figure A.26 Region of convergence for the two-sided "pulse" sequence. In this case, the Fourier transform exists because the mutual region of convergence contains the unit circle.

can be broken into a sum of two sequences for positive and negative values of n, which have regions of convergence as shown in Figure A.25. If $\beta < \alpha$, then there is a common region of convergence, as shown in Figure A.26. If, in addition, this common region of convergence includes the unit circle, then a Fourier transform of the sequence also exists.

We notice the similarity between the sequence in Figure A.25, which has the region of convergence in Figure A.26, and the existence of a Fourier integral transform of a pulse. A finite energy time pulse has a Fourier transform. We have a very similar situation here in that both the pulse and the sequence decay in a particular way for negative and positive values of the integer, which means that the sequence has a Fourier transform.

A.11 INVERSE Z-TRANSFORM

Let us now multiply both sides of Equation A.52 by z^{k-1}:

$$z^{k-1}X(z) = \sum_{n=-\infty}^{\infty} x(n)z^{-n+k-1}. \tag{A.53}$$

If we now integrate counterclockwise in a contour that encircles the origin, then, by the residue theorem, we obtain

$$\frac{1}{2\pi j}\oint z^{k-1}X(z)\,dz = x(k). \tag{A.54}$$

This is the inverse Z-transform, and it allows us to construct the sequence $x(k)$ if we know the Z-transform $X(z)$. The path of the contour integral in Equation A.54 must be within the region of convergence of the Z-transform (e.g., see Figure A.25). The region of convergence need not include the unit circle. The sequence, therefore, is defined by the set of poles lying within the region of convergence.

A.12 DISCRETE FOURIER TRANSFORM

We are especially interested in discrete Fourier transforms (DTFs) because they are the functions computed by digital spectrum analyzers and computer programs that do spectral analysis. To analyze sequences digitally, we must deal with a finite number of elements in the computation so that we convert a potentially infinite discrete sequence to a finite sequence of N values. We treat this finite sequence as periodic; that is, we assume that it occurs over an infinite range of integers but with the values that it has between the integers 0 to $N - 1$ repeated with period N.

We note that this periodic sequence does not have a Z-transform because it does not decay in either direction as the order goes to infinity. We therefore represent it as a Fourier series

$$\tilde{x}(n) = \frac{1}{N} \sum_{k=0}^{N-1} \tilde{X}(k) e^{j(2\pi k/N)n}, \tag{A.55}$$

where $x(n)$ is the periodic sequence and $X(k)$ is its DFT. If we multiply both sides by $e^{-j2\pi nr/N}$ and sum over n, we obtain

$$\begin{aligned} \sum_{n=0}^{N-1} \tilde{x}(n) e^{-j(2\pi/N)nr} &= \frac{1}{N} \sum_{n=0}^{N-1} \sum_{k=0}^{N-1} \tilde{X}(k) e^{j(2\pi/N)n(k-r)} \\ &= \tilde{X}(r), \end{aligned} \tag{A.56}$$

where we have used the relationship

$$\frac{1}{N} \sum_{n=0}^{N-1} e^{j(2\pi/N)n(k-r)} = \begin{cases} 1 & \text{for } k - r = mN, \\ 0 & \text{otherwise,} \end{cases} \tag{A.57}$$

where m is an integer.

Equation A.56 can be rewritten as

$$\tilde{X}(r) = N \sum_{n=0}^{N-1} \tilde{x}(n) e^{-j(2\pi/N)nr}, \tag{A.58}$$

and we notice that, on comparing this relation with Equation A.41, the DFT $\tilde{X}(r)$ can be regarded as the Fourier transform of the sequence $\tilde{x}(n)$, but now the Fourier transform is sampled at intervals in angle $2\pi/N$. Therefore, the Fourier transform that was continuous over the unit circle is now being sampled at frequency intervals $2\pi/N$ on the unit circle because we have now made the sequence periodic. In this case, then, we have developed a transformation between a discrete periodic sequence $\tilde{x}(n)$ and a discrete periodic Fourier transform $\tilde{X}(r)$, the DFT. This is the quantity computed by spectrum analyzers. The sampling is shown schematically in Figure A.27.

Just as frequencies in a time signal beyond $f_g - f_s/2$ will produce frequency aliasing when a signal is time sampled, discrete frequency representation implies that

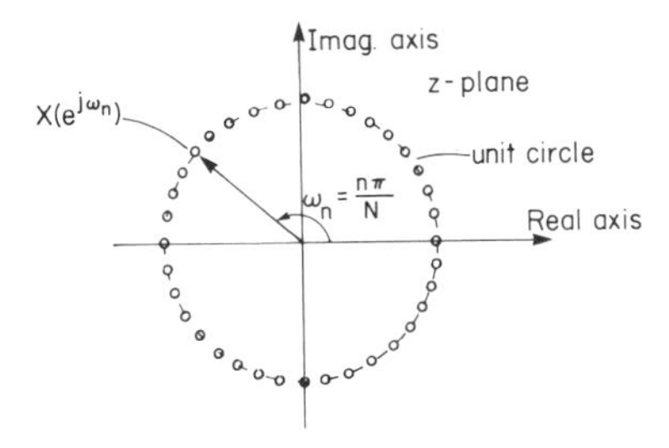

Figure A.27 The DFT is a sampling of the Fourier transform of a sequence on the unit circle.

constructed time signals will be aliased unless they are windowed in time. Thus, time signals are zeroed from $T/2$ to T after processing to eliminate such uncertainties by a process called "zero-padding."

A.13 HILBERT TRANSFORM

Most of the signals we are interested in are those that result from the excitation of structures. The response of a structure to an impulse $h(t)$ will vanish for $t < 0$, since there is no response before the impulse acts at $t = 0$. This also tells us that the poles of the system function $H(\omega) \overset{\text{F.T.}}{\longleftrightarrow} h(t)$ must lie in the upper half of the complex frequency plane. If $h(t)$ is real, then its Fourier transform has an even real part and an odd imaginary part. If these are individually transformed back into the time domain, we find that the impulse response can be divided into similar even and odd parts:

$$H = H_r + jH_i \overset{\text{F.T.}}{\longleftrightarrow} h_r(t) + h_i(t) = h(t), \tag{A.59}$$

where $h_r(t)$ is even and $h_i(t)$ is odd in t as shown in Figure A.28. Since the impulse response is zero for negative time and nonzero for positive time, this requires that $h_r = h_i$ for $t > 0$ and $h_r = -h_i$ for $t < 0$. This relationship is also illustrated in Figure A.28.

These conditions on the even and odd parts of the impulse response lead to conditions on the even and odd parts of the transfer function. For example, if we

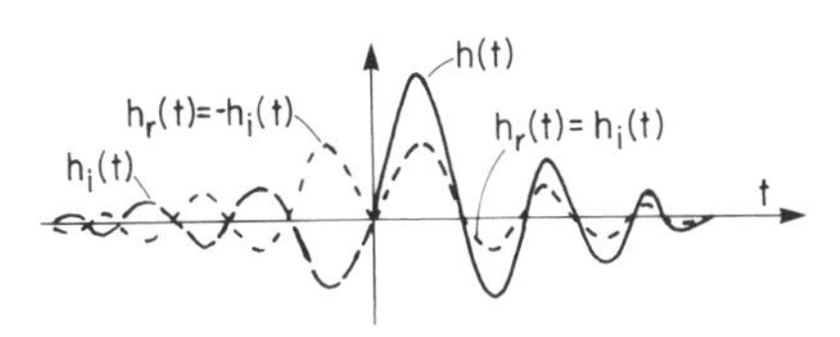

Figure A.28 Causality requires that an impulse response vanish for negative time, which puts a constraint on its even and odd parts that requires their Fourier transforms be Hilbert transforms of each other.

represent these components of the impulse response by their Fourier transforms, we get

$$h_r(t) = \frac{1}{\pi}\int_0^\infty H_r(\omega)\cos\omega t\, d\omega,$$
$$h_i(t) = \frac{1}{\pi}\int_0^\infty H_i(\omega)\sin\omega t\, d\omega, \tag{A.60}$$

$$H_r(\omega) = \int_0^\infty h_r(t)\cos\omega t\, dt = \int_0^\infty h_i(t)\cos\omega t\, dt,$$
$$H_i(\omega) = \int_0^\infty h_i(t)\sin\omega t\, dt = \int_0^\infty h_r(t)\sin\omega t\, dt, \tag{A.61}$$

since $h_r = h_i$ for $t > 0$; substituting into Equations A.61 for h_i and h_r, respectively, from Equation A.60, we get the Hilbert transform relationship

$$H_r(\omega) = \frac{1}{\pi}\int_0^\infty dt\cos\omega t\int_0^\infty d\omega'\sin\omega' t\, H_i(\omega'),$$
$$H_i(\omega) = \frac{1}{\pi}\int_0^\infty dt\sin\omega t\int_0^\infty d\omega'\cos\omega' t\, H_r(\omega'). \tag{A.62}$$

In Sections A.2 and A.3, we suggested that we might wish to represent a signal using positive frequencies only, but if we do so, then the signal becomes complex with real and imaginary parts. For example,

$$y(t) = y_l(t) + jy_h \leftrightarrow Y(\omega), \tag{A.63}$$

where $Y(\omega) = 0$ for $\omega < 0$, but it is a complex function. We then speak of $y(t)$ as being an "analytic" signal. The ordinary "laboratory" signal is $y_l(t)$, and $y_h(t)$ is a signal to be constructed by a Hilbert transform.

Since $Y(\omega)$ vanishes for $\omega < 0$, y_l and y_h are Hilbert transforms of each other. But since $Y(\omega)$ is complex, the Hilbert transform relations between y_l and y_h are different from those in Equation A.62. If the Fourier transform of $y_l(t)$ is $Y_l(\omega)$ with components $Y_{l,r}$ and $Y_{l,i}$, then, from Figure A.3, the negative frequency, or the clockwise rotating vector, is constructed from the counterclockwise or positive frequency components by reversing the sign of $Y_{l,i}$. The negative of this vector, which must be added to produce a conventional transform for $\omega < 0$, has real and imaginary components Y_i, $-Y_r$, as shown in Figure A.29. The real part of the transform jy_h is even, and the imaginary part is odd, as required.

A major use of y_h is to construct an "envelope" or magnitude of a laboratory time function $y_l(t)$. For each positive frequency, the envelope or magnitude is

$$|Y| = \sqrt{|Y_r|^2 + |Y_i|^2}, \tag{A.64}$$

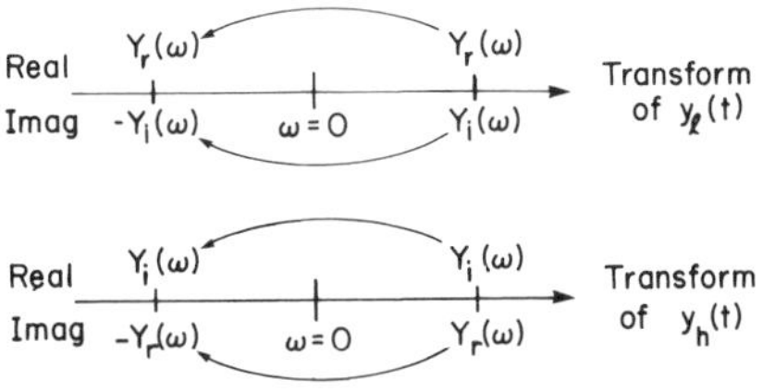

Real $Y_r(\omega)$ $Y_r(\omega)$ Transform of $y_l(t)$
Imag $-Y_i(\omega)$ $\omega = 0$ $Y_i(\omega)$

Real $Y_i(\omega)$ $Y_i(\omega)$ Transform of $y_h(t)$
Imag $-Y_r(\omega)$ $\omega = 0$ $Y_r(\omega)$

Figure A.29 Construction of the Fourier transform of $y_h(t)$, the Hilbert transform of $y_l(t)$. This procedure is particularly appropriate for constructing y_h by a Fourier analyzer.

so we can define a magnitude of the time function,

$$|y| = \sqrt{y_l^2 + y_h^2}, \tag{A.65}$$

a time-dependent phase

$$\phi(t) = \tan^{-1}\left(\frac{y_n}{y_l}\right), \tag{A.66}$$

and an instantaneous frequency

$$\omega(t) = \frac{d\phi(t)}{dt}. \tag{A.67}$$

A.14 POWER SPECTRUM AND THE CEPSTRUM

The power spectrum is derived from the Fourier transform

$$Y(\omega) = \frac{1}{T}\int_{-T/2}^{T/2} y(t)e^{j\omega t}\,dt. \tag{A.68}$$

If $y(t)$ is real, then there is also the constraint that

$$Y(-\omega) = Y^*(\omega), \tag{A.69}$$

which requires that Y_r be an even function of ω and Y_i be an odd function of ω:

$$Y(\omega) = Y_r(\omega) \quad [\text{even in } \omega] + jY_i(\omega) \quad [\text{odd in } \omega]. \tag{A.70}$$

Therefore,

$$|Y| = \sqrt{Y_r^2 + Y_i^2} \tag{A.71}$$

is also an even function of ω, and the phase

$$\phi_y = \tan^{-1}\left(\frac{Y_i}{Y_r}\right) \tag{A.72}$$

is an odd function of ω.

With this background, we examine the relationships shown in Figure A.30. In the upper left corner, we show a system with an impulse response $h(t)$ that is driven by a source $x(t)$ and produces a response or output $y(t)$. The output $y(t)$ is computed by convolving the input time record with the impulse response of the system to produce an output waveform. The result of this process is that the input and the system response are inextricably woven and folded together so that it is not possible to gain direct information regarding the input or the path from an observation of $y(t)$.

In the frequency domain, the transform of the output is the product of the Fourier transform of the input $X(\omega)$ and the system function $H(\omega)$. The magnitude

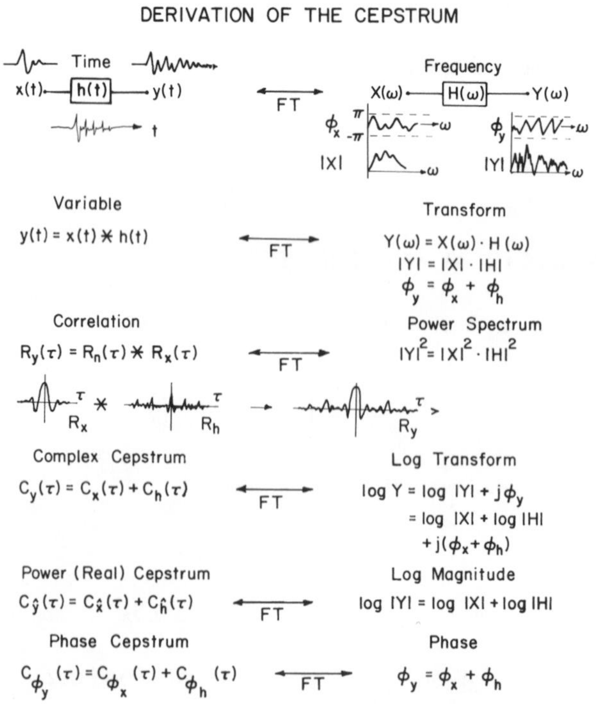

DERIVATION OF THE CEPSTRUM

Time — $x(t) \to h(t) \to y(t)$ — FT — Frequency — $X(\omega) \to H(\omega) \to Y(\omega)$

ϕ_x, $|X|$, ϕ_y, $|Y|$

Variable: $y(t) = x(t) * h(t)$ — FT — Transform: $Y(\omega) = X(\omega) \cdot H(\omega)$, $|Y| = |X| \cdot |H|$, $\phi_y = \phi_x + \phi_h$

Correlation: $R_y(\tau) = R_h(\tau) * R_x(\tau)$ — FT — Power Spectrum: $|Y|^2 = |X|^2 \cdot |H|^2$

$R_x * R_h \to R_y$

Complex Cepstrum: $C_y(\tau) = C_x(\tau) + C_h(\tau)$ — FT — Log Transform: $\log Y = \log |Y| + j\phi_y = \log |X| + \log |H| + j(\phi_x + \phi_h)$

Power (Real) Cepstrum: $C_{\hat{y}}(\tau) = C_{\hat{x}}(\tau) + C_{\hat{h}}(\tau)$ — FT — Log Magnitude: $\log |Y| = \log |X| + \log |H|$

Phase Cepstrum: $C_{\phi_y}(\tau) = C_{\phi_x}(\tau) + C_{\phi_h}(\tau)$ — FT — Phase: $\phi_y = \phi_x + \phi_h$

Figure A.30 The cepstrum has an even part (power cepstrum) and an odd part (phase cepstrum), each of which is additive in the time and frequency domains for source and path components.

of the output is the product of the magnitudes of the input and system functions, and the phase of the output ϕ_y is the sum of the phase of the input ϕ_x and the phase of the system function ϕ_h.

The magnitude relationship can be used to generate power spectra according to $|Y|^2 = |X|^2|H|^2$. Because of the product relationship of the power spectra, the correlation output is again a convolution of the input correlation function and the convolution of the impulse response. This relationship is also sketched in Figure A.30. Thus, correlation does not solve the problem of mixing between the input and the system response.

This separation problem can sometimes be alleviated, however, by going another step and forming the log of the transform so that, for example, $\log Y = \log|Y| + j\phi_y$. Clearly, the logs of the magnitudes of the input and transfer function add to produce the log of the output magnitude. The phase of the output is a linear sum of the phases of the input and of the transfer function. We now have a situation in which the transfer function and the input add their properties to produce an output. This frequency domain process may not result in an effective way to separate the source and transfer function; but if we take the inverse transform into the time domain, we generate the cepstrum, which may provide a way in some cases to separate the individual effects of source and propagation path on the output cepstrum, at least to a degree that is useful for diagnostic purposes.

Since the log magnitude is an even function of frequency and the phase is an odd function of frequency, their inverse transforms are real functions, so the complex cepstrum $C_y(t)$ is a real function of time. The Fourier transform of the log magnitude is called the power or real cepstrum, and in situations where the phase is unknown or ignored it may be a useful way to separate source and path effects. The inverse transform of the phase is the phase cepstrum, and the sum of the magnitude and phase cepstra is the complete complex cepstrum of the signal.

We note that if we are able to determine the input cepstrum $C_x(t)$, then by a Fourier transformation we could construct the log Fourier transform of x and by exponentiation recreate the Fourier transform itself. Having obtained the Fourier transform of x, both in magnitude and phase, we could use the inverse transform again and get back to $x(t)$. Thus, there is a unique and recoverable relationship between the complex cepstrum and the variable from which it is derived. However, it is not possible to recover the initial waveform from the power or real cepstrum because the inverse transform of the real cepstrum only allows us to compute the magnitude of the Fourier transform or the power spectrum. Inverse time transformation of the power spectrum reproduces the correlation function, not the initial waveform, because we have lost the phase of the signal.

APPENDIX B

Criteria for Machinery Noise and Vibration

B.1 INTRODUCTION

The criteria to be applied to machinery noise relate to two general situations. The first is the workplace. Machines used in the workplace may have to meet noise standards that include hearing loss hazard, the ability to communicate in the presence of the machine, and what we might call the quality of the acoustical environment produced by the machine.

Clearly, if the exposure of a worker to machinery noise during many years results in loss of the enjoyment of music, inability to understand conversation, or loss ofcontact with one's family, then a very serious situation exists. Hearing loss associated with such noise exposure should be avoided if at all possible. In the past, some workers have been proud because their hearing might be impaired by their work, and we hear stories of "boilermaker ear" as evidence of a life of good hard work. People are better educated today and not so proud of such loss of facility as they once seemed to be. Nevertheless, it still occurs that male workers feel that it is somehow less virile to protect their hearing. A newspaper plant, where workers were supplied with ear muffs as protection against the high noise levels in the press room, had sound levels of about 100 dB. Only 2 of about 20 pressmen present were wearing their muffs. The rest of the muffs were hanging on their storage pegs.

The issue of communication deals with a worker's ability to transfer information that might have to do with the job itself or with safety. Sound can interfere with communication at levels substantially below those that cause hearing loss. Concern about communication can occur in either a factory situation, in which speaking over short distances with a raised voice is considered acceptable, or in office environments, where one would expect to speak from one end of a room to the other at a normal or slightly raised voice level and be understood. The issue of communication not only relates to ability to communicate but also to the sense of privacy. Privacy is not concerned with machinery noise as such but is more related to the proper acoustical design of the spaces in which machines are placed. If you are in one office and can understand what the person in the next office is saying, then it is natural to assume that that person can also hear and understand you, and the sense of privacy is lost.

Quality of the environment means how acceptable the noise is in terms of the way it sounds. A background noise, whether produced by machines or other devices, that has too much high-frequency noise will sound "hissy." If it has too much low-frequency component, it will sound "boomy." The noise from office machines may be perfectly acceptable in terms of hearing loss or in terms of interference with speech, but still have an unpleasant quality that makes the machines undesirable in an office. Closely related to this is the idea of "appropriateness." An executive with a new computer terminal on his or her desk can be disturbed by even a very weak tone generated by a cooling fan.

We have been discussing the first situation in which machines are intended primarily for a factory or office. Machines sold to consumers have a different set of noise problems that are related to both disturbance and product quality. In some countries, products such as dishwashers, lawn mowers, vacuum cleaners, or snowmobiles are restricted by legislation in the amount of noise they may make. Products may have to be labeled according to the noise output. In this case, the product noise limits may be similar to those placed on products for the workplace, particularly as related to hearing loss.

Another product quality criterion, however, is "perceived quality," the impression of quality that the potential buyer or user infers from the sound of the product. The purchaser of an automobile slams the door to see what kind of sound it makes as a way of judging the quality of the automobile. Perceived quality is often more important than true quality in the decision-making process when making a purchase. Most engineers assume that such issues of perceived quality are not amenable to acoustical design, but it turns out that it is possible to develop

Table B.1 Considerations in the Development of Criteria for Machinery Noise and Vibration

Some effects of noise and vibration on people	Hearing damage Interference with speech communication Interference with work tasks Annoyance Violation of interroom privacy
Noise criterion	A physical measure A limit value of the physical measure
Searching for appropriate noise criteria consider	Variation in people's reaction to noise, depending on the situation Variation between individuals
Machinery noise criteria In the workplace	Hearing damage Speech communication Quality of noise environment
Product quality	Suitability in the workplace Perceived quality

procedures to allow us to be guided in the acoustical design of products by consumers' reactions to the various features or aspects of product sound.

The development of criteria for machinery noise and vibration involves many aspects, as indicated in Table B.1. We need to measure the various effects that the acoustical environment produces and the relation between these effects and a physical sound or vibration level. In doing this, we must somehow account for the great variability in people's sensitivities and reactions to noise, and the effect of this variability on how these criteria are to be applied to different situations.

B.2 HEARING LOSS

"Hearing loss" as a technical term defines the relative sensitivity of an individual's hearing to a population of young male adults. If one has the same hearing acuity as the average, then no hearing loss is presumed to exist. Such measurements are made at a sequence of frequencies, and the ones used for evaluating hearing loss are 500, 1000, and 2000 Hz. If, for example, one has a hearing sensitivity that is 4 dB less than the standard at 500 Hz, 8 dB at 1000 Hz, and 15 dB at 2000 Hz, then the hearing loss is defined to be the average of these three numbers, or 9 dB. No significant hearing loss is assumed to occur until this deficit is 25 dB. The reason is that as one loses hearing sensitivity, both the desired sound and undesired background noises are attenuated. The signal-to-noise ratio is not diminished until the background noise becomes imperceptible, presumed to occur at a hearing loss of about 25 dB.

A sample audiogram is shown in Figure B.1. This individual's hearing loss in the right ear at 500 Hz, 1000 Hz, and 2000 Hz is 10, 10, and 20 dB, respectively. The average of these is 13 dB, so this individual would be reported to have a hearing loss of 13 dB in the right ear.

The graph in Figure B.2 shows how a working population exposed to industrial noise tends to lose hearing acuity over years of exposure. There is quite a wide variation in sensitivity of different people to noise exposure, but hearing loss

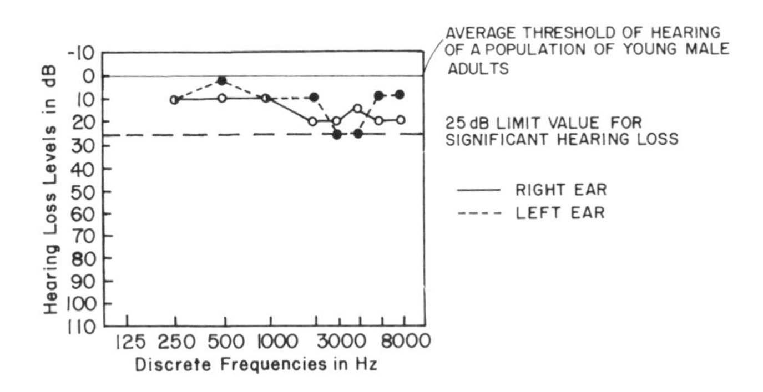

Figure B.1 A standardized recording of hearing loss for the two ears of a subject. The 25-dB hearing loss value is shown. The hearing loss for the right ear is 13 dB; for the left ear, 8 dB.

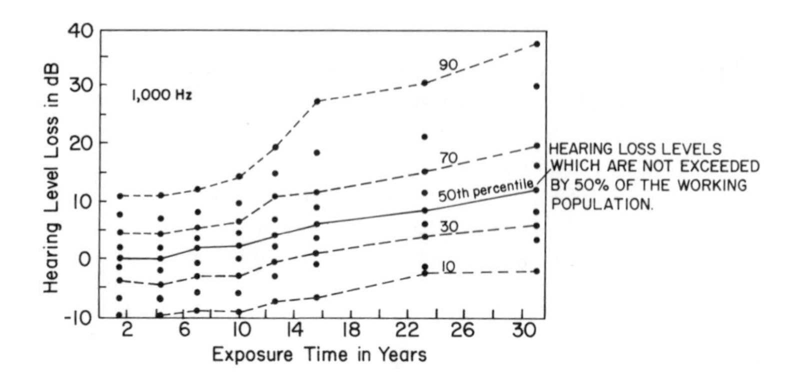

Figure B.2 Growth in hearing loss at 1000 Hz for population exposed to "ordinary" nonindustrial noises. Exposure time starts at age 20. Hearing loss first occurs mainly in the auditory frequencies above 2 kHz. In time the aural acuity is also affected at 500 Hz, 1 kHz, and 2 kHz.

generally increases for the entire population as the number of years of noise exposure increases.

Detailed studies have tracked hearing loss against various levels of industrial noise exposure, and the results of these studies are shown in Table B.2. At a sound level of 80 dB, there is no percentage increase in hearing loss versus years of exposure that we can associate with noise, since the entire population tends to lose hearing acuity with age in the absence of such exposure. This age-related hearing loss, therefore, is assumed to be unrelated to noise. Any increase over this amount of loss is then associated with the noise. For example, if the noise exposure is 90 dB, then after 15 years of exposure the total percentage at risk is 12%, but only 2% would be expected without any noise exposure. Therefore, the amount due to the noise is assumed to be 10%. It is clear that noise levels of 100 dB, which are not uncommon in some industries, will cause a significant fraction of the population to be at risk within 10 to 15 years of exposure.

The consideration of the effects of noise exposure on hearing loss has led to the Walsh–Healy criteria, which limit noise exposure in the workplace environments according to the chart shown in Table B.3. There is a 5-dB tradeoff between doubling of exposure duration and noise level, so for a 4-hr duration a noise level of 95 dB is allowed, whereas only 90 dB is allowed for 8 hr of exposure.

As shown in the table, noise-dose exposure for a varying environment is computed by combining fractional exposures. Suppose that in a workplace the sound level is 85 dB for 2 hr during the day, 90 dB for 4 hr during the day, and 95 dB for 2 hr during the day. Four hours at 90 dB is a fractional exposure of 0.5, 2 hr of exposure at 95 dB is a fractional exposure of 0.5, and 2 hr at 85 dB is 0 exposure. The fractions add to 1, so we have a 100% exposure. Thus, the limit is not exceeded. If the exposure is greater than unity, then personal hearing protection must be provided and engineering noise control solutions must be initiated.

Table B.2 Percentage Risk of Developing a Hearing Loss in Excess of 25 dB

Age (years)	20	25	30	35	40	45	50	55	60	65
Exposure years (re age 20)	0	5	10	15	20	25	30	35	40	45
80 dBA exposure level										
Total risk	0.7	1.0	1.3	2.0	3.1	4.9	7.7	13.5	24.0	40
Due to noise	No increase in risk at this level of exposure									
85 dBA exposure level										
Total risk	0.7	2.0	3.9	6.0	8.1	11.0	14.2	21.5	32.0	46.5
Due to noise	0.0	1.0	2.6	4.0	5.0	6.1	6.5	8.0	8.0	6.5
90 dBA exposure level										
Total risk	0.7	4.0	7.9	12.0	15.0	18.3	23.3	31.0	42.0	54.5
Due to noise	0.0	3.0	6.6	(10.0)[a]	11.9	13.4	15.6	17.5	18.0	14.5
95 dBA exposure level										
Total risk	0.7	6.7	13.6	20.2	24.5	29.0	34.4	41.8	52.0	64.0
Due to noise	0.0	5.7	12.3	18.2	21.4	24.1	26.7	28.3	28.0	24.0
100 dBA exposure level										
Total risk	0.7	10.0	22.0	32.0	39.0	43.0	48.5	55.0	64.0	75.0
Due to noise	0.0	9.0	20.7	30.0	35.9	38.1	40.8	41.5	40.0	35.0
105 dBA exposure level										
Total risk	0.7	14.2	33.0	46.0	53.0	59.0	65.5	71.0	78.0	84.5
Due to noise	0.0	13.2	31.7	44.0	49.9	54.1	57.8	57.5	54.0	44.5
110 dBA exposure level										
Total risk	0.7	20.0	47.5	63.0	71.5	78.0	81.5	85.0	88.0	91.5
Due to noise	0.0	19.0	46.2	61.0	68.4	73.1	73.8	71.5	64.0	51.5
115 dBA exposure level										
Total risk	0.7	27.0	62.5	81.0	87.0	91.0	92.0	93.0	94.0	95.0
Due to noise	0.0	26.0	61.2	79.0	83.9	86.1	84.3	89.5	70.0	55.0

Note: It is assumed that the effects of noise and age on people's hearing are additive, and that there is no interaction between the two effects; both effects are "nerve deafness" or "perceptive deafness."

[a] Acceptable noise exposure when percentage risk ⟨10%⟩.

Table B.3 Walsh–Healy Criteria for Industrial Noise Exposure—Tradeoff of Level Against Allowed Exposure Time (Permissible Noise Exposures per 8-hr Working Day)

Duration per day (hours)	*Noise level dBA slow response*
8	90
6	92
4	93
3	97
2	100
1.5	102
1	103
0.50	110
0.25 or less	115 max

Two main forces are at work to reduce industrial or occupational noise exposure in the United States. The first of these is the Department of Labor's obligation under the Walsh–Healy Act, which is to conduct surveys and to issue citations in the event of excessive worker exposure to noise. Surveys are run, and citations for excessive noise conditions are made by the Department of Labor.

The other impetus for control of noise in the workplace is the Workmen's Compensation Act, which requires that employers carry insurance against industrial hazards of all kinds, including exposure to noise. The trade unions and the courts have together been able to develop the precedent that if the Walsh–Healy criteria are exceeded in the workplace, then there is probable cause to expect that the worker's hearing has been damaged by industrial noise and that compensation is appropriate. This system is, therefore, driven by legal action between worker, insurance company, and employer, and the government regulators are bypassed in the process.

B.3 LOUDNESS OF MACHINERY NOISE

"Loudness" is the psychological attribute of an acoustical signal that relates to the magnitude or intensity of the signal. It is related to the physical magnitude or intensity of the sound waves and to the duration, spectrum, and amplitude of the acoustical signal as well. As a result of psychoacoustic experiments, we know that on average subjects tend to give reasonably consistent answers when asked to judge the relative loudness of two sounds or the changes in loudness due to a combination of sounds, although there is a fair amount of scatter in the data.

The simplest loudness judgment that we can make is to compare the loudness of two tones. The results of the experiment in which subjects are presented with a 1-kHz tone at a known pressure level and a second tone with its level adjusted until the subject judged it to be equally as loud as the 1-kHz tone are shown in Figure B.3. For example, suppose that a subject is presented with a 1000-Hz tone at a sound pressure level of 40 dB. According to Figure B.3, when the subject is instructed to adjust a second tone at 100 Hz until it is as loud as the 1000-Hz tone, then on average he or she will increase the sound pressure level to about 65 dB.

If the frequency of the second tone is continuously adjusted over the entire spectrum, a locus of sound pressure levels equally loud to the 1000-Hz tone will be generated. Since these tones are all judged to be as loud as the 1000-Hz tone (and as loud as each other), they are said to be equal in loudness level. The value of the loudness level (the unit is the phon) is equal numerically to the sound pressure level of the 1000-Hz tone. The contour so developed is the 40-phon contour.

Similarly, if we now change the sound pressure level of the 1000-Hz tone and do the experiment repeatedly, the set of contours in Figure B.3 will be generated. We see that the ear is somewhat more sensitive in the frequency range from 2000 to 6000 Hz than it is at 1000 Hz and that generally below 500 Hz and above 7000 Hz the ear is substantially less sensitive. In Figure B.4, we show data taken by other investigators using bands of noise. These curves have a slightly flatter shape than the curves of Figure B.3, but the general similarity is apparent.

The curves in Figure B.3 are the source of certain standard weighting spectra used in sound level meters. The A-weighted filter of the sound-level meter approximates the inverse of the 40-phon curve, as noted in Figure B.5. The A-weighted curve is widely used and has been shown to correlate well with a variety of subjective responses to noise.

A second experiment carried out with tones is concerned with growth of loudness as the physical intensity of the signal is changed. In this experiment, the

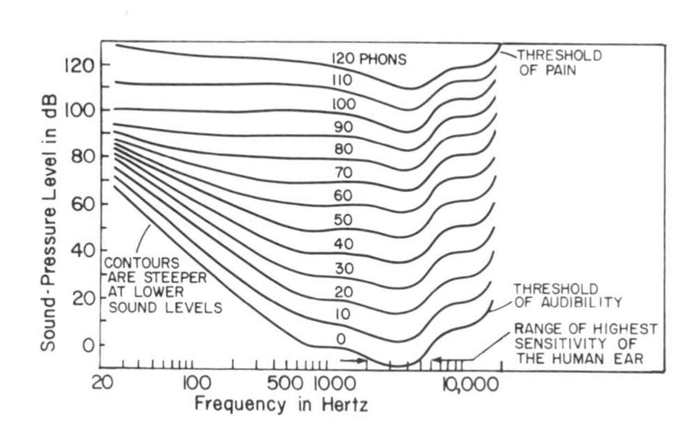

Figure B.3 Contours of equal loudness for pure tones. Note the peak in sensitivity at about 3 kHz.

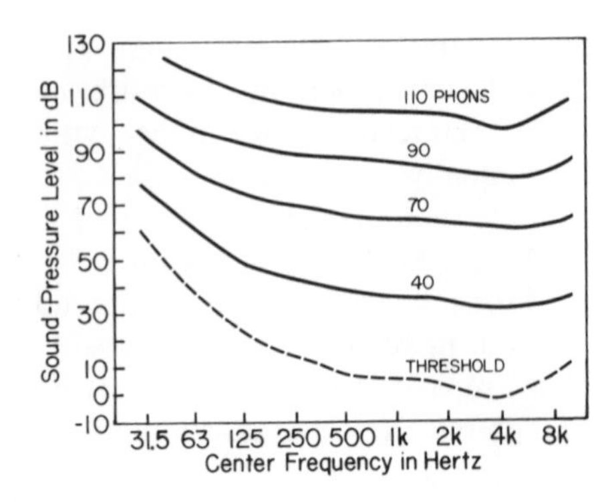

Figure B.4 Equal loudness curves for octave bands of noise are nearly parallel, flatter, and less steep compared with the contours determined for pure tones.

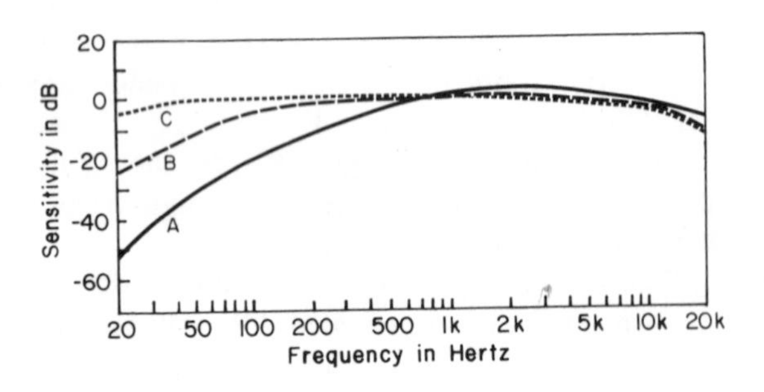

Figure B.5 Applications of equal loudness contours to weighting filter.

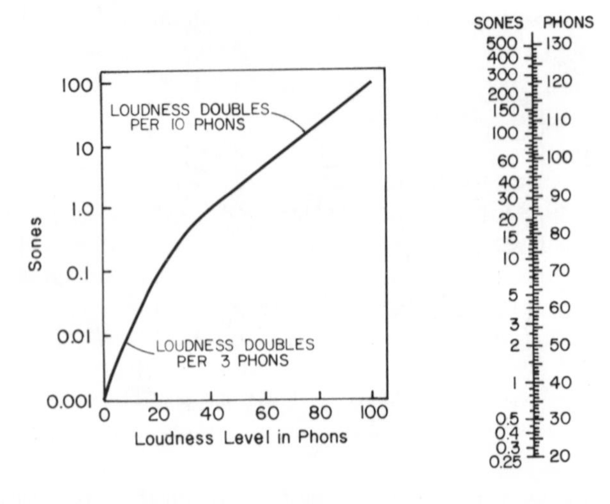

Figure B.6 Plot of function relating loudness in sones to loudness level in phons.

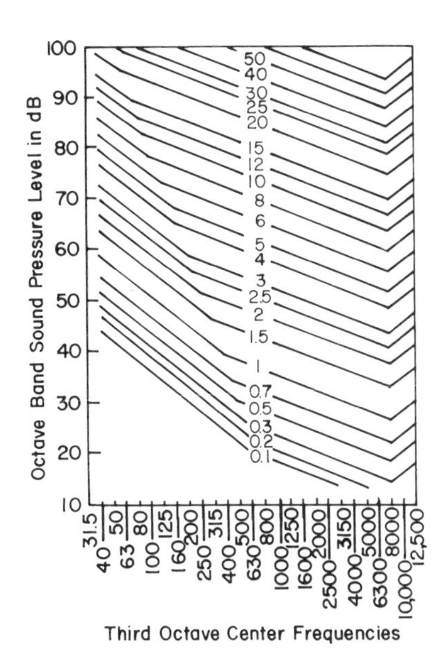

Figure B.7 Standardized third-octave and monograph center frequencies.

subject is asked to change the sound pressure level of the tone until it is twice as loud or half as loud as the original tone. The quantity of loudness is arbitrarily taken to be unity at the 40-phon loudness level. A sound that is "twice as loud" has two loudness units, for example. The loudness units are called "sones." The result of these experiments, shown in Figure B.6, is a subjective impression of loudness doubling with an increase of 10 phons above a level of 40 phons. Below a level of 10 phons, however, the loudness doubles with an increase of 3 phons. Thus at very low signal levels, the rate of change of loudness with amplitude is quite strong, but this rate is diminished with higher sound amplitudes.

The increase of loudness level for third-octave bands of noise is shown in Figure B.7. We can calculate the loudness in the number of sones corresponding to sound pressure level for each band in a complex noise spectrum. Unfortunately, we cannot find a total number of loudness units by simply adding up the loudness of each band. The reason for this is *masking*; that is, the loudest band will reduce the subjective impact of the other bands so that they affect the overall loudness less than they would if that loudest band were not present. Psychoacoustic experiments indicate that we should count the loudest band at full value and take 15% of the loudness for the remaining bands. Once the overall loudness in sones is calculated, we may use the chart in Figure B.7 to calculate the overall loudness level in phons.

An example of this calculation considers the spectrum of third-octave-band sound levels in a shop area. The band levels and the associated loudness values are shown in Table B.4. The total loudness is 21.2 sones.

Table B.4 Example of Calculation of Loudness Level in a Workshop

Band Frequencies		Sound Level		Loudness Index		
1/3 Oct.	Oct.	L_p(1/3 OB)	L_p(OB)	I_n(1/3 OB)	A-Weighting	L_p^A(OB)
100		76		7.0		
125	125	75	80	7.0	−15	65
160		74		6.8		
200		74		7.5		
250	250	73	78	7.5	−8	70
320		71		6.6		
400		69		6.5		
500	500	67	72	6.0	−3	69
640		65		5.8		
800		63		5.5		
1000	1000	61	66	5.2	0	66
1250		59		5.0		
1600		57		4.7		
2000	2000	55	60	4.4	1.2	61
2500		53		3.9		
3200		50		3.6		
4000	4000	47	52	3.2	0.9	53
5000		44		2.8		

$S_1 = I_{max} + 0.15(\Sigma I - I_{max}) = 7.5 + 0.15(91.5) = 7.5 + 13.7 = 21.2$ sone.
$L_1 = 84$ phon; i.e., the loudness of the noise equals the loudness of a pure tone of 84 dB at 1 kHz.
$L_p^A = 74$ dB
PSIL = 66 dB.

B.4 CRITERIA FOR COMMUNICATION IN THE WORKPLACE

If we record the sound levels of someone speaking through a third-octave-band analyzer, as shown in Figure B.8, then in different third-octave bands the level versus time of the speech signal will appear as shown in Figure B.9. It is well established that the intelligibility of speech depends on a quantity called the *articulation index* (AI), which in turn depends on this dynamic range and energy spectrum of speech.

The AI is a measure of the speech amplitude relative to background noise. The signal energy in the speech bands is determined by analysis of the spectrum into so-called articulation bands. In addition, speech ranges up to 30 dB between its minima and its peaks, and the various intervals of speech levels also contribute more or less equally to the AI.

Thus, plotting the speech level against frequency, as shown in Figure B.10, and then plotting the background noise spectrum, as shown in this curve, we can

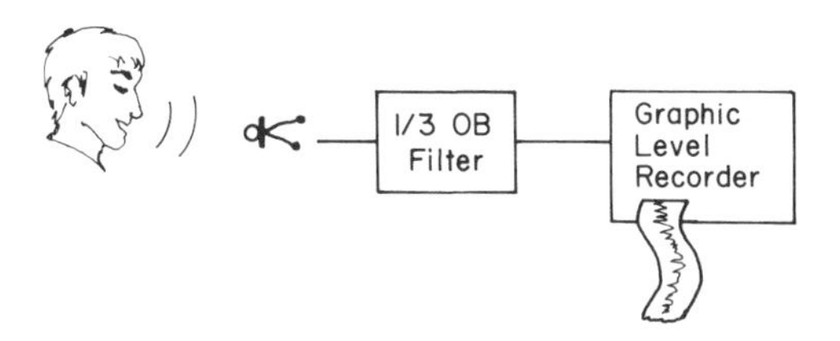

Figure B.8 Recording of third-octave-band filtered speech levels shows dynamic range of speech signal and its energy content as a function of frequency.

determine the fractional amount of area of the speech range that is above the background noise. This fractional area is the AI and determines how well speech will be understood in this noise background. If the entire speech area (i.e., the range of frequency and levels) lies above the background noise, then the AI is unity and the speech can be perfectly understood. On the other hand, if 50% of the area is covered by the background noise, then the AI is 0.5, and we expect less speech understanding. The relationship between AI and fractional amount of speech understood is shown in Figure B.11. This shows that for complete sentences an AI of only 0.2 can lead to between 70% and 80% of speech understood, and an AI of 0.5 results in nearly perfect understanding.

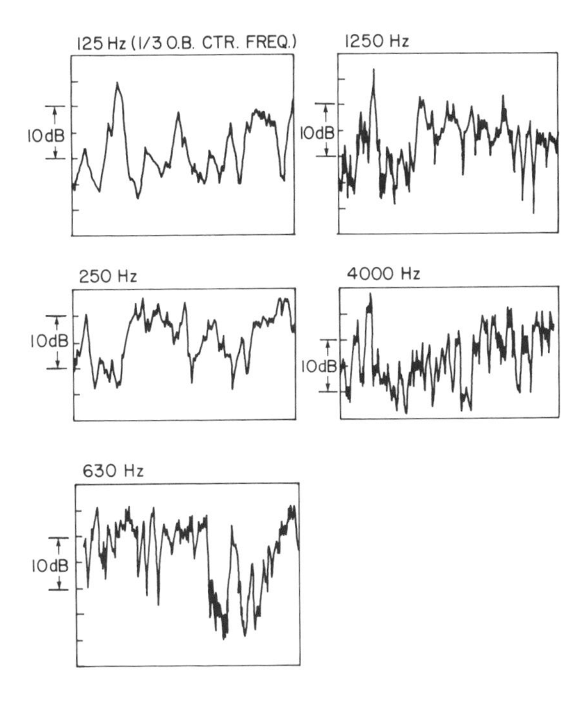

Figure B.9 Typical data for the speech levels in selected third-octave bands. The range of speech energy is about 20 to 30 dB.

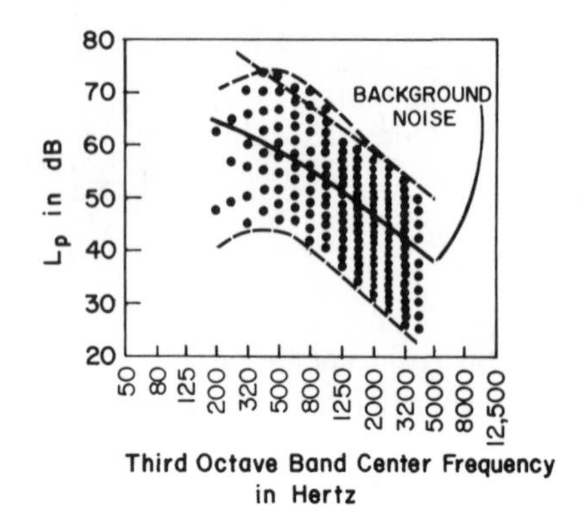

Figure B.10 The articulation index (AI) is the fractional amount of the frequency-level area that is not covered by background noise. A dot-counting procedure is a convenient way to calculate AI.

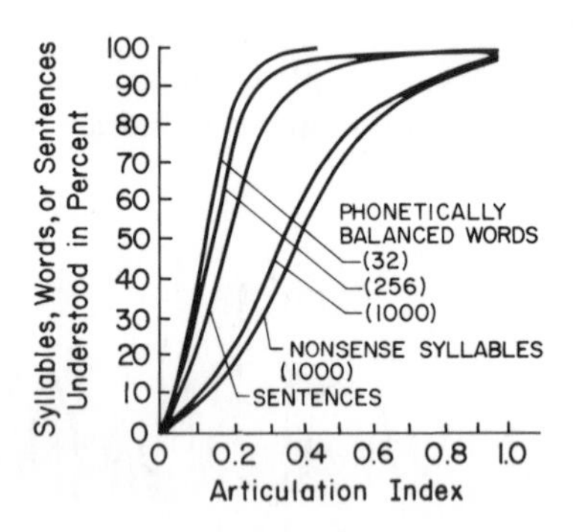

Figure B.11 The percentage of words understood for a given value of articulation index depends on contextual clues. A significant fraction of connected speech can be understood for even small values of AI.

Figure B.10 shows a level of background noise that gives an AI of 50%. If this were the ratio of background noise to direct speech in an open area, such as an aircraft cabin or restaurant, for example, we could understand the connected speech of a neighbor without difficulty. The background noise, however, can be used as a mechanism for privacy in such open spaces. For example, if the distance between you and the partner in conversation is 0.5 m and the distance to the next table or to nearby seats in the aircraft is 2 m, then the direct sound level for the people in the next area is down about 12 dB from that of your neighbor. Since the articulation range is 30 dB, this effectively decreases the speech levels by about 12 dB or 40% of the range. Therefore, if the AI was 50% for your neighbor, it will be about 10% for the people in the next seating area, and this will result in a drop in the fraction of speech understood from something approaching 90% to something under 20%. This drop in speech intelligibility is probably sufficient to give a feeling of privacy in such open public areas.

The combined criteria of loudness, speech interference, and hearing loss are of central importance in establishing limits on allowable noise production by machines. We can see that the loudness contours, the region of greatest speech energy, and the sensitive frequencies for hearing damage all reflect the importance of the 500-, 1000-, and 2000-Hz frequency bands. It is therefore most important that the noise radiated by machines in these frequency bands be reduced.

B.5 COMBINED NOISE CRITERIA

Machines used in office environments, such as printers, copiers, typewriters, and computer terminals, are not likely to damage one's hearing, but they may cause speech interference or produce an unpleasant background noise. We can say that such criteria are intensity related, and we discuss criteria in this section related to such measures.

We must note, however, that the magnitude of the sound, whether expressed in physical or subjective terms, is not the only quality that a sound has. Sound carries meaning, and we make judgments on the sounds that products make and on the products themselves, based on the interpretation we make of that sound. Thus, issues of the appropriateness of a sound, the acceptability of the sound, and the quality judgment that is made of the machine (perceived quality) must also be addressed. We shall deal with these latter issues in the next section.

The criteria for noise in office environments is based on speech interference level and loudness. As noted in the last section, the speech interference level of a background noise is defined as its average sound pressure level in the 500-, 1000-, and 2000-Hz bands. These are the bands that contain most speech energy and, therefore, the sound level average in these bands is a reasonably good measure of the ability of that background noise to reduce the AI. Noise criteria curves for which the speech interference of the background noise is balanced against its loudness have been established for background noise of offices. It has been determined experimentally that if the loudness level of the background noise is more than 22 units greater than the speech interference level, then the background noise becomes too loud. If the loudness level of the noise is less than 22 units above the speech interference level, then speech interference becomes the governing criterion for background noise.

A preferred noise criterion (PNC) curve is an octave-band spectrum level for which these two criteria are balanced so that the loudness level is exactly 22 units greater than the speech interference level. Thus, if we plot a noise spectrum as in Figure B.12, then the highest-valued PNC curve that is tangent to the background noise spectrum defines the PNC rating of that background noise. If the noise peaks in the low- or very-high-frequency ranges of the spectrum, then the PNC rating is loudness dominated, but if it peaks in the midregion, the noise rating is determined primarily by the speech interfering power of the background noise. The PNC curves are labeled by their speech interference level values.

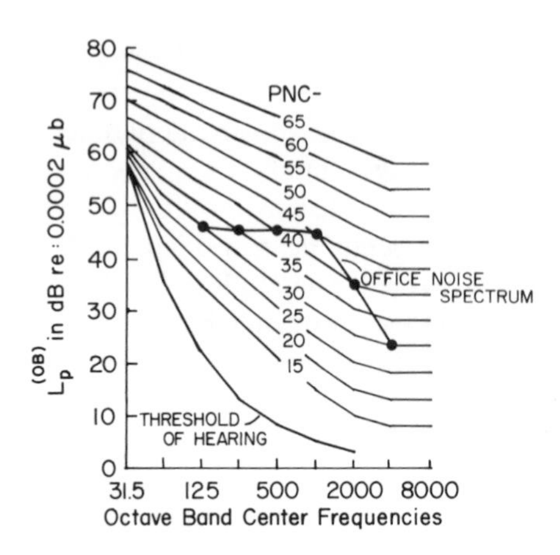

Figure B.12 The PNC curves are used as criteria for background noise in offices, balancing speech interference against loudness.

Preferred noise criterion curves have been used by suppliers of office equipment to express noise criteria to their design groups. They have also been used by companies purchasing office equipment to specify allowable noise levels for equipment. The PNC rating is an environmental criterion and is not directly a measure of noise emitted by the machine. To make the conversion between radiated sound power from the machine and ambient sound levels, we use the relationships developed in Chapter 5.

B.6 CRITERIA BASED ON PERCEIVED QUALITY

Several years ago, a new electric typewriter was about to be brought to the marketplace. The conventional typewriter, a "key-bar machine," had a rather distinctive sound. This new kind of typewriter used a ball element and the striking of the ball against the paper and platen produced a sharp impulse of sound similar to that of the traditional key-bar machine. However, the positioning of the ball element—its rotation, orientation, and translation—produced a low-level sound that was absent in the key-bar machine. Typists trying out the new electric typewriter felt the machine sounded more "rattley" and not as solid as the traditional machine. In this instance, it was necessary to reduce these extra sounds, not because they contributed to the overall loudness of the machine or interfered with speech, but simply because they resulted in a lower perceived quality of the product due to an "inappropriate" sound.

A second example involves the noise produced by a shop-type vacuum cleaner in which a small, high-speed motor with a vertically oriented shaft drives a centrifugal fan. This fan draws air from the interior of a tanklike cavity; attached to the tank is a vacuum hose. Debris and water and other objects can be pulled into the tank, and the air is extracted from the top of the tank by this motor/vacuum unit. The lower end of the motor armature is supported by a self-aligning sleeve bearing, and the upper end has a captured ball bearing that sits against a thrust pad and supports the weight of the armature shaft. The entire assembly is contained in a plastic housing.

The company determined that the operating noise of the unit was not a problem because that sound was expected by the user. The problem was that on rundown (i.e., when the machine was turned off and allowed to coast to a stop) the vibration had an unpleasant rising and falling modulation, sometimes accompanied by rattling of the assembly. The customer, owner, or user of the machine interpreted this as a defect in its manufacture and was concerned about its quality. The problem was to find out why this occurred on some machines and how to eliminate it.

It turned out that the unpleasant noise was due to a nonlinear stiffness of the armature-supporting ball bearing. When this spring was "linearized" by placing a rubber ring between the thrust washer and the support structure, the undesirable vibration was eliminated. The major point to be made here, however, is that the actual operating noise, which is relatively loud, is not considered by the company to be a problem, but the perception of low quality, triggered by a relatively weak sound, was a major problem.

It might appear that these subjective reactions to machinery noise are not susceptible to engineering procedures. It is possible, however, to use psychometric methods to determine the disturbing aspects of the complex sound of a machine. If those disturbing aspects can be attributed to particular mechanisms, then the designer knows which mechanisms to concentrate on for noise reduction.

This procedure has been used to determine the objectionable noisemakers in a sewing machine. The machine and some of the noisemaking mechanisms are shown in Figure B.13. The noise produced by each mechanism was recorded on a single track of multitrack tape so that the composite sound from all mechanisms together could be presented to a listener. But also, the sound of each mechanism could be increased or decreased to see the effect of noise reduction (or increase) on the judgments.

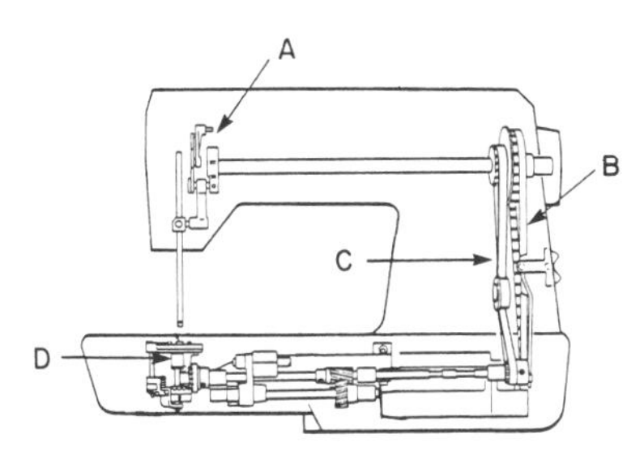

Figure B.13 Sketch of sewing machine showing many of the noise-producing mechanisms. A: takeup; B: motor belt; C: timing belt; D: bobbin case.

Table B.5 Consumer Panel Information

Total number	20 people with normal hearing according to ISO standards
Age	18–48 years
Gender	12 women 8 men
Sewing experience	16 consumers sew: 6 considered themselves to be novices, 7 intermediate, 3 advanced The consumers who do not sew have had exposure to sewing machine noise through mothers, sisters, friends Four of these consumers sell or repair sewing machines.
Brands of machines used	12 Singer users Elna and Kenmore were also mentioned specifically

The judgments of machine sounds were made by a jury described in Table B.5. These listeners were instructed in a psychometric method of evaluation called "magnitude estimation." When psychometric methods are used, it is important that the procedures be carefully supervised by workers experienced in the techniques, just as it is important that engineering measurements be made by experienced workers.

The results of jury tests on the perceived quality of the sewing machine ("How good a machine do you think made this sound?") are presented in Figures B.14 to B.17. Judgments in which the takeup noise was varied are shown in Figure B.14. At its natural level, the average judgment is between slightly and moderately good quality, but this improves about one category as the takeup noise component is

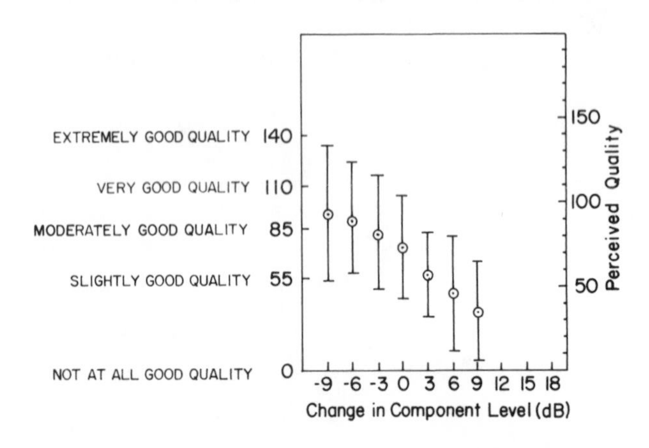

Figure B.14 Judgments of the perceived quality of the sewing machine vs. change in sound level of takeup.

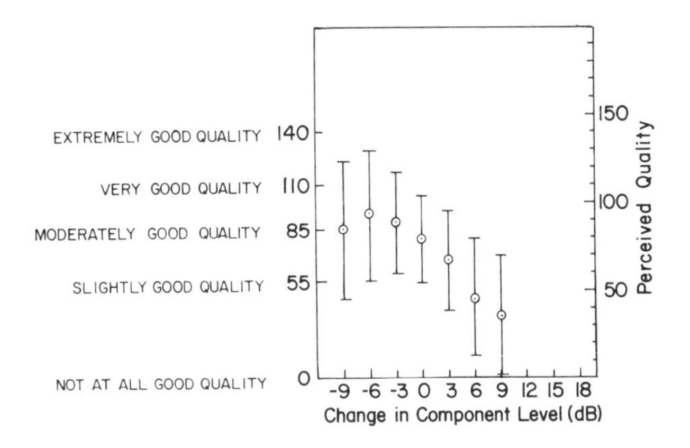

Figure B.15 Judgments of the perceived quality of the sewing machine vs. change in sound level of motor belt.

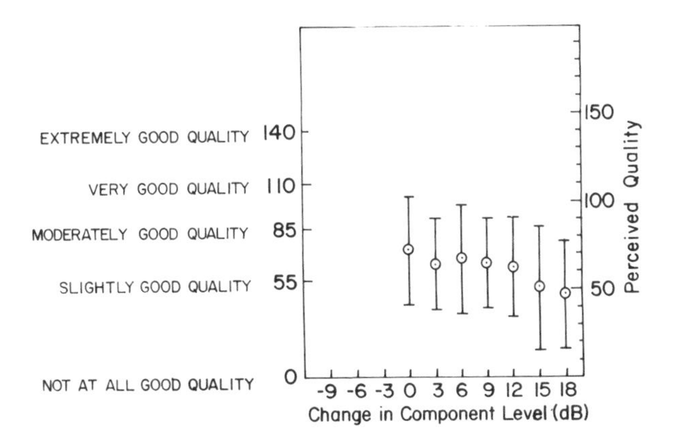

Figure B.16 Judgments of the perceived quality of the sewing machine vs. change in sound level of timing belt.

reduced by 9 dB. The perceived quality drops rapidly as the takeup noise is allowed to increase. The vertical lines at each level indicate the range of judgments for the sounds, indicating the great variability that is common in these procedures.

When the motor belt level is changed, we get a slightly different result, as seen in Figure B.15. There is improvement as the motor belt level is reduced, but perceived quality deteriorates. On the other hand, the judgments are almost completely insensitive to the timing belt noise shown in Figure B.16. These results strongly suggest that the timing belt noise could be allowed to increase without affecting perceived quality.

An interesting example is shown in Figure B.17, in which a rattle is introduced in the bobbin case. This is a common situation in that this part of the machine often

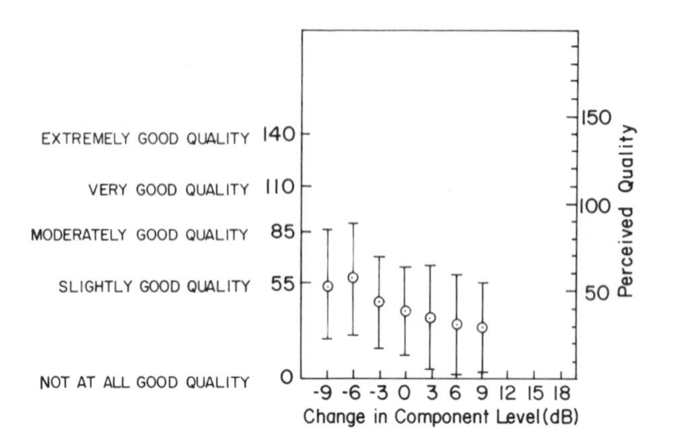

Figure B.17 Judgments in the perceived quality of the sewing machine vs. change in sound level of bobbin case rattle.

rattles if the bobbin is not secured. Note that the judgments are not greatly affected by the amplitude of the signal, but the mere presence of the rattle drops the perceived quality by a full category. Whenever a rattle in a machine is detectable, even if weak, we can expect a drop in perceived quality.

B.7 CRITERIA FOR VIBRATION

Vibration criteria for machinery have also been developed based on the various effects that vibrations have on the human body and nervous system. The curves in Figure B.18 are the established vibration criteria for vertical motion. Very low frequencies are determined by reaction to gross motions, which result in motion sickness or other discomfort due to disorientation from the large displacements. In a frequency range from about 1 to 20 Hz, the body and major groups of organs have resonances that result in internal forces in the body and feelings of discomfort. At very high frequencies, about 30–50 Hz, the primary effects of vibration on the body are very local and can damage skin capillaries and nerve endings. Workers who use vibrating equipment are subject to a disease called "white fingers," which is a destruction of the capillaries and the loss of blood circulation to the surface of the hands and fingers, in particular.

Criteria are expressed in terms of detectability, reduced comfort, reduced proficiency, and exposure limits. The International Standards Organization (ISO) has published suggested criteria levels for sinusoidal vibration, and these levels are shown in Figure B.18. The vertical motion criteria curves above 20 Hz tend to be constant-velocity curves; between 2 and 20 Hz they tend to be constant acceleration; and below 2 Hz they tend to be constant jerk, the time rate of change of acceleration. The transition from reduced comfort to reduced proficiency is an

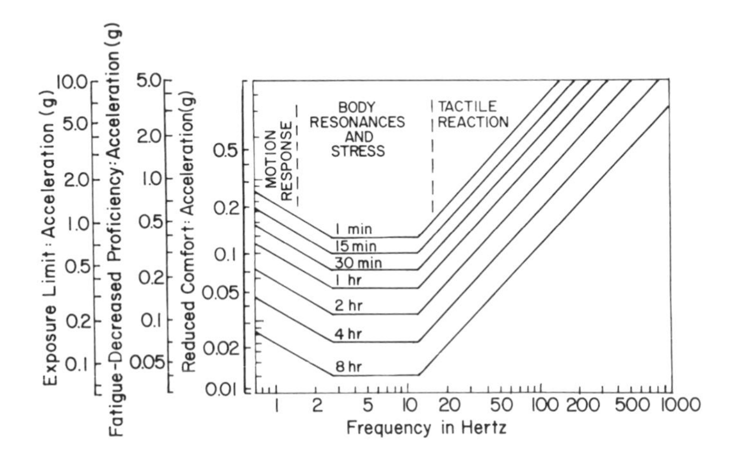

Figure B.18 ISO criteria for vertical vibration. The dynamic range between perception and discomfort for vibration is much smaller than it is for sound.

increase of about 10 dB. The acceptable level also depends on the time of exposure to the vibration. For example, considering the octave band of spectrum of vibration plotted in Figure B.18, we can see that it corresponds to a reduced comfort criterion if a time limit of about 1 or 2 hr is set.

APPENDIX C

Determining Transfer Functions from Measured Data

C.1 INTRODUCTION

We have discussed transfer functions throughout this book as though they are known exactly, by calculation from first principles, for example. But much of the time, our information is based on experimental data and is therefore imprecise. The purpose of this appendix is to indicate how noise affects measured quantities and how we can evaluate noise contamination of data.

C.2 CONSTRUCTION OF INVERSE FILTERS BASED ON MEASURED DATA

Let us now consider the situation in Figure C.1(a), where a transfer mobility Y relates an input force L to an output velocity V. The mobility is defined by the ratio of the velocity to the force, $Y = V/L$, but if we calculate the mobility by using the Fourier transforms of the velocity and the force directly in this way we would get a very poor representation of the transfer mobility. This is because truncation effects in the evaluation of the velocity and force transforms cause relatively large variations from one computation to the next. Measurement errors and noise will also affect the calculations. Let us call the combination of these noise and error components δV_n and δL_n, so in the presence of noise the velocity and force are

$$V_n \to V_n^t + \delta V_n, \qquad L_n = L_n^t + \delta L_n, \tag{C.1}$$

where V_n^t and L_n^t are the "true" values of the force and the velocity in the absence of

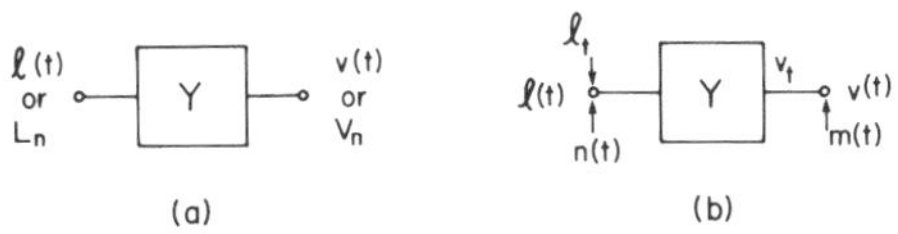

Figure C.1 A linear two-port system with and without additive noise at input and output.

noise. With these introduced, we would get

$$Y = Y^t + \delta Y = \left(\frac{V_n^t}{L_n^t}\right)\left(1 + \frac{\delta V_n}{V_n^t}\right)\left(1 - \frac{\delta L_n}{L_n^t}\right)$$

or

$$\delta Y = Y^t\left(\frac{\delta V_n}{V_n^t} - \frac{\delta L_n}{L_n^t}\right). \tag{C.2}$$

Since the relative errors in measuring the velocity and force may be considered independent, the relative variance in the transfer mobility equals the sum of the relative variances in the measured velocity and the input force:

$$\frac{\sigma_Y^2}{|Y|^2} = \frac{\sigma_V^2}{|V|^2} + \frac{\sigma_L^2}{|L|^2}. \tag{C.3}$$

Let us now multiply the numerator and denominator of the transfer mobility by the complex conjugate of velocity and then multiply mobility by the complex conjugate of force. If we now make N measurements and estimate Y by averaging over N, we get

$$\begin{aligned}(Y)_{\text{est}} &= \frac{\langle V_n L_n^* \rangle_N}{\langle (L_n)^2 \rangle_N} = \frac{S_{vl}(\omega)}{S_{ll}(\omega)} \qquad (= H_1),\\ &= \frac{\langle |V_n|^2 \rangle_N}{\langle L_N V_n^* \rangle} = \frac{S_{vv}(\omega)}{S_{lv}^*(\omega)} \qquad (= H_2),\end{aligned} \tag{C.4}$$

where the two possibilities generate two estimates for the transfer functions. In this equation, S_{vl} is the cross spectrum between the velocity and force, S_{ll} is the auto (or power) spectrum of the force, and S_{vv} is the auto spectrum of the velocity.

Since noise at the system input where the force is applied will affect the auto spectrum of the force more than it affects the cross spectrum between input force and output velocity, we would use H_2 when the measurement is expected to be contaminated by input noise; by a similar argument, we would use H_1 if noise in the output were expected. This situation can change frequency by frequency. At system resonances, the drive point force will become smaller, so noise at the input will be more important at frequencies near system resonances. At the same frequencies, the output velocity will tend to be large. At or near zero in the transfer function, however, the output velocity is small, even if the input force is large, so the measurement error of the output velocity will tend to be larger near transfer function minima. In such regions, we would tend to use H_1. Thus, the choice between the use of H_1 and H_2 tends to be a frequency-by-frequency matter, but it is possible to write a simple computer program that will take the measured data and use the more appropriate value of H_1 and H_2, depending on the spectral magnitudes of the input force and the output velocity.

A quantity that is closely related to the functions H_1 and H_2 is the *coherence*. Whenever we measure a transfer function, the coherence is also calculated to test the quality of the data as they are taken. Consider the situation as shown in Figure C.1(b), where noise $n(t)$ is added to the input and noise $m(t)$ is present at the output. The definition of the coherence γ^2 is

$$\gamma^2 \equiv \frac{|S_{lv}|^2}{S_{ll}S_{vv}} = \frac{H_1}{H_2}. \tag{C.5}$$

The measured input force and the output velocity are then

$$v(t) = v_t + m(t), \qquad l(t) = l_t + n(t). \tag{C.6}$$

If there were no noise in either the input or output, we would have

$$\gamma_t^2 = \frac{|S_{lv}|^2}{S_{ll}S_{vv}} = \left(\frac{S_{lv}}{S_{ll}}\right)\left(\frac{S_{lv}^*}{S_{vv}}\right) = \left(\frac{H_1}{H_2}\right) = 1. \tag{C.7}$$

Since the input and output noise are uncorrelated, the cross spectrum is not affected by the presence of input or output noise, but the auto spectra for input and output are, so the coherence with noise present is

$$\begin{aligned}\gamma^2 &= \frac{|S_{vl}|^2}{S_v S_l} = \frac{|S^2_{v_t l_t^2}|}{(S_{v_t} + S_m + S_n|Y|^2)(S_{l_t} + S_n)} \\ &= \left\{\left(\frac{1 + S_v}{S_{l_t}}\right)\left(\frac{1 + S_m}{S_{v_t}} + \frac{S_n|Y|^2}{S_{v_t}}\right)\right\}^{-1} \leq 1.\end{aligned} \tag{C.8}$$

The result in Equation C.8 shows how input and output noise or both reduce coherence. Another way of reducing coherence is to have frequency components at the output that are generated within the mechanical system and are not present in the input. This usually results from some sort of nonlinear behavior of the system, so a reduction of coherence can also be a good indicator of nonlinearity in the system.

Note, however, that coherence does not discriminate against correlated noises present at both the input and output or noises that are individually correlated with either the input or the output. Thus, we cannot use correlation to distinguish between different sources in a machine that have phase coherence. Since many machines have a periodic operating cycle, all the sources of vibration and excitation are likely to be coherent. Coherence analysis cannot easily discriminate against or determine the contribution of various noise sources in such machines.

Let us explore this last part further by referring to Figure C.2. Here we have two sources in a machine: a valve impact and a combustion pressure, both producing vibration pulses at an accelerometer. We have sketched the time

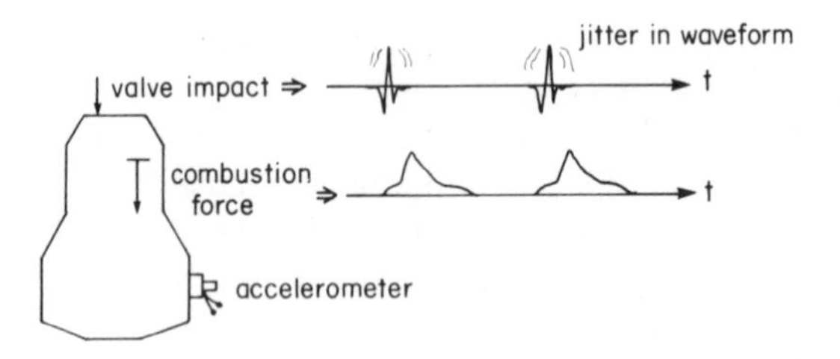

Figure C.2 Vibrations due to different sources that have a fixed time relation to each other will be coherent. A temporal uncertainty (jitter) in one of the sources produces loss in coherence.

waveforms of the two signals in the figure, and we note that since these two signals occur every cycle at the same time relative to each other they are perfectly coherent. Therefore, any combustion pressure signal that arrives at the accelerometer cannot be distinguished from a valve impact signature on the basis of coherence. Let us suppose, however, that there is a jitter in the time of valve impact due to uncontrolled irregularities and that this jitter is about 0.1 msec. Since 0.1 msec corresponds to a period of oscillation of 10 kHz, we might expect that at frequencies above 2.5 kHz, where the 0.1-msec jitter is a quarter of a period, this jitter will cause a loss in coherence between the valve impact and the combustion pressure. Thus for frequencies above 2.5 kHz, coherence analysis would allow an identification and separation of the sources, but below this frequency the signals will be coherent and cannot be separated by coherence analysis.

BIBLIOGRAPHY

BOOKS

Structural Excitation, Response, and Sound Radiation

1. L. Cremer, M. Heckl, and E. E. Ungar, *Structure-borne Sound*, Springer-Verlag, New York, 1973.
2. M. Junger and D. Feit, *Sound, Structures, and their Interaction*, MIT Press, Cambridge, Massachusetts, 1972.
3. F. Fahy, *Sound and Structural Vibration*, Academic Press, London, 1985.
4. R. Lyon, *Statistical Energy Analysis of Dynamical Systems*, MIT Press, Cambridge, Massachusetts, 1975.

Diagnostics, Condition Monitoring, Failure Detection

5. J. H. Williams, *Transfer Function Techniques and Fault Location*, Research Studies Press, Letchworth, England, 1985.
6. R. A. Collacot, *Mechanical Fault Diagnosis and Condition Monitoring*, UKM Publications Ltd., Leicestershire, England, 1982.

Signal Processing

7. A. Oppenheim and R. Schafer, *Digital Signal Processing*, Prentice Hall, Inc., Englewood Cliffs, New Jersey, 1975.
8. J. S. Bendat and A. G. Piersol, *Random Data: Analysis and Measurement Procedures*, Wiley-Interscience, New York, 1971.

JOURNALS AND MAGAZINES

1. *Noise Control Engineering Journal*, Institute of Noise Control Engineers, Poughkeepsie, New York.
2. *Mechanical Systems and Signal Processing*, Academic Press, London, New York.
3. *Vibrations*, Vibration Institute, Clarendon Hills, Illinois.
4. *Sound and Vibration*, Acoustical Publications, Inc., Bay Village, Ohio.
5. *Journal of Acoustical Society of America*, American Institute of Physics, New York.
6. *Journal of Sound and Vibration*, Academic Press, London.

INDEX

Page numbers in italics indicate figures.